T0093027

The Ecology of the Trees, Shrubs, and Woody Vines of Northern Florida

UNIVERSITY PRESS OF FLORIDA

Florida A&M University, Tallahassee
Florida Atlantic University, Boca Raton
Florida Gulf Coast University, Ft. Myers
Florida International University, Miami
Florida State University, Tallahassee
New College of Florida, Sarasota
University of Central Florida, Orlando
University of Florida, Gainesville
University of North Florida, Jacksonville
University of South Florida, Tampa
University of West Florida, Pensacola

THE ECOLOGY OF THE TREES, SHRUBS, AND WOODY VINES OF NORTHERN FLORIDA

Robert W. Simons

UNIVERSITY PRESS OF FLORIDA

Gainesville / Tallahassee / Tampa / Boca Raton

Pensacola / Orlando / Miami / Jacksonville / Ft. Myers / Sarasota

26 25 24 23 22 21   6 5 4 3 2 1

Library of Congress Cataloging-in-Publication Data
Names: Simons, Robert W., author.
Title: The ecology of the trees, shrubs, and woody vines of northern
  Florida / Robert W. Simons.
Description: 1. | Gainesville : University Press of Florida, 2021. |
  Includes bibliographical references and index.
Identifiers: LCCN 2020056303 (print) | LCCN 2020056304 (ebook) | ISBN
  9780813066929 (hardback) | ISBN 9780813057835 (pdf)
Subjects: LCSH: Woody plants—Ecology—Florida. | Trees—Ecology—Florida.
  | Shrubs—Ecology—Florida. | Climbing plants—Ecology—Florida. | Woody
  plants—Florida—Identification.
Classification: LCC QK154 .S46 2021 (print) | LCC QK154 (ebook) | DDC
  582.1609759—dc23
LC record available at https://lccn.loc.gov/2020056303
LC ebook record available at https://lccn.loc.gov/2020056304

The University Press of Florida is the scholarly publishing agency for the State
University System of Florida, comprising Florida A&M University, Florida Atlantic
University, Florida Gulf Coast University, Florida International University, Florida
State University, New College of Florida, University of Central Florida, University
of Florida, University of North Florida, University of South Florida, and University
of West Florida.

University Press of Florida
2046 NE Waldo Road
Suite 2100
Gainesville, FL 32609
http://upress.ufl.edu

To David W. Hall
For contributions beyond measure

CONTENTS

1. GYMNOSPERMS

2. ANGIOSPERMS: MONOCOTS

3. ANGIOSPERMS: DICOTS

ACKNOWLEDGMENTS

The following people assisted me in obtaining the knowledge that enabled me to write this book.

I am most indebted to my dear wife Erika. We have spent a lifetime exploring the natural world together, both here in Florida and elsewhere around the world.

I am also indebted to many friends and associates who have given freely of their time as we explored the natural world in each other's company. A partial list includes Jim, Mary, and Cayley Buckner; Archie and Marjory Carr; Judith Hancock; Helen Hood; Thomas Staley; Tom Forcee; Dana Griffin; Francis E. (Jack) Putz; David W. Hall; Richard Franz; Cathryn H. (Katie) Greenberg; Susan W. Vince; Robert K. (Bob) Godfrey; Angus Gholson; Walter S. Judd; Daniel B. Ward; John J. (Jack) Ewel; Paul E. Moler; Carol Lippincott; Buford Pruitt; Tom Morris; Ken Alvarez; Reed Noss; John Hintermister; Rex Rowan; Mike Manetz; and Michael Meisenburg.

I am also indebted to the people of Florida, who have voted on several occasions to tax themselves in order to preserve and maintain the many wonderful natural areas where many of these explorations have taken place and where many of the plants in this book live.

INTRODUCTION

Ecology is about the relationships of organisms with one another and with their physical environment. *The Ecology of the Trees, Shrubs, and Woody Vines of Northern Florida* is about the relationships of these plants with other plant species and with other organisms such as bacteria, fungi, insects, birds, and mammals, plus their relationships with soil, water, weather, sunlight, and fire.

In this book, these relationships are explored and explained as fully as possible within the scope of a single volume. A complete treatment, if that were even possible, would take an entire encyclopedia, with multiple volumes for each species. But we must start somewhere. Some species have been studied thoroughly enough to date to give us a good idea of where they fit into the natural world and also into our altered, human-dominated world.

I have spent the last seventy years exploring, studying, and learning about the natural world, driven by a passion for nature. I have not done this in isolation. Indeed, I have enjoyed the great privilege of exploring together with and learning from some exceptional naturalists, botanists, and zoologists, including Archie F. Carr, Francis E. Putz, Dana Griffin, David W. Hall, Walter S. Judd, Daniel B. Ward, John J. Ewel, Paul E. Moler, Richard Franz, Cathryn H. (Katie) Greenberg, Reed F. Noss, Susan W. Vince, Carol Lippincott, Buford Pruitt, Tom Morris, Ken Alvarez, James Buckner, John Hintermister, Rex Rowan, Michael Meisenburg, and others.

While I focus in this book on individual plant species accounts, it is helpful to keep in mind important overarching aspects of plant ecology. One of these occurs underground in the soil. The roots of most plants have an association with a group of fungi known as mycorrhizal fungi. These fungi form symbiotic associations with a plant's roots. This association of plant roots with an extensive fungal network functions to greatly expand the reach, surface area, and nutrient gathering capacity of each plant's root system. Much of the uptake of essential plant nutrients such as phosphorous, potassium, calcium, magnesium, and organic nitrogen is done by the fungi, using active transport. This comes at a cost to the plant, which supplies energy to the mycorrhizal fungi in the form of sugars transported down from the plant's leaves.

Another overarching aspect of plant ecology is photosynthesis, the process whereby plants obtain energy from sunlight, which enables them to convert carbon dioxide and water into glucose and oxygen.

The transport systems of the plants should also be taken into account. After plants obtain water and dissolved minerals in their roots, the water is translocated upward in vessels within the living tissue of the stems (sap wood in the case of gymnosperms and dicots). This upward transport is driven by the evaporation of water from the crown of the plant, primarily the leaves, and is often called "passive transport." The other main transport system in plants occurs within the living part of the bark of gymnosperms and dicots known as the "phloem." It is within the phloem that the sugars created by photosynthesis in the leaves are transported to all the living parts of the plant, including back down to the root system.

Finally, there are the various mechanisms by which plants grow to keep in mind. The growth of the bark and wood of woody plants, for instance, involves the cambium layer, from which new wood is laid down from the inner surface and new phloem (inner bark) is laid down from the outer surface of the cambium. (The growth of grasses and palm trees is quite different.)

The details of these and various other topics in plant physiology and anatomy are beyond the scope of this book, but a basic understanding is helpful for understanding the ecology of plants.

Information about the general physical characteristics of each plant described herein is provided in the species accounts, but this book is not intended to be a field guide or taxonomic text for the identification of species. There are many such books available, and it is likely that employing one or more of these in conjunction with this book will be helpful.

The main purpose of *The Ecology of the Trees, Shrubs, and Woody Vines of Northern Florida* is to explore the ecological relationships of these plants. Hopefully, this will stimulate readers to do their own additional exploring out in the natural world.

BIOLOGICAL COMMUNITIES OF NORTH FLORIDA

The vast scientific jargon of botany, ecology, and related fields has been avoided as much as possible in this book. However, the use of terms such as marsh, swamp, bayhead, pine flatwoods, sandhill, and hammock is necessary to give a proper description of the native habitat of each of the species discussed here. This can be a bit problematic, given changes in the meaning of these terms over time. For two hundred years, from the mid-1700s to the mid-1900s, the use of these terms remained fairly consistent among natural scientists. Unfortunately, their time-tested, well organized, and often-used system of biological community classification has now been largely replaced by a more complex and less organized classification system developed by the Florida Natural Areas Inventory. In order to give readers some sense of where each species naturally grows, I have chosen to use a simplified version of the older classification system based on the authoritative work of such botanists, naturalists, ecologists, and conservation biologists as Roland M. Harper, Archie F. Carr Jr., Albert M. Laessle, Ronald L. Myers, John J. Ewel, and Reed F. Noss, as follows.

Scrub Ancient Florida scrub communities on old sand dunes (both coastal and inland), often containing varying extents of sand pine, myrtle oak, chapman oak, sand live oak, crookedwood, Florida rosemary, scrub palmetto, and/or saw palmetto and supporting such species as the Florida scrub jay, scrub lizard, and sand skink. This community is quite variable, with some variants referred to as sand pine scrub, oak scrub, and rosemary scrub.

Sandhill The most xeric of the high pine communities, characterized by sandy soil, frequent ground fires, and an overstory of scattered longleaf pine, a midstory of scattered turkey oak, and a ground cover of wiregrass and many other grasses, legumes, and wildflowers.

High Pine A broad classification that includes sandhill as well as the more fertile longleaf pine upland forests that contain southern red oak, post oak, and mockernut hickory.

Hammock A term used for centuries to describe the patches of hardwood forest that occur across peninsular Florida. William Bartram used "hammock" for the hardwood forests he encountered in northern Florida, and most ecologists, botanists, and biologists since his time have used the term for some of the hardwood forests on the Florida peninsula. To the north and west of northern peninsular Florida, in the Florida Panhandle, the Carolinas, and in central Georgia and Alabama, this term is not used, even though the hardwood forests there are similar to the hammock forests in such places as San Felasco Hammock or Gulf Hammock. Thus, the term has always been somewhat colloquial and problematic.

    Hammocks in North Florida include mesic hammock (a variant of southern upland hardwood forest) in which species composition is not reduced significantly by drought, flooding, or lack of fertility; xeric hammock (dry hammock) in which species composition is restricted by drought and low fertility because of the low nutrient and water holding capacity of the soil; and hydric hammock (wet or lowland hammock) in which species composition is significantly restricted by periodic flooding. There are also areas of tropical hardwood trees in South Florida called tropical hammocks. Hydric hammock is a somewhat confusing identifier, given that other terms are often used to describe Florida's many hydric hammock variations, such as "bottomland hardwood forest," "floodplain forest," "prairie hammock," and "coastal hammock." Common tree species in the various North Florida hammocks include sweetgum, laurel oak, water oak, live oak, cabbage palm, pignut hickory, southern magnolia, persimmon, white ash, basswood, sugarberry, winged elm, black cherry, red cedar, loblolly pine, and spruce pine.

Pine Flatwoods Forests Pine forests on flat, poorly drained soil that historically burned frequently. The pines may be longleaf, slash, pond, or even loblolly in some places and on some plantations. (Most of these forests have been converted to pine plantations.) There is typically no midstory in pine flatwoods forests, but there is often a shrub layer consisting of saw palmetto and/or gallberry that usually contains a number of other shrubs, such as shiny blueberry, huckleberry, runner oak, and fetterbush (*Lyonia lucida*). Some variants of this community are known as "scrubby flatwoods," "mesic flatwoods," "wet flatwoods," and "pond pine flatwoods."

Bayhead Also called "baygall." This community may be either a seepage bog or a peat bog wetland that remains wet all year round but seldom floods and maintains a relatively constant water table near or at ground surface. Composed mostly of evergreen trees and shrubs, the characteristic species of bayheads include swamp red bay, sweetbay, loblolly bay, swamp tupelo, fetterbush, black titi, and white titi.

Bog Vegetated wetland that is constantly wet. This sogginess may be the result of hillside seepage or linked to a deep depression that is filled with peat that rises and falls as the water table rises and falls. Forested bogs are called "bayheads" or "baygalls." Some bogs are not forested and are covered with herbaceous vegetation. Unforested bogs are marshes, and some of them (the pitcher plant bogs of the Florida Panhandle and southern Alabama) are very interesting, containing such plants as pitcher plants, sundews, and terrestrial orchids. There are also bogs, dominated by shrubs such as fetterbush (*Lyonia lucida*), that are called "shrub bogs."

Swamp A wetland forest community composed mostly of deciduous trees in a periodically flooded wetland that has a variable water table. Common trees are pond-cypress, bald-cypress, red maple, swamp tupelo, water tupelo, ogeechee tupelo, pumpkin ash, green ash, and pop ash. Swamp variants include cypress domes (pond-cypress swamps), mixed cypress-hardwood swamps, bald-cypress swamps, and cypress strands.

Marsh An assemblage of various herbaceous plant communities in wetlands exposed to full sunlight (as opposed to their being shaded by trees). There are two main marsh types: freshwater marsh and salt marsh, each of which has many variants, including sawgrass marsh, cattail marsh, maidencane marsh, pickerelweed marsh, spatterdock marsh, black needlerush marsh, and smooth cordgrass marsh. Marshes are often patchy, with one plant species dominating one patch and another plant dominating an adjacent patch, and there are often areas where several species mix together. While marshes are typically dominated by herbaceous vegetation, there may be some woody vegetation scattered about, including a few trees.

Occasionally, it can be unclear whether to call an area a marsh or a swamp. The Big Cypress Swamp in South Florida provides a good example. Functionally, much of this area is a marsh, located in nearly full sunlight with a thin scattering of stunted pond-cypress trees. But there are enough cypress trees dispersed throughout the open marsh areas, along with some larger cypress trees in dense stands in scattered, deeper pockets and strands, to warrant identifying the region as a whole as a cypress swamp.

GLOSSARY OF TERMS

Acorn The fruit of an oak, composed of a nut and a cap.

Allelopathic The positive or negative effects of one plant on the growth of another plant or on the germination of seeds via the active release of chemicals into the soil.

Angiosperms Flowering plants whose seeds are contained within a fruiting structure. Examples: palms, grasses, oak trees, cherry trees, legumes.

Annosus Root Rot A fungal disease caused by the fungus *Heterobasidion annosum*.

Anther The pollen-bearing part of a stamen.

Anthracnose A disease caused by fungi in the genus *Colletotrichum*.

Apomixis Asexual reproduction in plants, resulting in offspring that are clones of the parent plant.

Aril An appendage growing at the point of attachment of a seed or fruit.

Bisexual Containing functional male and female structures.

Blade The expanded portion of a leaf, petal, or other structure.

Bract A reduced leaf or leaf-like structure, often subtending a flower or fruit.

Calcareous A descriptive term for a soil or habitat with a rich supply of calcium.

Calyx Outer floral envelope composed of the sepals that covers and protects the flower petals as they develop.

Cambium A layer of actively dividing cells between the bark and the wood of woody plants that produces both new wood and new living bark.

Capsule A nonpulpy fruit consisting of an outer shell containing seeds inside.

Catkins A flower spike with bracts but without petals. Examples include the flowering structures of willow, alder, and birch, plus the male flowers of oak and hickory.

Chert Associated with limestone formations, a hard, flint-like rock used to make stone projectile points.

Clonal Pertaining to groups of plants having arisen by vegetative means from the same parent plant and therefore being genetically identical.

Cotyledon The primary leaf of a plant within a seed or immediately after sprouting.

Deciduous Describing leaves or other plant structures that fall off at the end of a season, leaving the plant bare for a season.

Dicot One of the two subclasses of angiosperms, in which the newly sprouted plant has two cotyledons or primary leaves.

Dioecious Pertaining to a species with unisexual flowers, the male and female flowers occurring on different plants.

Drupe A fleshy fruit with a hard inner structure containing the seed or seeds.

Endemic Only occurring naturally in one particular area such as a state, one habitat type, or a specific biological community.

Entire Margin A leaf margin without teeth, serrations, or spines.

Ericaceous Indicating plants in the heath family, the Ericaceae.

Evergreen A perennial plant that has green leaves throughout the year.

Exotic A species living in a part of the world where it is not native.

Fall Line A narrow zone that marks the geological boundary between an upland region and a plain, distinguished by the occurrence of falls and rapids where rivers and streams cross it.

Glabrous Describing a surface without pubescence or hairs.

Glaucous Having a whitish, waxy coating.

Groundwater Podsol A sandy soil type on level terrain with the water table near the surface and often containing an organic hardpan between the topsoil and the subsoil.

Gymnosperm A vascular, nonflowering plant with seeds not contained within a fruit. Examples include pines, junipers, and cycads.

Hybrid Describing offspring that result from the crossing of two plants of different species or varieties.

Karst Topography Natural landforms and structures resulting from the dissolution of underlying limestone formations. Examples include sinkholes, caves, and karst prairies such as Payne's Prairie in Alachua County, Florida.

Lanceolate Lance shaped; wider at the base and tapering to an apex.

Laurel Wilt Disease Deadly tree disease caused by the fungus *Raffaelea lauricola*.

Mesic Pertaining to medium moisture conditions: not too wet, not too dry.

Midrib The main or central vein of a leaf.

Monocot One of the two subclasses of angiosperms, in which the newly sprouted plant has only one cotyledon or primary leaf. Examples include palms, grasses, lilies, and orchids.

Monoecious Describing a plant with unisexual flowers, the male and female flowers both occurring on the same plant.

Mucilaginous Having a viscous substance of gelatinous consistency.

Naturalized Describing a species that, though not native to an area, has escaped into the wild to reproduce and maintain or increase its population.

Nectary A plant structure (gland) that supplies or secretes nectar.

Niche An organism's position, role, or means of survival within a biological community.

Nodes The places on stems that bear leaves or other structures.

Ovate Egg-shaped in outline; broader below the middle and rounded at each end.

Palmate Describing a leaf with three or more leaflets originating from the same point.

Panicle A cluster of stalked flowers displayed on multiple, loose, irregularly arrayed stems.

Pectin Water soluble chemicals in fruits that gel when cooked.

Petiole The stalk that attaches a leaf to a stem.

Phloem Located outside the wood and cambium layer, the living inner bark of woody plants responsible for active transport of nutrients and water.

Pinnately Compound Having multiple leaflets attached to a linear central stalk (rachis).

Polyploid An organism with more than two sets of homologous chromosomes.

Prickles Sharp spines growing on the outer bark surface of a stem.

Pubescent Describing a surface covered with small hairs.

Raceme A flower structure with the separate flowers attached by short equal stalks at equal distances along a single central stalk, the flowers at the base of that stalk opening first.

Rachis The part of the leaf stalk to which the leaflets are attached.

Serrated Indicating a leaf margin with sharp, angled indentations (teeth) that point forward.

Sessile Without a stalk, such as a leaf attached directly to a stem without a petiole.

Shade Leaves Leaves that have grown in conditions of low light intensity. They are typically larger, thinner, and less distinctly lobed than sun leaves.

Spike An inflorescence with a single axis to which sessile flowers are attached.

Stamen The male part of a flower, including the stalk and the anther, which contains the pollen.

Stellate Hairs Hairs that branch in a star-shaped pattern.

Stalk A stem-like structure that attaches a leaf, flower, or fruit to a plant.

Sun Leaves Leaves that have grown in conditions of high light intensity. They are typically smaller, thicker, and more distinctly lobed than shade leaves.

Tendril A slender vine-like structure that twines or clasps as an aid in climbing.

Thorn A sharply pointed woody structure originating as a stem.

Umbel An inflorescence in which all the flowers arise from a common point.

Xylem The woody part of a stem, inside the bark and cambium layer, containing the vessels that transport water and dissolved minerals.

SPECIES ACCOUNTS

The organization of species accounts in this book is based on Robert K. Godfrey's excellent *Trees, Shrubs, and Woody Vines of Northern Florida and Adjacent Georgia and Alabama*. It served as one of my guides for selecting the species featured here, and I recommend that readers consider it a companion text as they seek fuller understanding of each plant species and variety.

The gymnosperms appear first. These are followed by the angiosperms, which are divided into two categories, the monocots and the dicots. The dicots are ordered alphabetically by family name, following Godfrey's method. For naming protocol, I relied on the Atlas of Florida Plants database, maintained by Richard B. Wunderlin, Bruce F. Hansen, Alan Franck, and the Institute of Systematic Botany. This resource is an invaluable guide for locating current and correct family, genus, species, and common names. As a final note, I occasionally include out-of-date plant names in parentheses following current names, when the dated name remains in use in particular areas such as the nursery trade or in older, pertinent literature.

1

GYMNOSPERMS

Atlantic White Cedar—*Chamaecyparis thyoides*

Cupressaceae

This evergreen tree has foliage reminiscent of southern red-cedar but much softer, not prickly, somewhat flattened, and generally oriented in one plane. Male and female flowers are monoecious, occurring on the same tree. Pollen, produced in mid-December, sometimes floats on the air like puffs of smoke. The fruits are small cones, about a quarter inch in diameter. The winged seeds are about an eighth of an inch in length and width. The bark is dark reddish- or grayish-brown with vertical, flat-topped ridges that often spiral around the trunk.

The growth habit of Atlantic white cedar is tall and straight, with a narrowly conical crown in youth that broadens somewhat as it reaches its mature height of about 80 feet with a trunk diameter of about 2 feet. The largest one reported from Florida, on Mormon Branch in the Ocala National Forest, was 87 feet tall with a trunk diameter of 3 feet and a crown spread of 31 feet (Ward and Ing 1997). Atlantic white cedar is slow growing and can reach ages of 200 years or more, with a maximum age of perhaps 1,000 years (Little and Garrett 1990).

The geographic distribution of Atlantic white cedar is unusual. It grows in a narrow strip along the Atlantic coast from Maine to the North Carolina–South Carolina border and in a narrow strip along the Gulf Coast from the Apalachicola River in the Florida Panhandle west to the Mississippi–Louisiana border (Little and Garrett 1990). In addition, there are three small disjunct populations along the fall line between the coastal plain and the Piedmont in South Carolina and Georgia. Finally, there are two small, disjunct populations on the peninsula of Florida, one in Putnam County and one in Marion County.

Throughout its wide range Atlantic white cedar grows in wetlands. In New Jersey and North Carolina, where it is most common, it often grows in shallow standing water in isolated wetlands. In North Carolina, it occurs most commonly in isolated wetlands known as "pocosins" or "Carolina bays." These unusual wetlands, each surrounded by a low ridge, are oval in shape and have a shallow wet center. The wetlands' oval shapes are all oriented in the same direction. These wetlands are thought to be the result of the impact of a cluster of meteorites. In all the isolated populations along the fall line and in the populations in Florida and along the Gulf coast, Atlantic white cedar grows in floodplains and seepage wetlands associated with streams.

Atlantic white cedar is dependent on an abundant and continually available water supply. It is not at all drought tolerant, nor is it especially flood tolerant. It will not grow in wetlands or next to wetlands where the water depth fluctuates widely and remains high for extended periods because the high water phases of these fluctuations will kill it. Due to its shallow root system, Atlantic white cedar is also susceptible to being toppled by high winds. It is especially vulnerable to such action when dense stands are thinned. The largest Atlantic white cedar in Florida, at Mormon Branch, was toppled in 1993 by the March superstorm called the "Storm of the Century." Atlantic white cedar has few insect or disease problems.

In Florida, the tree associates of Atlantic white cedar are primarily swamp tupelo, sweetbay, loblolly bay, swamp red bay, red maple, slash pine, and pond pine. Other associates include swamp laurel oak (diamond-leaf oak), water oak, water tupelo, pond-cypress, bald-cypress, and sweetgum. At Mormon Branch, cabbage palm is a common associate. In the other isolated population on the Florida peninsula, at Deep Creek in Putnam County, tulip tree is a common associate. Among both of these southernmost populations, Florida willow is another, very rare, associate, as is Grass-of-Parnassus and a large number of other unusual plants. One of these, climbing fetterbush (*Pieris phillyreifolius*), is an ericaceous vine-like shrub with evergreen foliage and pretty white flowers. When its stem grows upward, burrowing underneath the outer bark of pond-cypress and Atlantic white cedar trunks, it pushes through the bark at intervals, producing what appear to be branches coming out of the trunk of the host tree.

The wildlife value of Atlantic white cedar is moderate. Deer browse the foliage (Little and Garrett 1990), and the Hessel's hairstreak butterfly depends on it as its larval food plant (Glassberg et al. 2000). Up to nine million of the small, winged seeds can be produced per acre per year in a dense, healthy stand of Atlantic white cedar (Little and Garrett 1990), but whether or not these provide a food resource for birds or rodents has not been reported in the literature. At five sites in eastern North Carolina, an average of 7 percent of the seeds of Atlantic white cedar have been found to contain the larvae of the tiny wasp *Megastigmus thyoides* (Turgeon et al. 1997). From personal observation, the largest of Florida's Atlantic white cedars at Mormon Branch in the Ocala National Forest was, during its lifetime, repeatedly scarred by black bears that apparently used it as a territorial marker.

Atlantic white Cedar is rarely planted as an ornamental because of its exacting moisture requirements. However, if its needs can be met, it makes a handsome tree.

Eastern Red-Cedar—*Juniperus virginiana* var. *virginiana* and Southern Red-Cedar—*J. virginiana* var. *silicicola*

Cupressaceae

The common red-cedar of the eastern United States, including all of Florida, is now generally considered to be one species with two varieties: eastern red-cedar and southern red-cedar. Eastern red-cedar, located throughout most of the eastern United States, extends into Florida only on the extreme northern border of the eastern panhandle (and, where planted, elsewhere). Southern red-cedar extends along the coast of the southeastern United States, including most of the state of Florida (Wunderlin et al. 2020). The traits used to distinguish these two varieties are not very consistent and show considerable overlap (Adams 1986). Southern red-cedar has reddish bark, pendulous small branches, pollen cones 4 to 6 mm long, seed cones 3 to 4 mm long, and often a flat-topped crown at maturity, whereas eastern red-cedar tends to have gray bark, upright to horizontal branches, pollen cones 3 to 4 mm long, seed cones 5 to 6 mm long, and a conical crown at maturity (Wilhite 1990; Wunderlin et al. 2020).

Both eastern and southern red-cedar trees are medium-sized trees, normally reaching heights of 30 to 50 feet and having eventual trunk diameters in old age of 1 to 3 feet. The largest southern red-cedar in Florida is located in Archer. In 2014 it was 77 feet tall, had a trunk diameter of 5½ feet at breast height, and an average crown spread of 56 feet (American Forests 2015).

Southern red-cedar has evergreen foliage that is dense, fine textured, and prickly, with the juvenile foliage often much pricklier than the adult foliage. The trunk is often straight, but instead of being round in cross section, it is often composed of round bulges and folds. This growth form may provide some defense against being girdled by the antler rubbing of deer. The bark is thin, reddish-brown, and flakes off in long thin vertical strips. It has a conical growth form when young but is quite variable in shape when older. Some red-cedars, particularly open-grown ones near the coast, are very short with wide, irregular crowns. Others retain the straight trunk and conical crown shape until mature.

Southern red-cedar is most commonly found on calcareous soils in low-lying situations near the coast and along the St. Johns River. However, it is quite adaptable. It grows well on either clay or sand and on both excessively well-drained soils and poorly drained soils. It constitutes an important part of the coastal hammocks of Florida, Georgia, and the Carolinas in association with live oak and cabbage palm and extends westward from northern Florida along the Gulf Coast into eastern Texas (Wilhite 1990). It also occurs as a scattered tree in other upland and lowland forests, being especially common where limerock is near the surface. It will invade sandhills and other upland pine forests, pine plantations, and fencerows in the absence of fire.

Red-cedar can live a long time. Fowells (1965) reports a maximum age of about 300 years for eastern red-cedar. However, the age of large red-cedars can be deceptive. A southern red-cedar with the largest trunk diameter I have seen to date (well over 5 feet)—an ancient looking, gnarled, hollow tree between Micanopy and Payne's Prairie—appears to be only about 80 years old based on ring counts from the trunk above the hollow part, combined with an estimate of how long it would take to grow to the height where the ring counts were taken.

Southern red-cedar is moderately shade tolerant as a seedling but becomes shade intolerant as it grows older. It is highly drought tolerant and is one of the most salt tolerant of Florida trees, being roughly equal in this regard to cabbage palm and live oak. It is also somewhat tolerant of flooding and wet soils, being able to withstand, for instance, a week of deep flooding by brackish water as happened along the Gulf Coast north and south of Cedar Key during hurricane Elena in 1985. Even so, red-cedar does not normally occur in swamps or seepage areas. On the negative side, both varieties of red-cedar are quite susceptible to root rot fungi, are easily wind-thrown, often have their tops or large branches broken by high winds, and are easily killed by low intensity ground fires.

In the forest, red-cedar is often killed as a sapling by the antler rubbing of buck white-tailed deer. Of all the trees in Florida, red-cedar is the tree most preferred by bucks for this purpose. Once a cedar has been chosen by a buck, it may be rubbed repeatedly, sometimes for a period of several years if it survives that long. The end result for the tree is usually death by girdling. In some hammocks where red-cedar occurs as a scattered tree, most of the young cedars may be killed this way.

Cedar-apple rust sometimes damages red-cedar foliage and is considered a problem for apple growers farther north, where red-cedar is often exterminated near apple orchards. In Florida, where apples are not commonly grown, hawthorns and the southern crabapple serve as the alternate hosts for this fungus. The evergreen bag-worm also sometimes feeds on the foliage.

The wildlife value of red-cedar is moderately good. Its berry-like seed cones, which are only produced by the female trees (red-cedars being dioecious) are nutritious and remain on the tree from fall through winter, providing an important food source for cedar waxwings, robins, mockingbirds, blue birds, tree swallows, and yellow-rumped warblers (Martin et al. 1951). The cones are, to some extent, also eaten by other birds and some mammals. Aside from providing ideal antler rubbing posts, red-cedar is also used as a browse plant by deer. Finally, its dense foliage provides good nesting and roosting sites for various bird species.

The commercial value of red-cedar wood is high. Because of its beautiful red-brown heartwood, which contrasts sharply with the white sapwood, and because of its long-lasting aroma, it is used for making cedar chests, paneling, furniture, lamps, bowls, and all sorts of specialty items. The heartwood is highly rot resistant and was sometimes used for making fence posts. The wood, though moderately hard, is easily worked and ideal for whittling. It is evenly textured and smooth, moderately strong, but very easy to split. Historically, the most important use of red-cedar wood was for making pencils. Two pencil mills at Cedar Key were major world suppliers of pencil slats from about 1870 to 1896, when the mills were destroyed by a hurricane (Burtchaell 1949).

Red-cedar is a valuable landscape plant. It is adaptable and hardy and makes a handsome evergreen tree. It will grow in parking lots and other urban situations where many other trees do poorly. It is sometimes planted as a hedge to form a tall, dense visual screen, often on the border of a property.

Bald-Cypress—*Taxodium distichum*

Cupressaceae

Bald-cypress is, in many ways, an exceptional tree. It is potentially enormous, getting larger than any other tree in the eastern half of North America. It is also the longest-lived tree in that region, with one bald-cypress tree growing in a flood-plain swamp along North Carolina's Black River documented to be at least 2,624 years old (Stahle et al. 2019). David W. Stahle began working in this wetland area in the 1980s, documenting climate changes during the past two millennia by using tree ring analysis of cores from bald-cypress trees (Stahle et al. 1988). Bald-cypress is a conifer, yet it is also deciduous, losing its feathery branchlets in the fall. Together with the closely related pond-cypress, bald-cypress is more tolerant of flooding by freshwater than any other tree, sometimes living in locations where its root system is submerged beneath several feet of water for years at a time. Bald-cypress grows naturally on the Atlantic and Gulf coastal plains from Delaware into Texas, including all of mainland Florida, and up the Mississippi into the southern tip of Indiana (Wilhite and Toliver 1990). It has been planted well outside this area, including in southern Canada, where it withstands winter temperatures as low as -20 to -29 degrees Fahrenheit (Wilhite and Toliver 1990).

A point of confusion and contention remains over whether or not there are, in fact, two types of cypress trees in the southeastern United States: bald-cypress and pond-cypress. They are very similar in most respects. However, pond-cypress has much thicker bark than bald-cypress, doesn't get nearly as large, and grows in habitats that are usually less fertile, lower in soil pH, and much more likely to burn. They also leaf out at different times. At Gainesville, Florida, bald-cypress begins leafing out about the first of March, whereas pond-cypress begins leafing out around the first of April. The timing of pollination is different for each tree type as well.

Bald-cypress grows at the moderate rate of about 1 foot in height per year in most situations, though it can grow considerably faster under ideal conditions and very much slower under adverse conditions or when old. The trunk is typically tall and straight and the crown conical in youth, but after its first 100 years, the crown becomes flat-topped and the trunk continues to add girth.

The maximum size of bald-cypress was best represented by the "Senator tree" in Seminole County, Florida. Unfortunately, this tree was carelessly killed by a person who set the hollow inside of it on fire, starting a chimney fire that completely destroyed the tree. In 1946, the American Forestry Association reported the growth data for this tree: diameter at the base—17½ feet; height—138 feet. The diameter 4½ feet above ground was just over 11 feet, and the height just before it was killed was about 120 feet. The trunk was hollow and had an estimated total volume of 3,731 cubic feet (including the hollow), which is nearly twice the trunk volume of the other big bald-cypress trees that have been listed as state and national champions for this species. These other trees have been listed as champions because of enormous butt swell that greatly exaggerates the single trunk measurement used for ranking trees on these lists.

Bald-cypress usually grows on river edges and river floodplains, the margins of lakes that have streams flowing into and out of them, and in swamps that have streams that flow into and out of them. It sometimes forms pure stands but is more commonly associated with one or more species of tupelo (swamp tupelo, water tupelo, and Ogeechee tupelo), one or more species of ash (green ash, pumpkin ash, and pop ash), and red maple. Other trees sometimes growing with it include coastal plain willow, overcup oak, swamp laurel oak, sweetgum, sweetbay, water hickory, water locust, Florida elm, cabbage palm, and dahoon holly. Buttonbush is probably the most closely associated shrub. In the Florida Panhandle, black willow, silver maple, cottonwood, and sycamore can be added to the list of associates.

Bald-cypress prefers fertile soil with an abundant supply of moisture. It is replaced by pond-cypress on infertile sites with low pH and on sites subject to fire. It will grow well on sandy soils, but it prefers at least some clay and organic matter. It does not tolerate shade, drought, or fire very well, and it is only slightly salt tolerant, being unable to tolerate salt content in the water above 0.89 percent (Montz and Cherubini 1973). However, it can withstand most other hardships better than most other trees. Bald-cypress is very firmly rooted, even in swamps and other wet situations, and it is rarely wind-thrown even though tall, slender, forest-grown, pole-sized trees are sometimes broken off by high winds, and the tops and crowns of ancient trees often show damage by hurricanes. It is not bothered by insects or diseases to any significant extent, and it is most famous for its ability to survive extended periods of flooding. A related ability is that it can withstand soil accumulation on its root system better than any other tree with the possible exception of cabbage palm. It can also often survive being struck by lightning.

Bald-cypress is wind pollinated and reproduces almost entirely by seed, although it will grow vigorously from stump sprouts (Wilhite and Toliver 1990). Floodwaters are the most important means of seed distribution (Fowells 1965).

Bald-cypress grows conical woody structures called "knees," which rise above the soil from the lateral roots. The knees of bald-cypress are taller, more numerous, and more pointed than those of pond-cypress. They are usually from 1 to 4 feet tall and from 6 to 12 inches in diameter at the ground, but sometimes grow to over 6 feet tall and 2 feet in diameter at the base. The function of these knees has been a topic of speculation for centuries. The most popular idea is that they provide oxygen to the root system when it is underwater or in saturated soil. However, the knees also provide anchors for the lateral roots so that they do not pull through the soil easily, thus presumably providing bald-cypress a better anchored root system for overall tree support than it would otherwise have. Another function may be that the forest of knees surrounding the tree provides an efficient litter trap in floodplain habitats, enabling the tree to gather nutrient-rich organic debris about its root system during floods and to prevent flowing water from eroding soil away from the root system area. None of these ideas have been proven scientifically.

Whatever the function knees serve for the tree, they clearly help reduce flood water speed and trap litter, thus helping to slow and disperse flood peaks and helping to remove some of the litter and sediment load. The bald-cypress root system as a whole is more effective than that of most trees at reducing the erosion of river banks.

The economic value of bald-cypress timber is quite high, although the available supply is very small compared to the original supplies from the virgin forests of the nineteenth and early twentieth centuries. It is mostly used for lumber, with the reddish-brown heartwood bringing high prices. Pecky heartwood, now milled mainly from logs dredged from river bottoms or the muck of deep swamps, is especially valuable. The current, principle use of bald-cypress lumber is for paneling, though the heartwood is ideally suited for any structural purpose where rot and termite resistance is needed. (Although the heartwood is not bothered by either the common subterranean termite or the dry wood termite, there are two uncommon, native termite species that specialize in eating this wood.) Pole-sized timber is used in the construction of log cabin–style houses. Stands of young trees are sometimes harvested and ground up to make the cypress mulch used in landscaping, although pond-cypress stands supply most of this market. There are also local markets for cypress knees and for cross-section slabs cut from the fluted bases of cypress trees. These items are used decoratively for various home furnishings and landscape adornments.

The wildlife value of bald-cypress is moderate. The seeds are eaten in significant quantities by gray squirrels (personal observation), wood ducks (Terres 1980), wild turkeys (Brunswig et al. 1983), sandhill cranes (Martin et al. 1951), and, in southwestern Florida, by the Big Cypress fox squirrel (Humphrey and Jodice 1992). The extinct Carolina parakeet was reported to have fed extensively on bald-cypress seeds and to have used the hollow trunks for roosting and nesting (Sprunt 1954). Further north, for instance in the Carolinas, evening grosbeaks feed on the seeds (Brunswig et al. 1983), and mallards and gadwalls have been reported to eat them on occasion (Martin et al. 1951). Squirrels begin feeding on the seeds in October while the seeds are still in the green cones. Cranes, ducks, and wild turkeys eat them after they fall from the tree, from late October through December.

Mature bald-cypress are often used as roost, nest, or den trees due to their large size, long life, and wetland location. Huge, old, hollow cypress trees are some of the few trees that are ever large enough in Florida to provide a den site for a black bear or a roost tree for chimney swifts. Many other mammals, birds, and reptiles also use bald-cypress as den trees. I have seen wood ducks and both barred and barn owls using hollow bald-cypress trees for nesting. Large specimens are often used as nest trees by ospreys, and, although bald eagles prefer to nest in pines, I have seen two bald eagle nests in huge old bald-cypress trees, one beside Newnan's Lake, just outside of Gainesville, and one in the Lower Suwannee River National Wildlife Refuge. Stands of bald-cypress in the interior of remote swamps sometimes provide nesting sites for colonies of wood storks and roosting sites for wild turkeys.

Bald-cypress is often planted as an ornamental. It does well on upland sites as well as on wet sites if it is watered during establishment and droughts. It should be planted in the open, well away from fast growing large trees such as sycamore and laurel oak. Its feathery foliage and Christmas tree shape during the first half-century of its life are very attractive. It should not be planted near septic tanks, drain fields, or sewer lines, as its roots will grow into these vigorously. If planted too close to buildings, it will also grow roots beneath the foundations and lift them.

Pond-Cypress—*Taxodium ascendens* (*T. distichum* var. *nutans*)

Cupressaceae

Pond-cypress is a deciduous conifer closely related to bald-cypress. A smaller tree than bald-cypress, it can nonetheless reach a height of at least 115 feet with a trunk diameter of at least 6 feet. This growth report is based on measurements taken of one pond-cypress in Bradwell Bay in the Apalachicola National Forest in 1981 by me and Dale Allen of the Florida Trail Association. The ultimate size of pond-cypress is highly dependent on growing conditions, however. There are stands of dwarfed pond-cypress in Tate's Hell south of Tallahassee that are no taller than 4 feet with trunk diameters of only 6 inches. Typically, this tree reaches heights of 50 to 100 feet with trunk diameters of 1 to 3 feet at maturity. Like bald-cypress, pond-cypress can live a very long time, probably well over 1,000 years in exceptional cases.

Distinguishing pond-cypress from bald-cypress can be difficult, even for experienced botanists. Trees with well-developed bark can be distinguished by bark thickness. Pond-cypress bark, at least twice as thick as bald-cypress bark, is about half an inch thick with vertical ridges. Another difference between the two cypresses appears in the disposition of their foliage, although distinguishing the two by this feature gives many people trouble. The juvenile and shade "leaves" of pond-cypress (actually deciduous stems) are feathery, with the leaves sticking out at all angles from the deciduous stem. In bald-cypress, the leaves on the shade and juvenile foliage stick out in a horizontal plain as if they had been pressed flat in a book. The mature sun leaves of both species are often appressed along the stem, with the deciduous stems often growing in a vertical orientation on pond-cypress, which gives the small branches a distinctive appearance (hence the Latin name *ascendens*). Pond-cypress tends to have fewer knees than bald-cypress, and the knees are shorter, more rounded, and covered with thicker bark. Also, the growth form of young pond-cypress is more narrowly conical than that of bald-cypress, and the color of the foliage is somewhat more blue-gray green as opposed to the lime to emerald green of bald-cypress. Finally, in Gainesville, Florida, pond-cypress leafs out around April first, about one month later than bald-cypress. Pollen production also occurs later in pond-cypress.

Pond-cypress often grows in nearly pure stands in depression swamps scattered about the extensive pine flatwoods forests of northern Florida and southeastern Georgia. These pond-cypress swamps are sometimes called "cypress domes." They generally occupy from 10 to 30 percent of these pine flatwoods forests. Pond-cypress is the primary tree found in the Okefenokee Swamp in southeastern Georgia, in the Green Swamp on the central Florida peninsula, and in the Big Cypress Swamp on the southwestern Florida peninsula. It also occurs on the borders of ponds and lakes and along streams in the pine flatwoods regions on acidic, nutrient-poor soils. For instance, pond-cypress is the cypress tree that forms the forest border around Lake Santa Fe and Lake Alto in northeastern Alachua County. By contrast, the trees in the forest border around Newnan's Lake are bald-cypress.

Pond-cypress is not nearly as widely distributed geographically as bald-cypress. It occurs throughout Florida but is otherwise restricted primarily to coastal plain swamps from eastern North Carolina into southeastern Louisiana (Kartesz 2015).

The soils supporting pond-cypress are often very acidic and nutrient poor, and the areas where it grows, at least historically, were subject to occasional-to-frequent forest fires. Pond-cypress is highly fire tolerant, even more so than the pine trees often closely associated with it. (This tolerance distinguishes pond-cypress from bald-cypress, which is not fire tolerant.) Pond-cypress, similar to bald-cypress in its ability to withstand prolonged flooding, often occurs well out from shore into the shallow edges of lakes and ponds. The fact that bald-cypress sometimes occurs in the deep center of large pond-cypress swamps probably has more to do with the subduing of fire at these deeper, wetter centers than with any difference in flood tolerance. As for drought tolerance, neither species is at all well adapted to prolonged dry periods. During the droughts of the 1990s, large, old trees and many young trees of both species died due to lack of water. Nor is either species particularly shade tolerant, pond-cypress even less so than bald-cypress.

Although often growing in pure stands, pond-cypress is also commonly associated with slash pine and swamp tupelo. Other associates include pond pine, sweetbay, swamp red bay, dahoon, red maple, coastal plain willow, water oak, fetterbush, gallberry, Virginia-willow, waxmyrtle, Virginia chain fern, and maidencane. In transition areas from the flat, acidic, upper watersheds of streams where pond-cypress grows to the more fertile lower watersheds where bald-cypress grows, there are sometimes areas where the two species grow side by side. Corkscrew Swamp in South Florida provides a remarkable example of a swamp where a magnificent bald-cypress forest grows in the deeper water center of an extensive forest of pond-cypress.

The wildlife value of pond-cypress is similar to that of bald-cypress, although I have not seen as much use of pond-cypress by eagles and ospreys for nesting. I have observed wood ducks using partially hollow pond-cypress trees with old woodpecker holes for nesting. Other cavity nesters such as the prothonotary warbler, Carolina chickadee, and tufted titmouse no doubt do likewise. The seeds are eaten by gray squirrels (personal observation), Florida sandhill cranes (Martin et al. 1951), wood ducks (Terres 1980), wild turkeys (Brunswig et al. 1983), and, in southwest Florida, are a major food item for the Big Cypress fox squirrel (Kellam et al. 2016). Squirrels begin feeding on the seeds in October (as early as June in the Big Cypress Swamp) while the seeds are still lodged in the green cones. Cranes, ducks, and wild turkeys eat them after they fall from the tree from late October through December.

The timber value of pond-cypress is high, though not as high as that of bald-cypress. The heartwood of pond-cypress has similar properties to that of bald-cypress and is used for the same purposes. At the other end of the economic spectrum, one rather low-value use of pond-cypress is the making of cypress mulch, commonly sold as landscape mulch. This is currently the dominant economic use of this species, and vast acreages of pond-cypress have been cut for this purpose.

Pond-cypress is used to a limited extent as a landscape tree, often around water retention areas, along streams and ditches, and near other water features where its narrowly conical form and blue-gray green foliage can be quite attractive. However, it is not used as a landscape tree nearly as often as bald-cypress. This is the result of its being less well known, having a slower growth rate, and being less adaptable than bald-cypress to upland situations.

Longleaf Pine—*Pinus palustris*

Pinaceae

Longleaf pine is one of the best known and historically most important trees of northern Florida and the rest of the southeastern coastal plain of the United States. Prior to European settlement, longleaf pine dominated an open, fire-maintained forest with a grass-dominated ground cover extending from southeastern Virginia into eastern Texas on an estimated 60 to 92 million acres of land (Oswalt et al. 2012). As of 2012, only about 3.3 million acres of this forest remained intact (Oswalt et al. 2012). Longleaf pine is native to the Piedmont and coastal plain areas of the Carolinas, Georgia, Alabama, and Mississippi, plus some of Louisiana and adjacent eastern Texas; occurs throughout the Florida Panhandle; and extends into the Florida peninsula as far south as Lake Okeechobee (Boyer 1990). It also grows in the Valley and Ridge province and mountain zones of Alabama and northwestern Georgia, living at elevations of up to 1,970 feet above sea level on some of Alabama's mountain tops and hilltops (Boyer 1990). Longleaf pine has the longest needles, thickest twigs, and largest cones and seeds of any of the southern pines. The 6- to 12-inch-long needles, which grow in bundles of three, and the thick twigs are particularly noticeable when the tree is only a few feet tall. At this stage, it often has no or only one or two branches. As it gets larger, the overall shape and appearance of the tree more closely resembles some of the other pines. However, its twigs are always at least about half an inch in diameter, whereas the young twigs of all the other southern pines are much thinner.

Longleaf pine is a moderately large tree at maturity, somewhat smaller on average than slash, loblolly, or spruce pine, but larger on average than pond pine or sand pine. It commonly grows to a height of 60 to 100 feet and has a trunk diameter of 1 to 2 feet with a tall, straight trunk and an open crown of relatively few branches. The largest mature trees can be up to 125 feet tall and a little over 3 feet in trunk diameter (American Forests 1986). The crown of young trees is pyramidal in shape, but old trees develop a flat-topped crown.

Longleaf pine grows on several types of soil. It grows best and gets biggest on the well-drained sandy-loam or clay-loam soils of the red clayhills that extend into the northern Florida Panhandle from southwestern Georgia and on similar soils in western Alachua and Marion Counties and elsewhere. On these soils, and on sandy soils over clay subsoils, longleaf pine grew in association with southern red oak, mockernut hickory, post oak, bluejack oak, dogwood, sassafras, yellow haw, and chinquapin in an open forest strongly dominated by longleaf pine with a dense ground cover dominated by wire grass in mixture with a great variety of other grasses and wildflowers. I say grew, because there are almost no good natural examples of this forest left. The closest thing to it, and one of the finest stands of longleaf pine left in the world, is a 200-acre virgin stand in southwest Georgia near Thomasville on what is known as the "Wade Tract." Species of wildlife characteristic of this plant association are bobwhite quail, Bachman's sparrow, red-cockaded woodpecker, red-headed woodpecker, American kestrel, pine warbler, brown-headed nuthatch, gopher tortoise, fox squirrel, pocket gopher, pine snake, cotton-tail rabbit, cotton rat, Florida mouse, and diamondback rattlesnake. Other, generalist species that do well in this rather open longleaf pine savanna are red-tailed hawk, great horned owl, nighthawk, and Chuck-will's-widow. The majority of the original longleaf pine forest was of this type, but it has almost all been converted, long ago, to farmland, pasture, or other human uses.

Longleaf pine inhabits a second kind of soil as well: the very infertile and excessively well-drained deep sands of the sandhills. Here, its most abundant associates are turkey oak and wire grass. Indeed, these three plants often strongly dominate the sandhills, collectively forming sandhill forests, woodlands, or savannas. Animals which fare best in this habitat are fox squirrel, gray fox, oldfield mouse, Florida mouse, gopher tortoise, red-headed woodpecker, coachwhip snake, and gopher frog. Some sandhill forest remains in public and private ownership, because the soils are not conducive to agriculture. There are extensive examples of this forest type in the sandhill region of the Carolinas, in the Florida Panhandle, and on the northern half of the Florida peninsula.

The third category of soils inhabited by longleaf pine is the groundwater podzol soils and other soils of the pine flatwoods biological community. These soils occur on flat terrain and are poorly drained. They generally have a sandy topsoil containing varying quantities of organic matter, a strongly leached sandy subsoil, and then an organic hardpan. There may be either sand or clay beneath the hardpan, and in some cases there is limestone. Some pine flatwoods soils lack a hardpan. These soils are often strongly acidic, with a pH sometimes less than 4.0, and they are often very low in available phosphorus, a factor that limits plant growth on such sites. Another factor limiting plant growth on groundwater podsol soils is that during rainy periods they may be saturated with water for long periods while, conversely, during droughts, the water table often falls below the hardpan, out of reach of most plant roots. Longleaf pine's most common overstory associate in pine flatwoods forests is slash pine, although pure stands of longleaf pine were common in the virgin forest. There is usually no vegetation between the tall pine overstory and the ground cover vegetation. The ground cover is often a mixture or mosaic of saw palmetto, gallberry, and other waist-high evergreen shrubs intermixed with areas of much shorter vegetation that is a blend of wire grass, shiny blueberry, runner oak, dwarf live oak, bracken fern, and a wide assortment of other grasses, wildflowers, and dwarf shrubs. This plant association, when maintained with frequent fire, provides habitat for many species of wildlife including white-tailed deer, black bear, cotton-tail rabbit, cotton rat, eastern towhee, brown-headed nuthatch, pine warbler, wild turkey, southern black racer, dusky pigmy rattlesnake, diamondback rattlesnake, oak toad, pine woods tree frog, little grass frog, Florida chorus frog, and flatwoods salamander (now rare). Much of this type of forest has been converted to industrial pine plantation, which usually greatly alters and reduces its wildlife habitat values, especially if bedding (the practice of plowing alternating planting beds and furrows) is employed in establishing the plantations. National forests in Florida and other southern states contain extensive examples of pine flatwoods forests.

Longleaf pine depends on fire for its existence. It is better adapted to frequent ground fires of light to moderate intensity than any other Florida tree, with the possible exceptions of cabbage palm and pond-cypress. Its most notable adaptation for surviving fires is its seedling grass stage. Other pines (except for South Florida slash pine) begin height growth as soon as their seeds sprout. Longleaf pine, on the other hand, begins growing a strong root system and a dense tuft of needles while its stem and terminal bud remain at ground surface. This enables longleaf pine seedlings to survive fires because fire heat is less intense at ground surface than it is a few inches above the ground. During this grass stage, which often lasts two to four years but may last up to 20 years, the seedling stores energy in its long, thick, carrot-like taproot. Once the grass-stage seedling obtains a stem diameter at the root collar of about 1 inch, it begins height growth. Its growth is rapid then, and it grows with a thick stem covered with fairly thick bark. The thick stems of the young trees and the thick twigs of all longleaf pines are adaptations for withstanding the heat of fires.

One fascinating aspect of longleaf pine is that it provides fuel for the fires it needs for its own survival by growing long needles that are impregnated with resin before they fall, so that they are highly flammable and do not rot quickly. Their length and stiffness ensures that they will sit atop the ground cover and drape all the surrounding shrubs and small trees. These needles make such a flammable fuel that the ground cover beneath a longleaf pine forest will burn on almost any day it is not raining. By enabling the forest to burn almost any time, fire frequency is increased, and, because of this greater frequency and because fires can burn almost any time instead of just when the fuels are very dry, the average intensity of the fires is reduced. Wire grass in the ground cover also has these same fuel characteristics.

Another way longleaf pine provides fuel for the fires it needs is by accumulating bark flakes, pine needles, and pine cone fragments around the base of its trunk in what is called a "duff ring." This duff ring provides tinder for starting a fire if the tree is struck by lightning, which will surely happen to a few of the trees in any extensive longleaf pine forest every year, as the frequency of cloud-to-ground lightning strikes in the southeastern coastal plain ranges from 10 to 38 strikes per square mile per year (Noss 2018). This lightning frequency also provides for moderate fires at an appropriate time of year because most lightning comes when there is high humidity during thunder storms in late spring and summer. Fires are less intense when the humidity is high.

Beginning perhaps 12 to 15 thousand years ago, Native Americans began altering this natural fire cycle by starting fires themselves, and when European settlers began to settle in the southeastern part of North America, they also started fires. Cattlemen often started fires every spring on the open range of Florida from the time they settled there until the open range era ended in the middle of the twentieth century.

Where fires are frequent, the grasses and wildflowers in longleaf pine habitats grow luxuriantly. Where fires seldom occur, the turkey oak in the sandhills, the various oaks in the clayhills, and the tall shrubs in the flatwoods dominate so strongly that the short ground-cover vegetation is greatly reduced. If fire is kept out of an area for a decade or more, laurel oak, water oak, and other trees from outside of the longleaf pine habitats start invading and, eventually, given enough time, completely destroy these habitats. This process happens most quickly on the clayhill habitat, which is one reason why so little of this habitat is left.

Besides being fire tolerant, longleaf pine is also insect, disease, and drought resistant, is relatively wind-firm, and is moderately tolerant of flooding. One reason for longleaf pine's drought resistance is its massive taproot, plus the sinker roots that resemble little taproots on its extensive lateral root system. Another reason for its drought tolerance is its ability to drop many or even most of its needles during severe droughts. Longleaf pine is also able to live a long time. I have counted rings on stumps indicating ages in excess of 300 years on several occasions. The maximum age is around 500 years, as indicated by the growth rings on a cross section of a longleaf pine trunk on display at Eglin Air Force Base in the Florida Panhandle (Ward 2015). Longleaf pine is the least shade tolerant of our southern pines, requiring full exposure to sunlight to do well.

One of the main causes of mortality in mature longleaf pines is lightning, which strikes the earth here in the southeastern coastal plain on average about 10 to 38 times per square mile per year (Noss 2018) with the highest frequency of strikes on the Florida peninsula and the lowest in parts of eastern North Carolina. Perhaps because of its deep taproot, longleaf pine is more often struck than most other trees, and it usually dies if struck. Several surrounding pines will often also die after being attacked by bark beetles, which build up a large population in the struck tree and then attack the surrounding trees which have usually suffered root damage from the lightning strike. There are five species of bark beetles that attack southern pines, and each of them kills some pines every year. But only populations of the southern pine beetle occasionally explode into outbreaks that can kill hundreds of acres of pine forest.

Longleaf pine is a valuable wildlife tree and, in this regard, it is somewhat an exception to the general principle that the wildlife value of an individual tree is inversely proportional to its abundance in a stand. This is because many wildlife species have become adapted to the extensive, nearly pure stands of longleaf pine that once covered millions of acres in Florida and other parts of the southeastern coastal plain. Therefore, for instance, a mixed stand of longleaf pine, loblolly pine, southern red oak, laurel oak, sweetgum, and black cherry that one might find today on the clayhills near Tallahassee has less value for fox squirrels, red-cockaded woodpeckers, kestrels, quail, Bachman's sparrows, gopher tortoises, and many other animals than were the forest more open and consisted of longleaf pine intermixed with a few fire-adapted oak trees. This is mostly because the herbaceous ground cover, which is suppressed by too much shade, is a very important part of the wildlife habitat in these forests. Dense stands of longleaf pine have less wildlife value than open stands, but open stands of pine with a moderate mixture of the kinds of oak trees that belong in that habitat greatly enhance the habitat value for fox squirrels, deer, wild turkeys, red-headed woodpeckers, and many other species.

Longleaf pine is particularly necessary for the survival of the red-cockaded woodpecker (a federally listed endangered species), which makes its nest in living longleaf pines (and occasionally other pines) and requires extensive pine stands little encroached-upon by tall hardwoods. In order for the pines to be suitable for nesting, they have to be at least 60 years old and preferably much older. The southeastern American kestrel also prefers extensive open stands of longleaf pine. It nests in abandoned woodpecker holes in dead pines (or in living pines in old red-cockaded woodpecker holes that have been reworked by larger woodpeckers) and needs extensive low herbaceous ground cover for feeding on grasshoppers and lizards. Other animals almost entirely dependent on longleaf pine habitat are the Florida subspecies of the fox squirrel and Bachman's sparrow. Animals that are not quite as restricted to this habitat but do better here than anywhere else are the pocket gopher, gray fox, Florida mouse, screech owl, red-headed woodpecker, bobwhite quail, ground dove, night hawk, Chuck-will's-widow, eastern wood-pewee, eastern kingbird, brown-headed nuthatch, pine warbler, summer tanager, gopher tortoise, short-tailed snake, Florida pine snake, scarlet king snake, eastern diamondback rattlesnake, fence lizard, slender glass lizard, worm lizard, oak toad, pinewoods tree frog, and Florida gopher frog.

Finally, there are many other animals that benefit from the increased habitat diversity of having some fire-adapted longleaf pine habitat mixed in with other habitats in the overall landscape. Wild turkey is a particularly good example of this. The adult birds do well in hardwood forests, but young turkeys need a diet rich in insects, such as grasshoppers, which they find in the grassy ground cover of longleaf pine habitats. The adults also do better if they have the open longleaf pine habitats as part of their feeding range.

The wildlife value of an individual longleaf pine depends on its size, location, how many cones it produces, as well as whether or not it is a cavity tree. The seeds are a valuable food source for squirrels, which begin stripping them from the cones before they are fully ripe and as much as three months before the acorn mast crop is available to them. The seeds are very nutritious and, once they fall, many species of mammals and birds feed on them. Fox squirrels also feed on longleaf pine buds. Wild hogs will feed on seedlings, including the taproot (Wahlenberg 1946), and will also scar saplings with their teeth to induce a coating of resin, then use them as rubbing posts. The yellow-bellied sapsucker and other birds such as brown-headed nuthatches and some warblers and woodpeckers that search the bark for food seem to prefer longleaf pine to other pines. Once a longleaf pine becomes a den tree, either as a live tree or a snag, it is particularly valuable, because it is likely to last a long time and because cavities are at a premium in longleaf pine forests. Living longleaf pine cavity trees are able to survive for a very long time, even when in seemingly poor condition, and snags stand a long time if they are old enough to have developed the rot-resistant, resin-soaked, "lightered" heartwood for which this tree is famous. Animals of the longleaf pine forest which require cavities include flying squirrels, several kinds of woodpeckers (which are critically important because they are the ones that make the cavities), American kestrels, great crested flycatchers, screech owls, brown-headed nuthatches, tufted titmice, Carolina chickadees, and bluebirds. Another type of den supplied by longleaf pine is comprised by the old root channels under the lightered stumps that are left after old trees die or are cut. Many kinds of snakes and small mammals use these, and they are particularly important for indigo snakes.

The ornamental landscape value of longleaf pine is primarily a matter of individual preference. For people who like the unique appearance of longleaf pine, it is a superb tree. It does very well when left as a specimen tree during land development, provided the root zone is not overly disturbed, the landscape is not regularly irrigated with aquifer water or other water with a neutral or higher pH, and that no fast-growing large trees like laurel oak are planted close to it or allowed to grow up underneath it. Sometimes, it will even do well when the root zone is quite disturbed. If young pines are to be planted, longleaf is not as adaptable, fast growing, or healthy in most landscaped situations as are slash, pond, or spruce pine. Longleaf is also the most likely of the pines to drop resin on whatever is underneath it. On the other hand, it can be planted as a small, bare-root or containerized, one-year-old seedling and is the best of the pines for landscape planting on deep sandy soil. It requires full sun and acidic soil and is sometimes damaged either by fertilizing or frequent watering (or both) when these activities reduce the acidity of the soil. If its requirements are met, however, it is the most durable, dependable, and wind-firm of the pines. One note of caution: if longleaf pine is grown in a pot and then planted, the root ball must be totally disentangled and the taproot completely straightened, or the resulting tree may well fall over 10 or 20 or 30 years later due to a poorly developed root system.

The commercial value of longleaf pine is quite high. It produces high quality lumber and is also used to make utility poles, plywood, and pulpwood. When the virgin forest was being cut, longleaf pine was Florida's most valuable and abundant timber tree. It was also the main resource for the naval stores industry, which gathered the resin by cutting "cat faces" on the trunks of the trees, although it was not quite as valuable, per tree, as slash pine for this purpose. The resin-filled "lightered" stumps left from when the virgin forest was cut were also harvested to obtain the resin. Today, longleaf pine is still highly valued as a timber tree, but it is not nearly as abundant as it was in the virgin forest. Where pines are grown commercially, it has largely been replaced by slash and loblolly pines, which are easier to plant, easier to grow in a nursery, and which grow somewhat faster than longleaf pines. State and federal agencies promote the planting of longleaf pine, and some private landowners have been planting it in recent years, both for restoration purposes and for timber production.

The only place that longleaf pine is now beginning to hold its ground is on our public lands (state parks, state forests, and national forests). Even here, until the 1980s, longleaf pine was being replaced by other kinds of trees. Now, however, there is a strong effort to maintain the remaining longleaf pine forests by doing the necessary prescribed burning to meet the habitat's fire requirements and to even expand the area of longleaf pine forest by replanting longleaf pine on sites it used to occupy. The managers of these lands now realize, for the most part, that if we are to have longleaf pine forests and all the wildlife associated with them, it will have to be on these public lands. One unfortunate, but common, mistake some of these managers have made in this comeback effort is that they sometimes try to completely eliminate all of the longleaf pine tree's competitors, including natural associates such as turkey oak, post oak, sand post oak, southern red oak, and mockernut hickory through mechanical removal, mechanical girdling, herbicides, and firewood sales. What these managers fail to realize is that it is the longleaf pine ecosystem, not just the tree itself, that needs protecting, and that turkey oak, bluejack oak, blackjack oak, post oak, sand post oak, southern red oak, and mockernut hickory, where they occur naturally, are an important part of this ecosystem. The best way to manage longleaf pine forests is to burn them frequently with prescribed fires during the late spring and early summer and to let the fires determine the mix of species and the density of the forest. This is how the forests evolved in the first place.

The best places to see longleaf pine forests in Florida are on Eglin Air Force Base or Blackwater River State Forest in the western Florida Panhandle, in the Withlacoochee State Forest on the central Florida peninsula, and in the Apalachicola, Osceola, and Ocala National Forests. There are magnificent longleaf pine forests in Georgia on some of the older quail plantations such as Greenwood Plantation, the Wade Tract, and Ichauway Plantation, and in national forests elsewhere in the southeastern coastal plain from the Carolinas into eastern Texas.

Slash Pine—*Pinus elliottii*

Pinaceae

Slash pine is the pine tree most commonly seen in the vast pine plantations on commercial timber lands in northern Florida and southern Georgia. Because these plantations are usually less than 30 years old and, therefore, consist of rather small trees, people assume that slash pine is smaller than the other pines. Although smaller on average than loblolly pine, given time, slash pine grows to be larger on average than longleaf pine, typically achieving heights from 80 to 120 feet and trunk diameters of 2 or more feet at maturity. The largest reported slash pine is located in a small area of virgin forest on the east side of Newnan's Lake in Alachua County. In 2003, at 180 years old, it measured 130 feet tall, with a 4-foot trunk diameter 4½ feet above ground, and a crown spread of 60 feet. Slash pine can live to be over 300 years old.

The native range of slash pine includes every county in Florida, the southern half of Georgia, and small areas in the southern parts of South Carolina, Alabama, and Mississippi, along with a bit of southeastern Louisiana (Lohrey and Kossuth 1990). It has been planted extensively both within this native range and well beyond it. South Florida slash pine, *Pinus elliottii* var. *densa*, is the variety of slash pine on the southern third of the Florida peninsula. It grows as far south as Big Pine Key in the Florida Keys. A close relative of slash pine, Caribbean pine, *Pinus caribaea*, grows in the Bahamas, Cuba, and Central America.

Slash pine is the straightest growing of our pines, and it sheds its lower branches more quickly and completely and to a greater height than the other pines. At maturity, it usually has a tall, straight, clear trunk covered with bark consisting of large, flat, orange-brown plates. The bark on young trees is dark gray, furrowed, and scaly. The needles of slash pine, arranged both two and three per bundle, are usually 5 to 10 inches long, which is longer than on any other Florida pine except for longleaf pine. The cones are also larger than on any other Florida pine except longleaf, averaging 5 inches long by 3 inches wide when open. They are a shiny chocolate brown when mature, relatively fat, and drop from the tree shortly after the seeds are dispersed, so that the crown of a mature slash pine is not cluttered with old cones as many other pines are.

Slash pine originally grew most commonly in wet pine flatwoods forests, on the edges of pond-cypress swamps (cypress domes), and along creek drainages in the pine flatwoods forests. The soils here are either acidic, poorly drained groundwater podzol soils with an organic hardpan, or they are very poorly drained swamp soils that may range in pH from less than 4.0 to about 7.0. Slash pine also once grew on many other sites, including on old sand dunes along the Atlantic and Gulf coasts and in the interiors of extensive bogs and bayheads, sometimes on deep organic soils. It was particularly abundant along both coasts, often forming extensive pure stands in the pine flatwoods forests within 20 miles of either coastline. The most common tree associates of slash pine in these natural forests include longleaf pine, pond pine, pond-cypress, swamp tupelo, sweetbay, swamp red bay, and loblolly bay. The most common shrub associates are gallberry, saw palmetto, and fetterbush. Common ground cover associates include wire grass, bluestem grasses, silk-grass, shiny blueberry, runner oak, dwarf live oak, and bracken fern. The slash pine located right on the coast of the Florida peninsula is somewhat different genetically from slash pine in the interior forests, more closely resembling South Florida slash pine (*Pinus elliottii* var. *densa*). (The wood of South Florida slash pine is denser and the heartwood more densely filled with resin than the wood of any other Florida pine.) The best places to see natural slash pine forests (as opposed to pine plantations) are in the Withlacoochee State Forest, the Apalachicola National Forest, and the Osceola National Forest. Slash pine flatwoods forest occupies parts of the Ocala National Forest, and the Tosohatchee State Reserve east of Orlando has a large area of pine flatwoods forest containing a mix of North and South Florida slash pine and cabbage palm. Most of the extensive areas of pine flatwoods forest in private ownership are now young slash pine plantations.

Slash pine is a very adaptable tree. It can withstand salt spray and salt water impacts on its root system better than any other Florida pine and is similar in this particular tolerance to live oak, cabbage palm, and southern red-cedar. It can also grow in very acidic soil (at pHs of less than 4.0) or in the nearly neutral soils of coastal sand dunes. Most surprising of all, it can grow in both very dry, well-drained sandy soil such as in the dunes along the coast (although it sometimes grows poorly there or, on the poorest of these soils, fails completely) and in the middle of extensive, permanently wet bogs, bayheads, and swamps such as the Santa Fe Swamp in northeastern Alachua County and Impassible Bay in the Osceola National Forest. Among all pines, slash pine is second only to longleaf pine (and perhaps pond pine) in its ability as a mature tree to withstand fire. On the other hand, slash pine is not shade tolerant, requiring full sunlight to grow well and dying if overtopped by other trees.

Most mature, healthy slash pine trees will produce an abundant seed crop almost every year. This enables it to reproduce prolifically if a fire or other disturbance temporarily removes the litter and ground cover vegetation, exposing mineral soil. Young slash pines will outgrow all other trees on the poorly drained, acidic, nutrient-poor soils of pine flatwoods. However, on more fertile soils, loblolly pine and many kinds of hardwood trees grow faster than slash pine. Similarly, on deep sands, longleaf pine grows slightly faster and sand pine much faster than slash pine.

Slash pine is often thought of as a poor tree for wildlife because of the diminished wildlife habitat quality of extensive, pure, dense, slash pine plantations. However, its wildlife value—in situations where it does not dominate the forest so completely and where it is able to grow to maturity—is high. Individual slash pines vary widely in the amount of good seed they produce. The best trees produce an abundance of seed every year and are highly valuable as a food source for squirrels. The seeds are nutritious, and they become available at a critical time of year. Gray and fox squirrels begin stripping cones and feeding on immature seeds around the first of June, which is three months before acorns become available. The very earliest I documented such feeding was on May 14th in both 1996 and 1998. Squirrels continue feeding on the seeds until the cones open three or four months later. The ripe seeds fall in September, October, and November, supplying many other mammals and birds with a valuable food source. In addition, slash pine branches and trunks provide a feeding habitat for insects, spiders, scorpions, lizards, and birds such as pine warblers, several kinds of woodpeckers, brown-headed nuthatches, and brown creepers. Finally, the full crowns of large, mature slash pines are a favorite nesting site for bald eagles and several other raptors.

Even extensive slash pine plantations can provide habitat for many wildlife species. The habitat value varies according to the density of the pine canopy and the condition of the ground cover. Plantations on old farm fields are often devoid of habitat values after the pine canopy closes at about 10 years of age because the field weeds all die when the shade becomes too dense and there are no other plants to immediately replace them. However, plantations on former Bahia grass pastures or on native pine forest ground cover, even without wire grass, can support many native species of animals. The most common species on such sites include white-tailed deer, fox squirrel, cotton-tail rabbit, cotton rat, cotton mouse, wild turkey, downy woodpecker, red-bellied woodpecker, rufous-sided towhee, cardinal, blue jay, night hawk, Chuck-will's-widow, pine warbler, gopher tortoise, pocket gopher, black racer, coachwhip snake, diamondback rattlesnake, fence lizard, ground skink, southern toad, oak toad, and gopher frog.

Wider initial spacing of pines helps slightly to maintain habitat values, but the key to good wildlife habitat in these plantations is prescribed burning. In addition, an extensive pine plantation forest will have some stands that are new plantations. For the first few years, these open fields will be very good habitat for quail, rabbits, cotton rats, oldfield mice, ground doves, mourning doves, indigo buntings, blue grosbeaks, kingbirds, kestrels, red-tailed hawks, great horned owls, gray foxes, bobcats, coachwhip snakes, diamondback rattlesnakes, and many other species.

The commercial value of slash pine is second to no other tree in northern Florida where it is the most commonly planted timber tree, producing saw timber, poles, plywood, and pulpwood. Its clean, straight trunks contain high quality wood that is denser and stronger on average than longleaf or loblolly pine and considerably denser and stronger than pond, sand, or spruce pine. One benefit of using slash pine for pulpwood is that its wood contains abundant resin and turpentine, which are recovered as valuable by-products during the pulping process. Historically, slash pine was a major contributor to the turpentine industry. Another valuable product of slash pine plantations is pine straw, which is raked from the ground beneath the pines and sold as a landscape mulch. Slash pines are usually planted about 8 feet apart in rows about 10 feet apart. The pines are then either harvested at about age 20 to 25 or thinned at about age 12 to 15, again at about age 20, and finally harvested at about age 30 to 35. On public lands, slash pine is often grown on rotations of 60 years to produce high quality saw timber.

Slash pine is a common ornamental tree in Florida for several reasons. First, its small seedlings are abundantly available from forestry nurseries. Second, it is often already present on landscape sites. Third, it is highly adaptable and fast growing with a generally vigorous and healthy appearance. In fact, slash pine has the cleanest, brightest green and the healthiest aspect of all the Florida pines.

Drawbacks to using slash pine in the landscape include its susceptibility to fusiform rust, which kills some trees and damages others (loblolly and pond pine are also very susceptible), its dropping heavy green cones and pine resin which can damage cars parked beneath its crown, and its brittle branches that easily break during storms. Another drawback is that slash pine is overused. Overabundance of slash and loblolly pines creates an environment conducive to outbreaks of southern pine beetle, which can result in the death of thousands of pines in a short period of time. Substituting for slash pine some of the less common pines such as pond pine, shortleaf pine, longleaf pine, and spruce pine as well as other evergreens such as red-cedar or southern magnolia would result in a healthier urban forest.

One note of caution: pine trees do poorly in pots. It is much better to plant a small, bare-root or containerized pine seedling than to plant a slightly larger pine that has been growing in a pot for a year or two. Pines planted in the ground after life in a pot often fail to develop a proper root system, which frequently results in the tree's not growing well and eventually falling over. If you must plant a pot-grown pine, make sure to open up the root system, straighten the taproot, and plant the tree so that the taproot extends straight all the way down the planting hole and the lateral roots do not circle the root ball.

Loblolly Pine—*Pinus taeda*

Pinaceae

Loblolly pine is the most abundant pine tree in the southeastern United States, both in the wild and in pine plantations where it is grown for pulpwood and timber. In Florida, it comes in second to slash pine but is still abundant from Marion and Levy Counties north and west throughout much of northern Florida. It is the largest of the southern pines, commonly getting over 100 feet tall with trunk diameters from 2 to 3 feet. The tallest loblolly pine reported in Florida as of 1997 was 145 feet tall; the largest trunk diameter measured in a different tree was 4 feet 4 inches (Ward and Ing 1997). Loblolly pine foliage is not as bright a green as that of slash pine, and the crowns of mature trees often contain many old cones from previous years. The 4- to 8-inch long fairly straight needles occur in bundles of three, and the cones, which turn gray as they age, are narrow and prickly. The brown bark on mature trees takes the form of large plates.

Loblolly pine prefers fertile, moist, well-drained soil but is adaptable and can be found on most soil types, although it does not do well on the more acidic of the pine flatwoods soils or on the poorest of the deep, sandy soils. On fertile soils, such as in Gulf Hammock or San Felasco Hammock, it grows in association with hardwood trees such as laurel oak, live oak, sweetgum, and pignut hickory. Loblolly pine is a major component of upland hardwood forests, lowland hardwood forests (hydric hammocks), bottomland hardwood forests, and floodplain forests. The loblolly pine forests north of the Silver River in Marion County were described in 1915 by Roland M. Harper as a unique vegetation type in which loblolly pine and cabbage palm were the dominant trees in association with a scattering of hardwood trees such as sweetgum and water oak on flat, poorly-drained clay soils (Harper 1915). Part of this forest is now within Silver Springs State Park along Highway 40 east of Silver Springs, Florida.

The growth rate of loblolly pine on good soil is fast, and trees can be 120 feet tall and 2 feet in diameter in 40 years. Seeds are produced in abundance almost every year and are scattered widely by the wind, allowing loblolly pine to colonize old fields, clear-cuts, and forest openings. (One of the alternative names for this tree is old-field pine.) Its maximum age is about 300 years, with most trees getting not much over 100 years. One reason for loblolly pine's relatively short life span compared to longleaf, slash, or shortleaf pine is that loblolly is more susceptible to insect, disease, and mechanical problems than these other pines. It is one of the preferred hosts of the southern pine beetle whose periodic outbreaks kill thousands of pines. The numbers of loblolly pines in residential areas, natural areas, and plantations were reduced considerably in northern Florida by the southern pine beetle outbreaks that occurred around the year 2000. Besides the southern pine beetle, loblolly pine is sometimes attacked and killed by other bark beetles, such as the black turpentine beetle and various kinds of pine engraver beetles. As a tree dies, it will also be attacked by a whole set of other insects which hasten its destruction.

Another reason loblolly tends to die early compared to other pines is its high susceptibility to a fungal disease known as "fusiform rust." This disease, which has oak trees as its alternate host, forms a swelling on a young pine twig that grows larger with time. If the swelling is near the tree trunk, it will often eventually grow into the trunk, forming an open wound that continues to expand, eventually killing the tree if it completely girdles the trunk. About 10 percent of the large loblolly pines in Alachua County have wounds caused by this fungus and are thus doomed to gradually develop rot inside their trunks. (The Marion County strain of loblolly pine is somewhat more resistant to this fungus than other loblolly pines in Florida.)

The abundance of large loblolly pines in Tallahassee was greatly reduced by Hurricane Kate, which blew down many large pines in the fall of 1985 while taking out power lines, damaging buildings, and blocking roads. Loblolly pines were felled by this storm much more frequently than were longleaf or shortleaf pines, probably because of loblolly's shallower root systems and larger average height and crown size. Compared to other large trees, loblolly is intermediate in its ability to withstand high winds, being more wind-firm than laurel oak, water oak, swamp chestnut oak, and elm trees; about the same as sweetgum, maple, and ash; and less wind-firm than hickories, bluff oak, live oak, magnolia, beech, longleaf pine, and shortleaf pine. That being said, one of the main factors involved in an individual tree's succumbing to wind is the health of the base and lower part of its trunk. A tree with rot at or near the trunk base will fall much more readily than one whose trunk base is still solid.

The fire tolerance of loblolly pine is low for seedlings and saplings but increases with age until the mature trees are nearly as fire tolerant as longleaf pines. In some situations, mature loblolly pine is more fire tolerant than mature longleaf pine because it does not accumulate as deep and flammable a duff ring around its trunk base in situations where fire has been excluded for a number of years. Loblolly pine does not need fire for reproduction the way longleaf pine does because its smaller seeds can filter down into leaf litter more readily and also because the seedlings and saplings are more tolerant of shade and competition. Even so, loblolly pine generally benefits from fire because fires tend to reduce competition from hardwood trees.

The tolerance of this pine for flooding, wet soils, and salt is fairly high. It grows right beside salt marshes in some places along the Gulf Coast and grows in floodplains and the edges of bayheads, swamps, and wet prairies. It is common on Sanchez Prairie in San Felasco Hammock where, on occasion, it withstands flooding by several feet of water for up to a month during the growing season. However, flooding for more than two months will kill it.

As with the other southern pine species, being struck by lightning is a common cause of loblolly pine death. This is true whether the tree is part of a stand of pines or an emergent tree within a hardwood forest. Once a pine tree is struck, it will usually die unless the lightning exits the tree trunk before getting all the way to the ground, which sometimes happens in a hardwood forest when, for example, the trunk of a large vine provides an alternate pathway for lightning to reach the ground.

The wildlife value of loblolly pine is moderate. The crowns of mature trees are often up above the forest canopy, providing good sites for the nests of hawks and kites. In the 1980s, I observed a stand of six tall loblolly pines in Gulf Hammock used for nesting at least three years in a row by a pair of swallow-tail kites. Similarly in the 1980s, I noted a loblolly pine on the University of Florida main campus that supported a large osprey nest and was also used one year by a family of great horned owls. Though loblolly pine seeds are eaten by squirrels, mice, insects, and many bird species, they are not as eagerly sought out as the seeds of longleaf and slash pines. Loblolly pine occasionally becomes a den tree. I have seen two that harbored honey bee colonies. Finally, when a large pine dies, it provides resources for an abundance of wood boring insects, which, in turn, feed woodpeckers and other birds. The standing dead trunk (or snag) also often provides nesting sites for woodpeckers, and the old cavities they leave behind, in turn, often provide shelters for other cavity-nesting birds.

The economic value of loblolly pine is high. Commercially, it is the most important forest tree species in the southeastern United States, where it dominates about 29 million acres of natural stands and plantations (Baker and Langdon 1990). Loblolly pine is a primary source for utility poles, lumber, plywood, and pulpwood. That said, loblolly pine is not grown commercially to the extent that slash pine is here in Florida and southeast Georgia, mainly because it is not well adapted to flatwoods soils or deep sandy soils, is more susceptible to southern pine beetle attack, and is not as good for producing pine straw.

Loblolly pine is a common landscape tree but is not ideal for this purpose, primarily because of its susceptibility to bark beetles such as the southern pine beetle and because of the potential damage it can cause if toppled by high winds. It is usually present in subdivisions and other landscapes because it was there before development occurred. One problem with having a large loblolly pine near a house is that if it comes down, it is more likely to crash right through the house all the way to the ground than would an oak or other hardwood tree. These other trees have wide crowns that usually cushion the blow as the tree falls. Another landscape consideration is the likelihood of a tall pine being struck by lightning. Tall pines are much more apt to be struck by lightning than hardwood trees. This is not necessarily a bad thing. The lightning is going to strike somewhere, and having it strike a pine tree is preferable to having it strike a house.

Shortleaf Pine—*Pinus echinata*

Pinaceae

Shortleaf pine has the northernmost and largest area of distribution of the four major southern yellow pines (longleaf, loblolly, slash, and shortleaf pine), extending as far north as Pennsylvania and southeastern Missouri and occurring in Florida mostly on clay soils in the northern half of the panhandle. (A few minor exceptions include four mature trees growing on the east side of the Ichetucknee River floodplain in Columbia County.) Shortleaf pine is particularly common in and around Tallahassee. It has shorter needles and smaller cones than loblolly pine (in these features more nearly resembling spruce pine), with usually two, but occasionally three, needles per bundle. Shortleaf pine bark is fairly typical of the other yellow pines, forming large, flat plates on old trunks. There is one bark characteristic, however, that distinguishes this pine from the others: small, round, dimples 1 to 2 mm wide scattered about widely and irregularly on the surface of the flat plates. (One might have to peel away a few plates to find them.)

Another unusual characteristic of shortleaf pine is the crook at or just above the root collar of seedlings. Shortleaf pine begins growth from seed by growing sideways for about a half inch before growing upward, and it produces dormant buds in this crooked section of stem (Fowells 1965). This enables the seedling to resprout if the top is killed by fire, browsing, or mechanical injury.

Shortleaf pine is a medium- to large-sized tree, typically 60 to 100 feet tall with a trunk diameter of 2 feet at maturity. It is slower growing than the other southern pines but continues growing at a good rate for longer than the others. Its maximum recorded age is 500 years. Shortleaf pine has a larger and stronger root system than loblolly or slash pine and is both quite wind-firm and drought tolerant (Fowells 1965). In Florida, it occurs mostly on well-drained soils with significant clay content.

Shortleaf pine susceptibility to the southern pine beetle and other bark beetles, to tip moths, root rots, and pine sawfly larvae is similar to that of the other southern pines. It is much less susceptible to fusiform rust than the other southern yellow pines, but it is susceptible to little-leaf disease, which does not bother the other pines (Fowells 1965). Ranging from Virginia into Mississippi and along the Gulf Coast, little-leaf disease typically occurs in shortleaf pine stands that are 30 to 50 years old and located on poorly drained soils to which this pine is ill adapted. The disease causes a gradual decline in the trees' health and vigor that often culminates in the death of most of the stand (Lawson 1990). Shortleaf pine is primarily an upland tree and is particularly well adapted to well-drained, nutrient-poor, clay soils.

Shortleaf pine often grows in association with either loblolly pine, longleaf pine, or with upland hardwood trees, and it is often seen along the ecotone between upland longleaf pine forest and upland hardwood forest. It will invade old fields and cutover longleaf pine forest areas. Common associates in Florida are longleaf pine, loblolly pine, southern red oak, sweetgum, mockernut hickory, post oak, upland laurel oak, water oak, and live oak.

The economic value of shortleaf pine is high north of Florida, where it produces high quality lumber, utility poles, plywood logs, and pulpwood. In Florida, it is not planted in pine plantations and is not abundant enough to be significant in the timber market, but individual trees are as valuable as other southern pines of similar size and shape.

The wildlife value of shortleaf pine is similar to that of loblolly pine. It provides pine seeds for squirrels before the cones open and seeds for birds and mice after the seeds are released. It provides a tall platform for raptor nesting and, once the tree dies, a nesting structure for woodpeckers and other cavity nesters. It is second only to longleaf pine in being able to provide nesting habitat for the endangered red-cockaded woodpecker.

The landscape value of shortleaf pine is potentially high. It is the best pine for planting on the clay soils of the northern panhandle and throughout most of the rest of the southeastern United States. Shortleaf pine is very attractive as a large, reliable tree in Tallahassee and elsewhere. Unfortunately, in the fall of 1985, Hurricane Kate blew down numerous loblolly pines in Tallahassee, and many people overreacted after this storm by cutting down all the remaining pines including grand old shortleaf pines that had withstood the storm. In an odd twist of fate, shortleaf pines were sometimes planted in the 1970s and 1980s by people who thought they were planting spruce pines, which had become a popular ornamental. Nurseries ostensibly growing spruce pine sometimes mistakenly grew and sold shortleaf pine instead.

Pond Pine—*Pinus serotina*

Pinaceae

Pond pine occurs in scattered locations from the central Florida peninsula northward throughout the northern Florida peninsula, the Florida Panhandle, the southern half of Georgia, and then northward on the Atlantic coastal plain into Delaware and the southern tip of New Jersey, where it blends with its sister species, pitch pine (*Pinus rigida*). Pond pine is indistinguishable from pitch pine where the two occur together in Delaware and is similar enough to loblolly pine that these two species often hybridize where they are found growing together. Pond pine needles, which occur in bundles of three, are similar in length to those of slash and loblolly pine, but the cones are unique in appearance, being nearly as fat as they are long. They are egg shaped and remain closed for many years in the crown of the tree. The thick, rough, blocky bark of pond pine is dark gray and does not form large plates, even on the oldest and largest trunks. In addition to the normal position of foliage at the ends of twigs and branches, pond pine often has many small clusters of needles arranged along its trunk and branches.

Pond pine is a somewhat smaller and scrubbier tree than the four major southern yellow pines, often achieving not more than 1 foot in trunk diameter and only about 60 feet in height. However, on better soils, it can get large: up to 100 feet tall with a trunk diameter of up to 3 feet. It usually grows in wet pine flatwoods forests and seepage bogs that are subject to occasional fires. In fact, pond pine is one of the most fire-adapted of pine trees. A large pond pine can withstand fairly hot fires that scorch its entire crown by sprouting back from the trunk and branches, while smaller trees and seedlings can sprout back from the lower trunk or the base of the trunk at the root collar. Hot fires that scorch the crown will melt the resin that holds the cones shut, enabling these closed cones to dry, open, and release a large store of seed to repopulate the burned area. Pond pine is also very tolerant of wet soils and can grow on soils with odd pH combinations and chemical compositions that prohibit the growth of other pines.

Pond pine is not very common regionally but sometimes occurs in pure stands or mixed with slash pine, pond-cypress, and loblolly bay in wet pine flatwoods forests. These pond pine flatwoods forests often have either a dense shrub layer of large shrubs such as saw palmetto, gallberry, and fetterbush or sometimes a dense ground cover of grasses, wild flowers, and small shrubs such as shiny blueberry, hairy-laurel, dwarf live oak, and huckleberry. Examples of pond pine flatwoods can be found at the north end of the Ocala National Forest on either side of Highway 19 in Putnam County and at the intersection of County Road 241 and County Road 236 in the Mill Creek Nature Preserve in Alachua County. Pond pine also occurs in bogs, bayheads, and wet seepage areas in many other places both on the peninsula and in the Florida Panhandle in association with a wide assortment of wetland species such as loblolly bay, sweetbay, swamp red bay, swamp tupelo, gallberry, big gallberry, black titi, white titi, red maple, water oak, dahoon, highbush blackberry, elderberry, red chokeberry, and waxmyrtle. An associated tree at Mormon Branch in the Ocala National Forest is Atlantic white cedar.

The wildlife value of pond pine is probably similar to that of loblolly pine. However, because pond pine is not nearly as abundant as other pines and because it does not become as large, its value for wildlife activities such as raptor nesting and cavity nesting is not as great.

Although pond pine produces timber that is considered inferior to that produced from the four main southern yellow pines, it is sometimes logged along with them and mixed into the loads of logs sent to various mills for producing lumber, plywood, poles, and pulpwood. It is difficult for a logger to differentiate between pond pine and loblolly pine, especially as these two pines often hybridize in forests where they grow together. In any case, pond pine is of minor importance as a timber tree.

Pond pine is rarely used for landscaping. There are two pond pines performing very well at Epcot in Orlando and a few located here and there in Gainesville that are also doing well, but for the most part this has come about by accident. The late Noel Lake, former landscape designer for the University of Florida's main campus, planted a few pond pines that grew successfully there in places where he had tried other species of pines without success (personal communication). The ability of pond pine to grow where other pines will not grow provides some potential for its landscape use. It seems to be the most widely adaptable of the southern pines. When used in landscaped situations, pond pine has, on average, a shorter stature and a darker, denser crown of foliage than loblolly, slash, or longleaf pine. Because of its ability to sprout new growth from cut branches, pond pine can be shaped by pruning more easily than any of the other pines.

Sand Pine—*Pinus clausa*

Pinaceae

Sand pine is the scrub pine of Florida. It has short, somewhat twisted needles in bundles of two, and, particularly in the Ocala National Forest, it often has many closed cones stored on its twigs and branches. On some sand pines, particularly in the panhandle, the cones will open the first year and then remain on the tree as open cones. Sand pine has very thin, smooth bark on the upper trunk and branches and rather thin, scaly bark on the lower trunk. The trunk is often crooked and full of live and dead branches for most of its length.

Sand pine is exceptional in many ways. It is the only pine species almost entirely restricted to the state of Florida, occurring in two separate populations that have sometimes been recognized as separate varieties, one in the western panhandle and the other on the central Florida peninsula. Sand pine is further restricted, under natural conditions, to the oldest of Florida's habitat types: the Florida scrub. It is the smallest of the Florida pines, rarely getting over 90 feet tall or achieving more than 20 inches in trunk diameter. One of the largest sand pines on record (now long dead) grew at Wekiwa Springs State Park. It was 25 inches in diameter and 103 feet tall with a crown spread of 46 feet. The National Champs as of 2015 grow in Alachua County (80 feet tall with a 27-inch trunk diameter) and Osceola County (96 feet tall with a 26-inch trunk diameter) (American Forests 2015). Sand pine grows on the very poorest of the deep sands left over from old coastal sand dunes, and, on these deep sands, it outgrows all other pines by a wide margin for its first 20 years or more. It is very drought tolerant. On the other hand, it is also the shortest lived and least wind-firm of our pines. It rarely lives more than 70 years and lacks a strong taproot or even very substantial lateral roots. Sand pine normally begins growing a straight, vertical trunk, but, after a succession of storms over the years, during which it is tilted first one way and then another, it often becomes crooked with a slanted trunk.

The natural associates of sand pine belong to the community of plants and animals that make up what is called the "Florida scrub." The trees of this community, other than sand pine, are all small, normally not getting over 20 feet tall. Stunted sand live oak, myrtle oak, chapman oak, and crookedwood are the most common of these tree species. There are also many kinds of shrubs in this community including Florida rosemary, saw palmetto, scrub palmetto, scrub pawpaw, and garberia. Some of these shrubs and herbs are considered endemic to the scrub community (that is, completely restricted to it). There are several endemic animals as well, the best known of which is the Florida scrub jay.

The fire ecology of this community is dramatic. Scrub does not burn nearly as easily as longleaf pine forests do and, therefore, does not burn nearly as often. However, when the scrub does burn, it does so with such intensity that all vegetation is usually killed to the ground. The most rapidly developing wildfire in United States Forest Service history was the 1935 scrub fire in the Ocala National Forest. Starting from a single burning brush pile near the Ocklawaha River, it burned eastward across the 14-mile width of the forest to the shore of Lake George in three hours. The wind then shifted, blowing a 14-mile wall of flame southward. In the hour that followed, before the fire was extinguished by rain, about 30,000 acres of sand pine forest were consumed. Oaks, palmettos, and most other plants survive these fires by sprouting back from extensive root systems or underground stems. Sand pine, however, is completely killed by such fires. The way sand pine comes back is by its release of many years' supply of seeds that have been stored in the closed cones in the tree crowns. The cones dry out and open after the heat of the fire melts the resin that has kept them sealed shut.

The number of seeds released after a fire has burned through a sand pine forest that is more than 10 years old is enormous. These seeds are a mix of seeds that will sprout after the first rain and seeds that will remain dormant for varying lengths of time. Therefore, some seeds sprout shortly after the first rain following the fire, and others sprout from several months to a year or more later. This enables sand pine to regenerate even if the weather conditions are not favorable for seedling survival during the first month or two after the fire.

By far the largest forest of sand pine and the best place to see this species is the 200,000 acres of sand pine scrub in the center of the Ocala National Forest. This is also where the largest populations of many other scrub species occur including Florida scrub jay, scrub lizard, red widow spider, scrub palmetto, silk bay, garberia, Ashe's savory, and Florida bonamia.

Sand pine has been planted extensively, particularly in the panhandle, on deep sands that once supported longleaf pine and turkey oak. These off-site plantations are nearly pure stands of sand pine and, unlike natural forests of sand pine, are largely sterile, containing almost no wildlife or ground cover vegetation. They are much more destructive of wildlife habitat than are plantations of other pine species.

The wildlife value of a sand pine tree is fairly low. The seeds of sand pine are often stored for many years in closed cones. Because these cones are rarely chewed open by squirrels or other wildlife species, they do not appear to be highly valuable as a food source for wild animals. However, both gray and fox squirrels do feed on them to some extent. Once the cones open, the seeds are eaten by a wide variety of birds and mammals. Wildlife ecologist Katie Greenberg has seen female red-winged blackbirds feeding on the seeds of open sand pine cones on trees killed by fire in the Ocala National Forest (personal communication). I have observed brown-headed nuthatches probing the open cones of sand pine, probably looking for both insects and pine seeds to eat. Once on the ground, many species of birds, mice, and insects eat the pine seeds.

The rather low wildlife value of an individual sand pine tree should not be confused with the wildlife value of the sand pine scrub habitat, which is high, especially for the first decade following a fire or timber harvest. As an example of this habitat value timing, the Florida scrub jay lives in sand pine forests in scrub habitat only while the sand pines are mostly less than about 15 feet tall. If the trees form a fairly dense stand, when the trees get taller than that, the birds move to nearby sand pine stands that are younger, smaller, and/or less dense or to scrub habitat with few or no sand pines.

Sand pines are sometimes used as nest trees by crows, scrub jays, red-tailed hawks, ospreys, and even bald eagles. However, all of these birds more often nest in other kinds of trees. When sand pines die, bark beetles and various wood boring insects feed on them in large numbers, in turn providing food for woodpeckers. Hairy woodpeckers are particularly attracted to fire-killed stands.

Sand pine is harvested commercially for pulpwood and, to a limited extent, for low-grade plywood. It is sometimes grown for Christmas trees and is also harvested from the wild for Christmas trees. Sand pine is occasionally used as an ornamental tree on deep, well-drained, sandy soils where it will grow rapidly, but it has the twin drawbacks of blowing over easily and not living very long.

Spruce Pine—*Pinus glabra*

Pinaceae

Spruce pine is a short-needled pine with dense, dark green foliage. The dark gray bark on large spruce pine trunks lacks the plate structure of typical southern pines and is more like the tightly ridged and furrowed bark of a red oak, sweetgum, or northern white pine. Its twisted needles are bundled in twos and are about 3 inches long. The small cones, which average about 2 inches in length, open when ripe in the fall and then remain attached to the tree for many years.

Spruce pine grows rapidly and can become quite large, second only to loblolly pine among Florida's pines in average size attained at maturity. On its preferred habitat of moist, fertile slopes, it can outgrow most other trees, including other species of pine. In the wild, spruce pines usually attain 100 feet or more in height and from 1 to 3 feet in trunk diameter. Through personal observation and conversations with foresters, I am aware of several spruce pines with trunks in excess of 4 feet in diameter. As of 2018, the largest reported spruce pine, from Louisiana, is 122 feet tall with a trunk diameter of 4 feet 9 inches (American Forests 2018). Spruce pine seldom lives more than 100 years.

Spruce pine grows as a scattered tree or in small groups in hardwood forests on moist, moderately fertile soil from southern South Carolina across southern Georgia, Alabama, and Mississippi into southeastern Louisiana, ranging as far south on the Florida peninsula as Putnam and Alachua Counties (Kossuth and Michael 1990). Its common associates include sweetgum, pignut hickory, white oak, southern magnolia, beech, Florida maple, water oak, laurel oak, swamp chestnut oak, Shumard oak, winged elm, loblolly pine, and shortleaf pine. Some places to see spruce pines in the wild are San Felasco Hammock, the River Rise section of O'Leno State Park, Holton Creek Wildlife Management Area, Wakulla Springs State Park, Torreya State Park, and Florida Caverns State Park. The southernmost trees occur in Rock Springs State Preserve north of Orlando in Orange County, and the easternmost stand is in Jennings State Forest in Clay County. The University of Florida campus in Gainesville is a good place to see spruce pines used as landscape ornamentals.

Spruce pine is unique among southern pines in being moderately shade tolerant although it clearly grows best in full sunlight. Numerous slowly growing spruce pine seedlings often become established in the full shade of a hardwood forest in much the same way that some oak species establish advanced reproduction in these same forests. A few of these seedlings will eventually be able to grow into gaps in the canopy to become emergent trees with crowns above the adjacent hardwood trees.

Spruce pine is also unique in being the least drought tolerant of the southern pines. It is most often found in one of three situations: on slopes that derive some moisture by underground seepage from the adjacent upland at the slope top; on stream banks and floodplains; or at the edge of ponds or swamps. In spite of its need for water, spruce pine is not as tolerant of flooding as the other southern pines (with the exception of sand pine) although it will tolerate some short-duration flooding. Spruce pine is also not very fire tolerant and, as a combined result of fire and drought, seldom occurs in fire-adapted pine forests. Finally, spruce pine is the least able of the southern pines to grow in nutrient-poor or very acidic soils. Conversely, it is the best able of the southern pines to grow on calcareous soils.

Within its native range, spruce pine is free of fusiform rust cankers and is rarely bothered by tip moths. It is susceptible to annosus root rot and appears to be the most prone of the southern pines to attacks by bark beetles, being especially susceptible to the southern pine beetle. The most common cause of death seems to be lightning, to which it is especially vulnerable because it usually occurs as a single pine tree that is taller than the surrounding hardwood trees.

The wildlife value of spruce pine is low to moderate. Its cones are not chewed open by seed-seeking squirrels nearly as often as are the cones of the other southern pines. Once they fall, the seeds are no doubt eaten by birds and rodents to some extent. Large spruce pines sometimes provide nest sites for raptors, and pileated woodpeckers sometimes excavate nest cavities in the trunks of large, sound, living spruce pines. Several kinds of woodpeckers make holes in dead spruce pines although these snags do not stand as long as those of the other southern pine species.

The wood of spruce pine is lighter in weight and more brittle than the wood of the four main southern pine species and is not at all rot resistant. It is used for either rough lumber or pulpwood when harvested, but its short wood fibers make it less desirable for pulpwood than the wood of the other southern pines (Kossuth and Michael 1990). Spruce pine is used to some extent in Christmas tree plantations (Kossuth and Michael 1990), where its soft, dense, dark green foliage is advantageous.

Spruce pine is sometimes used as an ornamental. It is easy to grow and transplant, it grows rapidly, and it makes a uniquely beautiful tree with dense, dark green foliage that somewhat resembles northern white pine in appearance. When planted in the open, it retains its lower branches and develops a more crooked trunk and a lower, fuller, more compact crown than do the other native pines. Although normally a tall tree in the forest, spruce pine tends to be shorter, with a lower and wider crown than the other southern pines when planted in the open. It is less adaptable to poor soil conditions and inadequate moisture than the other pines, causing it to do poorly in such situations, but in most landscaped situations it does well.

Florida Yew—*Taxus floridana*

Taxaceae

Florida yew is a shrub or small tree with evergreen foliage of flattened, soft, needle-like leaves about three quarters of an inch long and arranged in a flat, horizontal plane. It grows with a single upright stem from seed, but older plants are usually multi-trunked and irregular in shape. The bark on larger stems is purplish-brown, sloughing off in thin plates (Godfrey 1988). Florida yew is dioecious, with separate male and female plants. Female plants sometimes produce a few, single-seeded reproductive structures (a seed surrounded by a red aril) on the underside of the twigs. The aril is about half an inch in diameter and turns a bright, translucent red when ripe in the fall.

Florida yew is a very rare plant (state- and federally listed as endangered) that is restricted to the bluffs and ravines on the east side of the Apalachicola River in the vicinity of Bristol, Florida. It grows mostly in clonal patches that originate as a result of the layering and rooting of horizontal branches that come in contact with the soil (Sullivan 1993). Florida yew is very slow growing and very shade tolerant but not drought tolerant. Its natural habitat is the moist fertile slopes of ravines underneath upland hardwood forests of American beech, southern magnolia, American holly, pignut hickory, laurel oak, and white oak, often in association with mountain laurel, sweetleaf, and tree sparkleberry (Sullivan 1993).

Though much remains to be learned about Florida yew, it is probably similar to the other yew species in being able to live a long time and having strong, hard, but flexible wood. The European yew (*Taxus baccata*), which provided the ideal wood for making the old English longbows, is thought to be able to live as long as 1,500 years (Hickman 2017).

The wildlife value of Florida yew is low due to its rarity. However, the red arils are attractive to birds; beavers sometimes eat the stems; and deer like to rub their antlers on the stems (Sullivan 1993).

The only human use of Florida yew is as an ornamental shrub, although it is rarely planted. One drawback is that all parts of the plant are poisonous. Another drawback is that Florida yew is not at all drought tolerant and requires watering to keep it alive during dry periods. Another potential commercial use for Florida yew (and a potential threat to wild populations) is to harvest the bark for the production of Taxol (paclitaxel), a compound that has cancer-fighting properties. However, Florida yew is so small and rare that it has reportedly never been harvested for this purpose. (The same cannot be said for some of the other eleven species of yew. Pacific yew was decimated in the rush to produce this medicine before a new way to produce Taxol semisynthetically from the foliage of the commonly planted European yew was discovered. Unfortunately, *Taxus contorta* in India, Nepal, Afghanistan, and Pakistan (Thomas 2011) and *Taxus chinensis* in China (Thomas et al. 2013) are currently being pushed to the brink of extinction by overharvesting for Taxol production and other uses.

Florida Torreya—*Torreya taxifolia*

Taxaceae

Florida torreya, occasionally called "gopherwood tree" and "stinking-cedar," is a small- to medium-sized evergreen tree in the yew family with flattened, needle-like, sharp-tipped leaves that give off an unpleasant odor when crushed (Godfrey 1988). It grows with a single, upright trunk and a pyramidal crown. Its native habitat is narrowly restricted to the bluffs and ravines along the east side of the Apalachicola River north of Bristol, Florida. Unfortunately, all of the original trees here have been killed to the ground by a fungal canker disease identified by Smith et al. (2011) and subsequently named *Fusarium torreyae* (Takayuk et al. 2013). The only plants left in its native range are sprouts from the base of the former trees.

There are a few individual trees that were planted outside of the native range that give some notion of what this tree was like at maturity. The largest of these is (or was) located in Warren County, North Carolina. At last report in 2008, it was 53 feet tall, had a crown spread of 40 feet, a trunk diameter of just over 3 feet (American Forests 2008) and was shaped like a large Christmas tree. It may have been planted there around the year 1830 (Godfrey 1988).

The native habitat of this tree is a diverse upland hardwood forest on moist, fertile soil that contains common trees such as tulip tree, American beech, pignut hickory, white ash, sweetgum, winged elm, Carolina basswood, Florida maple, southern magnolia, various oaks, and also includes some rare species such as Florida yew, Ashe magnolia, and pyramid magnolia.

Because Florida torreya is highly restricted in its range, very rare, and is being exterminated by *Fusarium torreyae,* it is listed as an endangered species by both the United States and Florida governments.

The wildlife value of this tree is low due to its rarity and the fact that there are no longer mature, fruit-producing female trees in the wild. Before the blight began, mature female Florida torreya trees produced seeds about 1 inch long that were edible for some wildlife species.

If a cure for the fungus blight could be found, Florida torreya would make an interesting ornamental. Several have been planted in Tallahassee and Gainesville, but most have eventually died. There is still one alive and healthy in my yard in Gainesville as of 2020. It is possible to reproduce this tree both by seed and by cuttings, but cuttings from branches produce plants that do not have the central trunk and upright growth form of a plant that originates from seed.

Coontie—*Zamia integrifolia* (*Z. pumila, Z. floridana, Z. angustifolia, Z. silicicola,* and *Z. umbrosa*)

Zamiaceae

This small shrub is Florida's only native cycad. Cycads are ancient, primitive, vascular plants that were widely distributed in the early Mesozoic Period, 200 million years ago (Laqueur and Spatz 1968). A coontie plant looks like a patch of stiff, leathery, pinnately compound, evergreen, palm-like or fern-like leaves coming out of the ground. The leaves, which are 1 to 3 feet long, are attached to a stout underground stem resembling a large tuber. The only similar plants are other species of cycads, some of which have been introduced into Florida as ornamentals, such as the sago palm.

Coontie plants are either male or female. The males produce a pollen-producing structure that looks a bit like a large, narrow, upright pine cone. The cone of the female is much shorter and fatter and, when ripe, falls apart to release seeds that have an orange, fleshy, outer covering. Pollination is facilitated by beetles and perhaps also the Florida pink scavenger moth (Hua et al. 2018).

In Florida, the native range of coontie extends from St. Johns, Suwannee, and Dixie Counties at the north end of the Florida peninsula southward into southern Florida and the Keys, with an isolated population in Jackson County in the central panhandle (Wunderlin et al. 2020). There is also an isolated population in southeastern Georgia and populations of the same or closely related species in the Bahamas, Cuba, and the Cayman Islands. The population in southwestern Alachua County, western Marion County, Levy County, Citrus County, and along the Gulf Coast has narrow leaflets and has been given the Latin name *Zamia floridana* by some botanists. The population in northern and eastern Alachua County, eastern Marion County, Putnam County, Flagler County, Volusia County, and along the Atlantic coast has broad leaflets and has been given the Latin name *Zamia umbrosa* by some botanists. There is a population of coontie in Putnam County consisting of plants that are considerably larger than average and a population consisting of plants in the sand pine scrub in the Ocala National Forest that are smaller than average.

Although coontie is more common as an ornamental than as a wild plant, it is considered a native species even though it may have been introduced to Florida long ago by Native Americans. Coontie occurs naturally, usually as an uncommon and widely scattered plant, in scrub, sandhill, xeric hammock, and coastal strand habitats. However, it is quite common on some of the islands near Cedar Key. Some common associates of coontie in scrub include sand pine, myrtle oak, sand live oak, Chapman oak, crookedwood, scrub palmetto, saw palmetto, Florida rosemary, and sandyfield beaksedge. Its common associates in sandhill include longleaf pine, turkey oak, shiny blueberry, deerberry, woolly pawpaw, and wire grass. Common associates of coontie in xeric hammock include sand live oak, saw palmetto, crookedwood, winged sumac, sparkleberry, huckleberry, and Cherokee bean. Its associates on the islands in the Gulf of Mexico are live oak, southern redcedar, cabbage palm, and saw palmetto.

Coontie is a very tough and long-lived plant. It can withstand severe fires and severe drought better than almost any other Florida plant. In fact, it benefits by these events, which reduce competition from other plants and remove insect pests. It is also very salt tolerant. Its main weakness is that it does not do well in dense shade. In habitats that used to burn regularly but are now protected from fire, it is slowly disappearing because of encroaching shade. Coontie is not especially tolerant of flooding, although it can withstand short durations of it.

The wildlife value of coontie is low. Dense clumps may sometimes provide good cover for small animals, and there are a few insects that feed on the foliage or other parts of the plant. Three of these, the long-tailed mealybug, the hemispherical scale, and the Florida red scale, can damage coontie plants (Culbert 1995). These pests are controlled in the wild by natural predators and by periodic fires. The caterpillars of the echo moth feed on the foliage of several kinds of plants including coontie. The beetles *Pharaxonotha zamiae* and *Rhopalotria slossoni* feed on the male cone and facilitate pollination (Tang 1987). At the southern end of the Florida peninsula, the brightly colored caterpillars of the beautiful Florida atala butterfly feed on the leaves of coontie and other cycads (Culbert 1995).

One reason for the low wildlife value of coontie is that all parts of the plant, including the seeds, are very toxic and carcinogenic.

The starch in the tuber-like stem of coontie was used for food by Native Americans and white settlers. The starch was removed from the woody tissue and processed by cooking and leaching with water to remove the poisons. The principle (although not the only) toxic chemical found throughout the plant and its seeds is cycasin, a cancer-inducing glycoside that is also highly toxic, causing liver and kidney damage. Symptoms of cycad poisoning include vomiting, bloody diarrhea, depression, paralysis, coma, and death (Perkins and Payne 1978).

Today, the main value of coontie is as an ornamental. It is popular as a specimen plant and for creating borders of dark, evergreen foliage. It is usually raised from seeds, which are obtained from the cones of female plants. The cones, which fall apart when ripe, contain perhaps two or three dozen orange, flesh-covered seeds, each of which contains a single, hard inner seed the size and shape of a small acorn. These seeds often exhibit delayed germination, sometimes not sprouting for one or two years. Growth is then very slow. Another way of propagating coontie is by dividing side shoots from the main trunk. These are very tough and can be left out on a shelf for several weeks prior to planting without killing them, although this is not recommended.

A note of caution: all cycads, including coontie and sago palm, are very toxic, mutagenic (capable of causing mutations in living cells), and carcinogenic (Laquer and Spatz 1968). It is recommended that rubber gloves be worn when handling any part of these plants including the seeds. Eating even a small piece of the fleshy covering of one coontie seed can cause severe illness or death in a human or a pet (Youssef 2008). Since the orange coontie seeds are produced well within reach of children at ground level, are attractive in appearance, and are extremely poisonous, it would seem that great caution should be used in deciding about using this plant as an ornamental, and yet it is frequently planted along public rights-of-way and in all sorts of other places accessible to the general public.

2

ANGIOSPERMS

Monocots

Yucca

Spanish Bayonet—*Yucca aloifolia*; Spanish Dagger—*Y. gloriosa*; and Adam's Needle (Beargrass)—*Y. filamentosa* and *Y. flaccida*

Agavaceae

The first two somewhat similar, upright-growing species of yucca—Spanish bayonet and Spanish dagger—are coastal dune specialists occurring naturally in the southeastern United States and occurring as landscape plants on a much wider scale. Of the two, Spanish bayonet is much more common, both naturally and as a landscape plant. These two yucca species have thick, bayonet-like evergreen leaves that can be anywhere from 12 to 30 inches in length and are armed with a sharp spine on the leaf tip. Compared to Spanish dagger, the leaves of Spanish bayonet are narrower, rougher, and stiffer with a stiffer and more dangerous spine on the tip. Spanish bayonet is also more rhizomatous, spreading both by underground rhizomes and by the trunks eventually falling over and taking root. Spanish dagger, by contrast, is less spreading, slower growing, and eventually grows a thicker and stronger trunk.

Spanish bayonet grows naturally along the coast from North Carolina southward around the Florida peninsula and then westward into Texas and Mexico. Spanish dagger is more restricted in range, growing along the coast from North Carolina into the northeast corner of Florida (Nassau County) (Godfrey 1988), with isolated occurrences in the central panhandle and in Pinellas County (Wunderlin et al. 2020). Both species are coastal dune specialists growing in association with other dune vegetation such as sea oats, saw palmetto, and prickly pear cactus.

Yucca filamentosa and *Yucca flaccida* share the same common names of Adam's needle or beargrass and are sometimes considered one species with the Latin name *Yucca filamentosa* (Wunderlin et al. 2020). Conversely, they are sometimes considered two distinct species (Ward 2012). (The original and widely used common name beargrass is problematic in that two Florida species in the genus *Nolina* are also called "beargrass" (Wunderlin et al. 2020). If considered two distinct species, then compared to *Yucca flaccida*, *Yucca filamentosa* has longer, narrower, stiffer leaves that are concave in cross section. Similarly, *Yucca filamentosa* has longer leaf margin fibers than *Yucca flaccida* and produces a taller inflorescence with branches that are not as wide spreading. Some detailed differences in the flowers of the two species have also been reported. Both of these forms have a trunk that rarely extends above the ground surface but that often branches underground to produce an apparent cluster of plants. This *Yucca filamentosa* species complex is native to fire-adapted pine forests on well-drained uplands in the southeastern United States including all of Florida, except the extreme south end of the Florida peninsula. It occurs in widely scattered patches as part of the ground cover of the sandhill and upland pine forests that still burn occasionally in association with wire grass, bracken fern, blazing star, summer farewell, sand blackberry, shiny blueberry, deerberry, woolly pawpaw, and many other upland pine forest species. It also grows on dunes along the coast in association with coastal dune vegetation such as sea oats, saw palmetto, and prickly pear cactus. It has underground stems and produces clumps of long-lived, sword-like leaves that have long fibers along the edges. These plants produce a 3- to 10-foot-tall woody bloom spike of showy white flowers in late spring or early summer.

The bisexual flowers of yucca are produced in showy, white clusters on spikes that grow from the tops of the plant stems (Godfrey 1988). They bloom in April and May and are pollinated primarily by one type of insect, the yucca moth. The moth lays eggs in the flower during pollination. When the eggs hatch, the larvae eat some of the developing fruit. This interaction provides an example of obligate mutualism (Brown and Cooprider 2012).

Florida's yucca species are very drought and salt tolerant, and Adam's needle is very fire tolerant, whereas the other two species are not. They all can withstand some shade, but do best in full sun. They are not picky about soil, other than needing it well drained. They are not flood tolerant, although they can withstand brief flooding, even by salt water. They are fairly pest free, although there is one insect that can greatly damage Spanish bayonet. It is the agave snout weevil, which uses its snout to puncture the base of the stem where it lays its eggs. This wound initiates rotting of the stem, which the weevil larvae eat as the plant dies (Brown and Cooprider 2012). This weevil larvae is the "worm" that is placed in bottles of tequila.

The wildlife value of yucca plants in Florida is not great, but they do provide escape cover and nesting sites for some animals, and they provide specific habitat for four insects. Besides the two insects already mentioned, the larvae of the yucca giant skipper and the Cofaqui giant skipper specialize on these and other yucca species (Glassberg et al. 2000). Pollinators visiting the flowers, in addition to the yucca moth, include the ruby-throated hummingbird (Miller and Miller 1999).

The economic value of these plants is primarily a result of their use in landscaping. They are well adapted for landscape use provided the spines are taken into account. Because of its stiff, sharp spines, Spanish bayonet should not be used in landscapes frequented by children or where regular foot traffic is expected. Because of its spines, Spanish bayonet is sometimes specifically planted for security purposes under windows or along property borders to discourage trespass by unwanted visitors. Adam's needle is suitable for planting in rock gardens. The white blooms are showy in April and May and are edible (Christman [1997] 2012).

Switch Cane—*Arundinaria tecta* and Giant Cane or River Cane—*Arundinaria gigantea*

Poaceae

Whether switch cane is one or two species is still being debated (Wunderlin et al. 2020). These are the plants that once formed large, dense stands of cane known as "cane breaks" along streams and rivers in various places throughout the eastern United States as far north as southern Ohio and Illinois and as far south as eastern Texas and Polk and Osceola Counties on the Florida peninsula (Godfrey 1988). Switch cane is a grass, with grass-like leaves and bamboo-like stems, which can reach 2 inches in diameter and 30 feet in height on fertile alluvial sites. Most *Arundinaria* populations consist of plants that are smaller than this, with canes that are one half to three quarters of an inch in diameter and 5 to 15 feet tall.

Cane breaks were much more common in the past. The decline of this ecosystem is thought to be the result of a combination of fire protection, cattle grazing, and clearing for agriculture. However, there are still some populations of cane here in Florida, often either on upland soils over limestone or along stream floodplains. It is common along Mill Creek and at the downstream end of Blues Creek, both in Alachua County.

Cane breaks in forested situations are the preferred nesting habitat for two species of songbirds in the southeastern United States: the hooded warbler (Chiver et al. 2011) and the Swainson's warbler (Anich et al. 2010). Cane breaks also provide escape cover for wildlife species such as the swamp rabbit (Miller and Miller 1999).

Arundinaria is grazed by cattle, but it is not used by people to any significant degree. The canes are too weak to make good cane fishing poles, and non-native species of bamboo are usually used for landscaping purposes or for making cane poles or other cane items. Potentially, *Arundinaria* can make a delicate and attractive addition to a garden. One drawback to switch cane's use as a landscape plant is that it spreads rather aggressively by underground rhizomes. Nonetheless, it is much easier to control than most exotic bamboo species: when my wife asked me to eliminate the switch cane I had planted in one area of our yard—a patch that had been there for forty years and had spread widely—it took only one cutting of all of the cane stalks to ground level to completely exterminate it.

Needle Palm—*Rhapidophyllum hystrix*

Arecaceae

Needle palm is a slow growing native palm, the trunk of which remains at ground level or, occasionally, grows to perhaps 1 meter tall on very old individuals. Trunks that remain at ground level sometimes grow horizontally along the ground, although this is evident only on very old plants. Some plants retain a single trunk while others divide into multiple trunks that remain together in a tight clump. The leaf blades are divided into 12 to 24 segments that are separate from each other for most of their length, the sum of which creates a blade that is roughly circular in overall shape on a long stalk. The stems of the leaves are long, thin, and unarmed. A healthy old plant may have as many as 55 leaves (Godfrey 1988). At the trunk, which is covered with brown fibers, there are long, very sharp needles that point upward. Flowers and fruit are produced on short flower stalks that come from the trunk. The plants are dioecious, with male and female flowers on separate plants. Reproduction is by seed.

Needle palm grows in moist areas on river bluffs, ravine sides, limestone outcrops, and in spring-fed bottomlands and fertile upland hardwood forests from South Carolina and southern Georgia southward into the central Florida peninsula and westward into the Florida Panhandle, southern Alabama, and southeastern Mississippi (Godfrey 1988). It is spotty in its distribution, being common only in a few widely scattered places that are particularly well suited for it. The largest population I've seen is in the bottomland forest along and to the south of Mormon Branch in the Ocala National Forest. There are also good stands in the Florida Panhandle in Torreya State Park and in various ravines elsewhere. It occurs as a scattered plant in San Felasco Hammock. Needle palm has a very wide range of plant associates including the common and uncommon plants growing in such species-rich places as Torreya State Park in the Florida Panhandle and the Mormon Branch Botanical Area in the Ocala National Forest in Marion County.

Needle palm almost always grows in the shade and is very shade tolerant.

However, it can grow in full sunlight if planted in such a situation. It seems to be fairly drought tolerant although it typically grows in moist situations in its natural habitats. It is neither fire tolerant nor tolerant of repeated defoliation by landscape maintenance crews. One reason for its lack of tolerance to leaf loss by fire or trimming is its very slow growth rate. If it has fifty leaves, fire or trimming might reduce them to ten leaves or fewer. It might then take the plant five or more years to regrow the leaves it has lost. If fire or trimming takes away all of a needle palm's leaves, it may not have the strength to recover within ten years, even if the fire did no damage to the main trunk and terminal bud. In comparison to saw palmetto, which also does not like to have its green leaves removed, needle palm responds to the removal of leaves much more slowly. Saw palmetto has a store of energy reserves that allows it to rapidly regrow missing leaves after a single removal; thus it is only severely damaged by repeated removals. Needle palm, on the other hand, seems to have no such energy stores in reserve.

One oddity of needle palm is that it is perhaps the most cold tolerant of the palm species. It is grown as an ornamental in places as far north as Long Island, New York, and Seattle, Washington, and is reported to withstand temperatures as low as zero degrees Fahrenheit without harm. It appears that, like bald-cypress, which can also withstand temperatures much colder than it experiences in its native range, needle palm is likely an ancient species that was once more widespread or endured a much colder climate within its native range than now occurs there.

Needle palm seems to live a very long time. I saw one plant on the east side of Gad's Bay in Levy County that had spread about 15 feet in three directions from its initial base with its horizontal stems lying on the ground. Since a needle palm does not even begin to spread in this way for at least one hundred years, and then does not grow in length more than perhaps an inch a decade, this individual might well have been several thousand years old. (Of course, it might be that this was an aberrant plant that grew in this way much faster than normal.) Even so, all of the plants I have observed seem to grow very slowly, continuing to grow gradually larger, and none of them seem to be approaching a declining state of old age.

The wildlife value of needle palm is not well documented, but the clusters of inch-long, hairy fruits are eaten by squirrels and perhaps other animals. The flowers are pollinated by beetles, and thickets of needle palm provide escape cover for animals such as black bear and wild hog.

Needle palm has good potential as an ornamental plant. It is already used ornamentally to some extent and is a great choice as a dwarf palm in a shady situation. It is virtually pest free and easy to plant and grow. Its only drawbacks are its slow growth, its dislike of hot, dry, sunny conditions, its inability to recover well from pruning, and the sharp needles hidden at the trunk. Its foliage is denser and more graceful than that of any other palm. Its main requirement is that it be left alone and not be pruned. Due to its exceptional cold hardiness, it is now popular far north of its native range. One drawback of such popularity is that this rare plant is sometimes stolen from the wild. As a result, it is listed by the state of Florida as a plant endangered by exploitation.

Saw Palmetto—*Serenoa repens* (including notes on Silver Saw Palmetto [*S. repens* forma *glauca*])

Arecaceae

Saw palmetto is generally an evergreen shrub with its 6-inch-diameter trunk growing horizontally along the ground. Occasionally, trunks grow upright to a height of about 5 feet, and, on rare occasions, to as much as 15 feet. The leaves, which number, on average, eight per stem (Abrahamson 1995), are fan shaped without a midrib, and the leaf stalk (petiole) on robust leaves has sharp teeth along the two outer edges. Flower stalks, about 2 feet long with small side branches, produce clusters of flowers in January and February. The fruits (drupes), which turn black when they mature in August through October, are larger than those of cabbage palm, bluestem palmetto, scrub palmetto, and needle palm, being about an inch in length and shaped like dates.

Saw palmetto forms dense shrub thickets made of numerous individual patches. The patches are often individual clones derived from the continual branching and vegetative spreading of a single original plant. Since the stem growth rate of this plant on infertile sites is very slow, generally from a quarter of an inch to 1 inch per year (Abrahamson 1995), large clones on such sites appear to be very old. Individual stems of old saw palmettos growing horizontally along the ground in a healthy manner at one end and dying and rotting away at the other end in a continuous process appear to have the potential of living forever. Based on stem length (both living and dead) and stem growth rate, Abrahamson (1995) has estimated the age of some of the longest such individuals to be in excess of 700 years. Since an unknown length of stem may have already rotted away, it is likely that some saw palmettos are much older than this. Indeed, researchers analyzing clonal patches of saw palmetto in ancient scrub communities on the central ridge of the Florida peninsula have come up with the estimate that many of these clones may be more than ten thousand years old (Takahashi et al. 2011).

Saw palmetto is abundant in the pine flatwoods forests of southeastern Georgia, the Florida Panhandle, the northern Florida peninsula, and the dry prairies and pine flatwoods forests of the southern Florida peninsula. In these habitats, it often dominates a dense shrub layer that may also contain gallberry, shiny blueberry, highbush blueberry, blue huckleberry, and several species in the genus *Lyonia*. Also, it is quite common in scrub and xeric hammock habitats, where sand live oak, myrtle oak, Chapman oak, crookedwood, and sometimes scrub palmetto are common associates. It is uncommon but usually present as widely scattered small clumps in longleaf pine sandhill habitat, where it also sometimes forms rings of dense vegetation around wet depressions. Finally, saw palmetto is often abundant on the dunes along Florida's coast.

On the dunes along the Atlantic coast, the typical inland form of saw palmetto with its bright green leaves is often less common than a coastal form of saw palmetto that is more robust and has leaves that are a chalky blue-gray green color. This coastal blue-gray form is called "silver saw palmetto" in the nursery trade. It maintains its distinctive appearance when planted together with typical saw palmetto in landscape situations, and these two forms have remained distinct while growing intermixed along Florida's Atlantic coast. Silver saw palmetto breeds true from seed, which plant nurseries have demonstrated as they produce new plants for sale. The late Noel Lake, former landscape designer for the main campus of the University of Florida, gave me two seedlings he raised from seed, which are now growing vigorously in my yard in Gainesville, Florida.

Saw palmetto has increased in size and abundance in many flatwoods and sandhill habitats due to changes in fire frequency. Consequently, in some pine flatwoods forests and dry prairies, the saw palmetto thicket is now so dense in places that it is nearly impossible to walk through. Prior to about 1930, these habitats burned frequently. Harper (1915) estimated the overall average fire frequency for high pine (sandhill and clayhills) in Florida to be about once every two years and for pine flatwoods to be once every few years. Fires since about 1950 have been much less frequent.

This is not to say that saw palmetto is not fire tolerant. It is one of the most fire tolerant of Florida's plants, and it rarely grows except in areas that experience occasional fires, except for dunes along the coast that are kept open and sunny by wind storms and salt spray. Saw palmetto is clearly fire adapted. It is abundant in scrub, a plant community that burns with very hot fires on long cycles, and it is abundant in pine flatwood forests that burn with lower intensity ground fires on a natural cycle of perhaps every two to four years. So, why is saw palmetto so uncommon in sandhill and clayhill habitats, where the soil and light conditions should be at least as favorable as in these other two habitats? The answer seems to be the difference in fire frequency. Saw palmetto can withstand occasional severe fires quite well, but it is progressively weakened by having its leaves killed back every year or two even by relatively mild fires. This has been demonstrated using prescribed burning. If a dense saw palmetto understory is burned every year for three years in a row, most of the palmetto stems will die, and those that survive will be greatly reduced in crown size and vigor. Saw palmetto seems to prefer a fire frequency of once every three to ten years and does not do as well if the fires come either less or more frequently than this.

Saw palmetto is a very tough plant. Not only can it withstand hot fires, but it is very drought and salt tolerant. It is also widely adaptable in its soil requirements, being able to grow in virtually any soil type in Florida that does not experience prolonged flooding. The three things that saw palmetto cannot withstand are prolonged flooding, prolonged deep shading, and frequent removal of the leaves (by fire, grazing, clipping, mowing).

Although saw palmetto is abundant in many areas, it is not a weedy plant. It grows slowly, has very low mortality rates (Abrahamson 1995), and does not invade new areas rapidly. Under ideal conditions, it takes a seedling at least two decades to attain the large crown size of a mature plant and to begin spreading along the ground.

The wildlife value of saw palmetto is very high. It is the most important plant in Florida for providing cover for wild animals. It is also one of the most important mast producers in Florida. The fruits (drupes), which resemble half-sized dates, are an abundant, nutritious, and widely used food resource for many species. They are heavily utilized by white-tailed deer, black bear, and wild hogs (Halls 1977). According to Halls (1977), 21 pounds of saw palmetto fruit were taken from the stomach of one black bear in Florida. Raccoons, opossums, armadillos, foxes, wild turkeys, and gopher tortoises also benefit greatly from this mast crop, which ripens beginning in August in Alachua County, with the main crop ripening in September and October and with a few dried fruits still available in November. The flowers bloom in January and February, providing food for bees and other insects at a time when nectar and pollen sources are scarce. One consequence of the winter flowering period is that hard freezes sometimes kill the flower stalks, thus preventing fruit production. A hard freeze on January 19, 1997, caused a complete crop failure that year on the northern and north central Florida peninsula.

Saw palmetto "heart of palm" is also eaten by black bears and is a significant part of their diet. Bears feed on the bud by biting the newly forming leaves and pulling them out, bringing part of the bud out with them, which does not remove enough of the bud to kill the plant or stop it from growing. Bears sometimes selectively feed on saw palmetto buds in areas that have recently burned, apparently benefiting from the rapid and tender growth that has been stimulated by the fire.

In the distant past, Native Americans in Florida ate saw palmetto fruits as a staple, although distasteful, part of their diet (Bartram [1791] 1928). More recently, the human dietary use of saw palmetto has primarily been realized through the harvest of game animals fattened, in part, on the mast crop. An economically important use of saw palmetto flowers is the overwinter foraging of bees from commercial bee hives, which are often brought to Florida from farther north.

The past forty years or so has seen the development of an economically important use of the fruit. Saw palmetto fruit, and extracts from it, have become popular herbal medicines in Europe, Asia, and the United States, being especially valued for the treatment of prostatic hypertrophy. It is now among the top ten herbal remedies sold in the United States, thus creating an active market for the fruits, which are gathered from the wild (Florida DEP 1995).

The landscape value of saw palmetto is potentially high. Evergreen, attractive, trouble free, very tough, and very adaptable, palmetto is one of the best native shrubs for general landscape use. It is greener and more attractive in landscaped situations than in the wild, and, once established, it needs no care at all although finicky landscapers and gardeners are prone to pruning away the old, dead leaves as they do with all palm species. It never needs watering, spraying, or fertilizer and can live virtually forever. It does not spread rapidly or get out of control, and its roots do not thicken with age, making it trouble free around sidewalks and buildings. The two main reasons it is not more widely used ornamentally, besides the prickly spines along the leaf stem, are that people were prejudiced against it in the past because it was a common native plant known for sheltering wildlife in wild places and because it is slow to grow from seed and difficult to transplant from the wild. However, when grown in a container from seed and sold in a nursery, it is very easy to plant and grow.

The blue-gray coastal form, silver saw palmetto, has recently become a popular landscape shrub. Compared to regular saw palmetto, it is more vigorous and faster growing in urban landscape conditions, and it provides a striking contrast in color to regular saw palmetto and most other vegetation.

Cabbage Palm (Sabal Palm)—*Sabal palmetto*

Arecaceae

The official Florida "State Tree" is normally 30 to 80 feet tall at maturity with a trunk about 1 foot in diameter, give or take 5 or 6 inches. Once the trunk is formed, it does not grow any thicker, although the trunk of large, old individuals sometimes has an enlarged base consisting of a solid mass of root growth. The crown consists of a single stem (a few have been found with trunks that have branched into two or more stems) topped with a bud that produces very large, partially folded, fan-shaped leaves that have a central rib running from the leaf stalk (petiole) all the way to the tip of the leaf blade. The total length of the leaf blade and its stalk varies from about 6 to 12 feet with the blade being about 3 to 5 feet long and 2 to 4 feet wide. The trunk of a young cabbage palm is clothed with the remains of leaf bases (often called "boots"), but as the tree ages, these eventually rot and fall away. The resulting clean trunk is gray and somewhat fibrous in appearance. As it ages further, the vertical fibrous texture becomes more apparent; these fibers are sharp and stiff, so that a person trying to climb an old cabbage palm can end up with many splinters deeply imbedded in their skin.

Cabbage palm does best on moist soils with a high calcium content, but it is very adaptable. It is abundant in hammocks along the Atlantic and, especially, the Gulf Coast and along the St. Johns River, where it sometimes grows in pure stands but is more commonly intermixed with live oak and southern red-cedar. Along the Atlantic and Gulf coasts, cabbage palm grows naturally as far north as the southern tip of North Carolina and as far west as the coast of Louisiana (Kartesz 2015). Further inland, it is rare in the wild north of Gainesville, Florida, but common south of Gainesville on the Florida peninsula in floodplain swamps, around the edges of prairies, and in mesic and hydric hammocks. Its most common hammock associates include live oak, laurel oak, sweetgum, and hornbeam. Its most common swamp associates include red maple, green ash, pumpkin ash, pop ash, swamp tupelo, and bald-cypress. Near the Ocklawaha and Silver Rivers in Marion County it grows on poorly drained uplands in association with loblolly pine, live oak, water oak, and sweetgum. On the central and southern peninsula of Florida, cabbage palm is also a common tree in many pine flatwoods forests where its most common associate is South Florida slash pine. (Unfortunately, the nuisance exotics melaleuca and Brazilian pepper have now also become abundant in many of these forests in South Florida.) Cabbage palm reproduces and grows very well in urban and suburban situations. Locations supporting large populations of cabbage palm include all of the cities and towns of Florida, St. Marks National Wildlife Refuge in Wakulla County, Waccasassa Bay State Preserve in Levy County, Chassahowitzka National Wildlife Refuge in Citrus and Hernando Counties, Tosohatchee State Reserve in Orange County, and the Big Cypress Swamp in Collier County.

The largest cabbage palm on record occurred at Highlands Hammock State Park in Highlands County. It was 90 feet tall with a 45-inch trunk circumference (14-inch diameter) and a 14-foot-wide crown. The largest one in Florida in 1997 was 60 feet tall with a trunk circumference of 69 inches (22-inch diameter) and a crown spread of 14 feet (Ward and Ing 1997). The crowns of these old ones, as with most very old palms, were not very big. A young cabbage palm in my yard in Gainesville has a 25-foot crown spread, which is probably pretty close to the maximum for this species.

Cabbage palms reproduce by seed, starting life with a single, grass-like blade a few inches tall. After several years, if growing well, the leaves begin to assume a fan shape. As more years pass, the leaves become more numerous and larger. During these first few years, the trunk actually grows downward with the bud making a sharp bend at the bottom to send its leaves to the surface. This allows the trunk to become firmly anchored in the ground and achieve a large trunk diameter before growing above the ground. The crown of leaves reaches nearly adult size before the palm finally begins to grow upward. This stemless stage of cabbage palm lasts at least ten years and often more than twenty years. Once height growth starts, the tree may grow about 1 foot in height per year, but growth is much slower in adverse situations. The average growth rate for young individuals is about 6 inches per year (Moyroud 1996). As the tree gets older, growth slows even further.

Cabbage palm can apparently get very old. It is difficult to tell how old because palms produce no growth rings for us to count. However, some of the old palms in Gulf coastal hammocks that were logged around 1900 bear the marks from the logging cables that pulled out the timber, indicating that these trees were there with fully developed trunks at that time. It would appear from this, and from the very slow growth of the old trees, that cabbage palm can reach at least 200 years old.

The cabbage palm is Florida's toughest, most durable tree, able to survive more severe environmental stresses of more varied kinds than any other tree. It is our most fire-tolerant tree, being able to withstand very intense fires. It also withstands the effects of cattle grazing better than any other tree. Severe fires in mixed live oak, red-cedar, and cabbage palm hammocks often result in pure cabbage palm forests, and cattle grazing has greatly increased the abundance of cabbage palm in many places. Until very tall, cabbage palm is also more wind-firm than most other trees. It is more salt tolerant, both of salt spray and flooding by brackish water, than any of Florida's trees, except the mangroves. (Live oak and red-cedar are nearly as salt tolerant). Cabbage palm also tolerates freshwater flooding and continuously wet soil conditions better than most trees, though not as well as bald-cypress and pond-cypress. Cabbage palm is fairly resistant to most insect and disease problems with one major exception. Unfortunately, a new disease called "Texas Phoenix palm decline," which can kill cabbage palm and many other palms, has been introduced into the Tampa Bay area and is now spreading widely (Harrison and Elliott 2007). As of 2019, it appeared that this disease was attacking only an occasional cabbage palm. However, it is continuing to kill individual cabbage palms in Gainesville, Florida. As of June 1, 2020, in a row of 48 cabbage palms planted in the median of Depot Avenue several years ago, 15 have died or are dying of what appears to be this disease. Recently planted trees seem more susceptible. None of the 26 cabbage palms in my yard, which have been there for decades, have died, whereas all three of the recently planted ones in a neighbor's yard have died.

Cabbage palm is only moderately shade tolerant. It can survive for a long time in the shade of other trees but will eventually die in the dense, full shade of a closed hardwood canopy. Also, it does not reproduce well in such situations. However, it reproduces prolifically in the hammocks where it grows in the wake of periodic disturbance such as logging, cattle grazing, or fire. Near the coast, the reduction of competition caused by salt stress on its hardwood associates greatly benefits cabbage palm.

Now that sea level is rising due to the increased greenhouse effect caused by rising carbon dioxide levels in the atmosphere, huge numbers of cabbage palms are being killed along the Gulf Coast as the depth of flooding and the salt concentrations become too great for them to withstand. Because cabbage palm is the most abundant and noticeable tree along much of this coast, because it is usually the last tree species standing, and because its dead trunks last as visible reminders for a long time, this large-scale palm massacre is highly conspicuous.

Cabbage palm provides the primary habitat for three species of epiphytic ferns. The large epiphytic fern known as the "golden polypody" ("goldfoot fern" or "rabbit's foot fern") starts growing in the detritus-filled boots on the trunk, continues to grow upwards in and along the boots, and finally becomes established in and just below the crowns of mature cabbage palms. When it has been growing there for many years, a large, dark mass of organic matter and fern roots and rhizomes occurs around the trunk just below the crown. This organic mass is much more noticeable than the fern itself and serves as an easy way of finding the fern. The hand fern is a much rarer fern that grows almost exclusively in detritus-filled boots of cabbage palm (Godfrey and Ward 1979). Finally, shoestring fern occurs on the open trunk or on the boots, often only a few feet above the ground. All of these ferns are killed by hard freezes and are therefore more common on the central and southern Florida peninsula, reaching only about as far north inland as Ocala.

The wildlife value of cabbage palm is moderately high. Like most other trees, the value of a single one is proportional to the scarcity of the species in the forest. In pure stands of cabbage palm, one palm has less value than it does if the palms are scattered about in a mixed stand of many tree species.

The fruits (drupes) of cabbage palm are an important food source for a wide assortment of birds and mammals, most of which benefit the palm in turn by widely distributing the seeds. The fruits begin ripening in September and remain on the tree until late March if not eaten before then. In addition, more fruits are produced in the spring, so that fruit is available until at least the end of May. (There is some variation in the timing of fruiting from year to year, place to place, and between different trees.) The fruit seems to be eaten most commonly in December, January, and February, when other foods are scarce. Raccoons and robins (Martin et al. 1951) and fish crows and common grackles (Sprunt 1954) eat great quantities of them. Halls (1977) says the fruits are a major winter food for wild turkeys in Florida and are also eaten by songbirds, bobwhite quail, raccoons, squirrels, white-tailed deer, and black bear. Along the coast, flocks of ring-billed gulls feed on the fruits while in flight (Broun 1941). Other species that eat the fruit include mockingbird, Eurasian collared dove, rusty blackbird, cedar waxwing, yellow-rumped warbler, pileated woodpecker, and red-bellied woodpecker (personal observation). The fruits are edible by people, tasting somewhat like dates, but they are so dry and have so little flesh as to be nearly useless for we humans.

The bud (heart of palm) is eaten by people. The only wild mammal that commonly eats the heart of the cabbage palm is the black bear. It is an important part of their diet in places where it is available, such as the Ocala National Forest and the Big Cypress Swamp. There is a large native beetle, the palmetto weevil, the larvae of which sometimes infest the bud of cabbage palms weakened by transplanting or excessive pruning of green leaves (Giblin-Davis and Howard 1989; Weissling and Giblin-Davis [1997] 2013). When the bud is taken out by a human or invaded by palmetto weevil larvae, the palm dies. Death may also occur when a bear tears out the heart of a palm that is near the ground. However, black bears will climb tall cabbage palms, sit on top of the crown, and pull the heart out to eat it. When this happens, the palm usually survives. I have seen the evidence of this on a number of occasions. Apparently, enough of the bud remains in the tree to produce new growth.

The fibers on the palm leaves are a good source of nest-building material for some songbirds, and Halls (1977) points out that gray squirrels often nest in the crowns of cabbage palms and that the caracara in south central Florida almost always uses a cabbage palm for its nest site.

Kilham (1989) observed that American crows at the Hendrie Ranch on the south central Florida peninsula frequently used cabbage palms as foraging sites. One or two crows would shake the large leaves by moving about or jumping on them while the other crows stood on the ground below to catch whatever small creatures were dislodged and fell to the ground. Examples of food items captured in this way are large insects, spiders, tree frogs, snakes, and lizards. The crows would also flip over dead palm leaves lying on the ground to capture similar prey hiding beneath.

Cabbage palm leaves provide an ideal site for paper wasps of various kinds to construct their nests. The large, tough, long-lasting leaf blades provide a roof over the paper nest. The only animal that I have seen raiding these nests is the blue jay, although I would not be surprised if crows did this as well. Twice, I have seen blue jays attack the nest from the top side of the leaf, in relative safety from the wasps, stabbing their beaks through the leaf until the nest drops to the ground. They then swoop down and fly off with the nest to feed on the wasp larvae.

Another kind of wasp, the southern yellowjacket, which normally nests in holes in the ground or in hollow logs, will sometimes construct an enormous, liberty-bell-shaped nest about 4 feet in both height and diameter on a 30-to-50-foot-tall cabbage palm trunk just below the crown of leaves. I have seen this twice on the property that became Silver Springs State Park just south of State Road 40 in Marion County and have heard of it happening in other parts of the state. These yellowjacket nests are so large that any major disturbance, such as knocking the palm down during a logging operation, could result in someone being stung to death.

Human uses of cabbage palm have changed over the course of history. Native Americans used the leaves for making thatched roofs and the fruits and buds for food. The early white settlers sometimes used the trunks to construct forts. In the nineteenth century, fibers extracted from cabbage palm trunks at a factory at Cedar Key were used to make brushes (Burtchaell 1949). The trunks themselves were often used for pilings, especially in salt water, where they resist marine borers better than other timbers.

Although heart of palm is still used extensively for food, the main value of cabbage palm today is its landscape value. Since this tree can be transplanted when 10 to 40 feet or more in height, many thousands of wild ones are transplanted into landscaped situations each year. All of the leaves and roots are cut away, leaving the tree looking much like a log when it is planted. The trunk serves as a water and energy reservoir, allowing the tree to survive and begin growing new roots and leaves in the first several months following planting. Cabbage palm is very abundant in the coastal hammocks from which most of the landscape-bound palms come, and as it reproduces prolifically in these areas, this practice probably does little harm to the overall size of the wild population in most cases. In forests where the palm is uncommon, removal is probably detrimental to wildlife habitat values.

If the trees are watered frequently for the first year after transplanting, their survival rate is high. Once established, they need almost no care. Landscape maintenance people often trim off the dead and dying leaves each year, along with a varying number of the green leaves, and this is often overdone. Removing the healthy green leaves is detrimental to the tree, reducing the amount of photosynthesis it is able to do and thus the amount of food reserves it is able to produce. It also interrupts the tree's conservation of phosphorus and potassium, which the tree removes from old leaves as they die and moves to growing leaves. Excessive annual pruning of green leaves can result in the decline and eventual death of the palm, as has been demonstrated on the University of Florida campus and in many other overly manicured landscapes. Trimming should remove only the dead and clearly declining leaves; all the healthy green leaves should be retained to support the life functions of the tree (Broschat [1993] 2013). In situations where the leaves do not have to be trimmed, it is better for the tree to do no trimming at all.

Cabbage palm is an excellent landscape tree. It is very hardy and can grow right next to buildings, sidewalks, and streets without doing any damage to the structures with its roots. This is because, like its trunk, its roots do not increase in thickness with age. Cabbage palm excels at withstanding the stresses of urban life, such as soil disturbance, increases in pH or salt content of the soil, construction or paving next to the trunk, heat radiation from paved areas, and air pollution. It is much tougher and lives much longer than any of the exotic palms that are used for landscaping. The main downside to planting cabbage palms is the possibility of the trees dying from the Texas Phoenix palm decline disease.

Bluestem Palmetto, Dwarf Palmetto—*Sabal minor*

Arecaceae

Bluestem palmetto is a dwarf palm that grows in scattered locations beneath hardwood forests throughout most of the coastal plain of the southeastern United States including the panhandle of Florida and most of the northern and central Florida peninsula (Kartesz 2015). In Florida, it rarely if ever grows a trunk above ground. Its leaves are a dark green color, the leaf stems are smooth without spines, and the slender stalks that bear the flowers and fruit are often tall enough to reach above the top of the foliage. The fruits occur in clusters, are small and round, and turn from green to black as they ripen. The leaf blades do not have a central rib the way those of scrub palmetto and cabbage palm do and are thus much flatter in shape.

This little palm is found mostly in rich, fertile, southern hardwood forests on ecotones between well-drained soil and soil that is moist to wet. Bluestem palmettos can be abundant in seepage spots that remain moist to wet but have good drainage downslope such that they do not flood. In places where this palm does especially well, such as seepage areas along the edge of the Silver River floodplain in Marion County and in San Felasco Hammock in Alachua County, the height to the top of the foliage can be 6 to 8 feet. In drier places, the plants are usually half that tall. Unlike saw palmetto, this palmetto does not spread by vegetative growth, but it does reproduce by seed. The growth rate is very slow, and individual plants can live for hundreds of years.

Although normally growing in moist environments, this palm has good drought resistance once established. It is quite shade tolerant but can also grow in full sun and seems to have few insect or disease problems. Whether the newly introduced Texas Phoenix palm decline disease will impact it remains to be seen. Because its trunk is below ground, this palm can survive occasional fires. However, it does not do well if its leaves are removed often and is rarely found where fires are frequent or where cattle grazing is intense.

The wildlife value of this palm is good. The fruits are eaten by raccoon, white-tailed deer, gray squirrel, yellow-rumped warbler, and various woodpeckers (Miller and Miller 1999). Other birds such as cardinals, mockingbirds, gray catbirds, and wild turkeys also eat the fruit (personal observation). The flowers attract small, native bees and other pollinators. Thickets of this little palmetto provide cover for various species, including white-tailed deer and wild hogs.

The landscape value of bluestem palmetto is good. The foliage is a dark, shiny, blue-green, and it is hardy and care free once established, making a great shrub or tall ground cover. It is almost impossible to transplant but is easy to grow from seed. It is more cold tolerant than other palms. One advantage it has over the blue-gray forms of saw palmetto used in landscaping is that it does not have spines along the leaf stem. Also, it is smaller and slower growing, thus requiring less maintenance once established. One important note, however, is that it does not like to have its leaves trimmed. Most landscape maintenance personnel like to remove palm leaves, including healthy green ones. This is completely unnecessary for this palm and causes considerable damage. (Removing healthy green leaves is also at least somewhat harmful for all other kinds of palms.)

Scrub Palmetto—*Sabal etonia*

Arecaceae

Scrub palmetto is a small palm with an underground stem and with leaves similar in shape to those of the cabbage palm but much smaller. The leaf shape, with its central rib, is even more curled than is typical for the leaf of a cabbage palm. Scrub palmetto inhabits deep, excessively well-drained sands on the Florida peninsula from Columbia County south to just north of Lake Okeechobee on the central ridge of the peninsula and from Volusia County south into Dade County along the Atlantic coast (Godfrey 1988). (Most of the coastal plants have long since been removed by coastal development.) The largest single population of this species is in the 200,000-acre central scrub area of the Ocala National Forest. Although on shorter stalks, the flowers and fruits are similar to those of cabbage palm.

Scrub palmetto is endemic to the Florida scrub biological community, a relict community left over from a time, long ago, when the highest parts of the central ridge of the Florida peninsula were reduced to a group of islands or a sliver of peninsula during much higher sea levels. Scrub palmetto is fire adapted, as all Florida scrub plants are, surviving occasional intense fires by having its trunk and bud below ground. It grows in association with saw palmetto, crookedwood, garberia, myrtle oak, sand live oak, sand pine, and Florida rosemary and is attended by scrub wildlife including Florida scrub jay, scrub lizard, sand skink, red widow spider and a wide assortment of other Florida scrub plants and animals. It has a single, unbranched trunk underground and therefore does not form thickets the way saw palmetto does.

Scrub palmetto grows horizontally, with its trunk a few inches below the surface, growing forward at the growing end, with the trunk at the other end eventually dying and then gradually rotting away. In this manner, scrub palmetto can continually grow, slowly traveling across its habitat. Age estimates for saw palmetto, which grows in a similar manner to and in the same habitat as scrub palmetto, range from 700 years for individual plants to over 10,000 years for clones derived from individual plants (Takahashi et al. 2011). Presumably, scrub palmetto could reach similar ages, for, although it does not branch the way saw palmetto does to form large patches, it can still grow in one direction for a very long time.

Scrub palmetto is very fire tolerant, being able to withstand the hottest wildfires. Like saw palmetto, scrub palmetto benefits from occasional fires that prune back competing vegetation but probably does not do well if the fires are more frequent than every three or four years. It is also very drought tolerant and seems to have few insect or disease problems although the Texas Phoenix palm decline disease may eventually cause trouble. Scrub palmetto does best in full sun although it also grows well in partial shade.

The wildlife value of scrub palmetto is high. It produces fruits in October and November that are eaten by a wide variety of wildlife species. It also provides cover for some species. The structure of the leaves at the top of scrub palmetto (and, on the Lake Wales Ridge, also saw palmetto) provides the main habitat for the red widow spider that occurs in the Florida scrub community on the Florida peninsula from Marion County southward into Highlands County.

The landscape value of scrub palmetto is rather low, as it does not form dense patches the way saw palmetto does and as it is a somewhat smaller and an even slower-growing plant. It is also not as broadly adaptable as saw palmetto or as shade tolerant as bluestem palmetto. However, it still has some potential, especially on well-drained sandy soil. One advantage scrub palmetto has over saw palmetto is that it lacks spines on the leaf stalks (petioles). Another is that, according to botanist and ecologist Carol Lippincott, the blossoms are more sweetly aromatic than those of the other native palms (personal communication).

Greenbriar—*Smilax* spp.

Smilacaceae

There are as many as 12 species in the genus *Smilax* occurring in northern Florida. They are all vines with tough, hard stems that are green when young, sometimes turning black with age. Greenbriars are monocots, and, therefore, the individual stems and roots do not grow thicker once they are fully formed. The maximum thickness of the largest stems of the largest species is about half an inch. All but one species are tall-growing vines with thorns (technically prickles) and tendrils. All form underground tubers with considerable variation in tuber size, shape, and thickness from one species to the next. The leathery leaves are simple and, usually, either evergreen or semievergreen. The flowers and berries occur in clusters in the leaf axles, with male and female flowers on different plants (dioecious). In all but two species, the berries are black at maturity. Below are descriptions of some of the more common species.

Earleaf Greenbriar—*Smilax auriculata*

Earleaf greenbriar is a medium-sized thorny vine with glabrous, evergreen leaves that is commonly found on poor, sandy soil in sunny situations throughout Florida, parts of southeastern Georgia, and northward along the coast into the Carolinas (Kartesz 2015). It often forms dense tangles on fences, bushes, small trees, the trunks of pine trees, and sometimes on itself in the open. It can be weedy along fencerows and in pine plantations.

Saw Greenbriar—*Smilax bona-nox*

Saw greenbriar is a tall-growing thorny vine with slightly scruffy semievergreen leaves. It is widely adapted and grows throughout Florida and the rest of the southeastern United States (Kartesz 2015).

Sawbriar—*Smilax glauca*

Sawbriar is a tall-growing and very thorny vine with slender but very tough stems that turn from green to black with age. Its semievergreen leaves start out chalky white underneath, becoming blotchy with age. It is quite adaptable, being most common in moist to wet pinelands, seepage areas, bayheads, and sunny wetland edges, but it is sometimes also abundant in pine plantations on well-drained sandy soil. It spreads aggressively by underground runners and often forms large clonal patches. It can be very weedy and hard to control in residential landscapes and pine plantations. It is common in the southeastern United States and throughout the northern parts of Florida, ranging as far south on the peninsula as Sarasota (Wunderlin et al. 2020).

Laurel Greenbriar—*Smilax laurifolia*

Laurel greenbriar becomes a large, very thorny, tall-growing vine with thick, leathery, evergreen leaves. It inhabits wetlands of all sorts, especially bayheads and bogs, but also cypress and hardwood swamps throughout Florida and the rest of the Atlantic and Gulf coastal states from Virginia to eastern Texas (Kartesz 2015).

Sarsaparilla Vine—*Smilax pumila*

Sarsaparilla vine is the least typical of the greenbriars. It does not have thorns, and the vines are rarely more than 2 feet long. Its berries are orange-red at maturity. It grows as a ground-cover plant in mesic to xeric upland hardwood forests from South Carolina to eastern Texas, getting down into Florida as far south as Highlands County (Kartesz 2015). It is infrequently used as an ornamental.

Roundleaf Greenbriar—*Smilax rotundifolia*

Roundleaf greenbriar is a tall-growing thorny vine with large, heart-shaped semi-evergreen leaves. It occurs throughout most of the eastern United States and is more common north of Florida, occurring sparingly in the central panhandle of Florida with only a few isolated records for Alachua, Bradford, Clay, and Volusia Counties (Wunderlin et al. 2020). It is widely adapted, but prefers well-drained to moist upland clay soils, often adjacent to wetlands or water.

Lanceleaf Greenbriar—*Smilax smallii*

Lanceleaf greenbriar becomes a large, tall-growing vine with shiny evergreen leaves and with stout thorns on the lower part of its stems. It often grows as high as 50 to 80 feet into the crowns of hardwood trees. The female plants produce abundant crops of black berries that ripen over an extended period from March to May. It grows on fertile soil in mesic upland hardwood forests, in slope forests, and occasionally on high, fertile river floodplains. It occurs from North Carolina to eastern Texas as well as in the Florida Panhandle and some of the counties on the northern and central Florida peninsula (Kartesz 2015). It is common in Alachua County.

Bristly Greenbriar—*Smilax tamnoides* (*S. hispida*)

Bristly greenbriar is a medium-sized thorny vine with semievergreen leaves. The thorns are a shiny, dark brown to black and needle-like (Godfrey 1988). This greenbriar is quite adaptable, occurring mostly in upland situations throughout the eastern United States and most of Florida (Kartesz 2015).

Coral Greenbriar—*Smilax walteri*

Coral greenbriar is a small- to medium-sized, slender, sparingly thorny vine with bright green, deciduous leaves and berries on the female plants that turn bright red at maturity, usually after most of the leaves have fallen. It grows in the wettest situations of any of the greenbriars, being largely restricted to swamps, marshes, and wetland edges. It occurs in the Atlantic and Gulf coast states from southern New Jersey to eastern Texas, ranging as far south on the Florida peninsula as Highlands County (Kartesz 2015).

The wildlife value of greenbriars is fairly high. This is mostly because of the berries, which are available for birds and other animals to eat throughout most of the year. Because of the abundance and wide distribution of the various greenbriar species, together with the nearly continuous availability of the fruits, a large number of wildlife species make some use of them. Animals that use them for more than 2 percent of their diet, as estimated by Martin et al. (1951), are wild turkey, catbird, mockingbird, fish crow, hermit thrush, black bear, raccoon, and wood rat. I have seen cedar waxwings eating the fruit in Gainesville in March and April, and have seen greenbriar seedlings sprouting by the hundreds under a shade tree in one of my neighbor's yards where the cedar waxwings roost before and after feeding in a red mulberry tree nearby. It has been estimated that white-tailed deer obtain from 5 to 10 percent of their diet by browsing on the foliage and young stems of greenbriar (Martin et al. 1951). Dense tangles of greenbriars provide escape and nesting cover for several species and are a preferred nesting site for catbirds (Martin et al. 1951).

Human use of greenbriar by the aborigines and by wild food enthusiasts include using both the growing tips of the vines and the large underground tubers for food. The tubers are very woody and therefore not easy to make edible, but the growing tips can be eaten raw, tasting rather like raw green beans (sawbriar is the least tasty), or they can be cooked. However, great care must be taken to be sure the growing tips that are gathered are indeed greenbriar, because the tips of some other vines, notably yellow jessamine, are quite poisonous.

There is little current use of greenbriar as an ornamental. Indeed, it is generally considered to be a highly undesirable weed. However, the foliage and the fruits of some species are quite pretty. Anyone interested in attracting and feeding wildlife might plant several species along a back fence or allow them to climb into a shrub or tree in order to provide berries and nesting places for birds.

3

ANGIOSPERMS

Dicots

Black Mangrove—*Avicennia germinans*

Acanthaceae

Over 90 percent of the mangrove forests in Florida occur in the four counties at the south end of the Florida peninsula (Wikipedia 2020, "Florida Mangroves"). However, black mangrove, being the most cold tolerant of the mangroves, reaches as far north as St. Augustine on the east coast and Cedar Key on the west coast. After a series of warmer winters, seeds washed northward along the Gulf Coast often establish seedlings along the coast in the Florida Panhandle, and there is a substantial population of black mangrove shrubs along the coast of Louisiana south of New Orleans. Black mangrove is a very distinctive bush or tree, living on the shallow, intermittently flooded edge of salt water. It has opposite, smooth margined, evergreen leaves that are shiny green on top and covered by gray hairs (pubescence) on the lower surface. The small flowers, produced at any time of year, are white with a gold center. The most distinctive feature of this tree is its production of abundant, thin, woody structures called "pneumatophores" that stick up 2 to 6 inches above the surface of the ground from the lateral root system.

Black mangrove is not cold tolerant and is killed by sustained temperatures below minus four degrees centigrade (25 degrees F) (Zielinski 2013). However, as the earth's temperature gradually warms, this species has been slowly moving north along Florida's coastlines (Zielinski 2013). The best places in northern Florida to see it are at Cedar Key and along the Bella Vista Trail at Washington Oaks Gardens State Park beside the Matanzas River south of St. Augustine. This far north, black mangrove is a shrub or small tree, but, at the south end of the Florida peninsula, it can grow 60 feet tall with a trunk 2 feet in diameter (Godfrey 1988). It has dark, rough bark and produces very heavy, dense, black wood.

Black mangrove is remarkable in its ability to grow in salt water. Its pneumatophores provide oxygen for the root system, and the leaves exude the excess salt it gets from the salt water (Hill 2009). The sap inside black mangrove is about ten times as salty as the sap inside red mangrove (and most other trees) because red mangrove blocks salt from entering at the root surface, whereas black mangrove allows salt to enter and then exudes any excess beyond what it can tolerate (Hill 2009). (It may be this high internal salt concentration that allows black mangrove to survive cold weather better than the other mangrove species.) Black mangrove grows where salt water floods the root system at high tide but retreats at low tide. It also grows a bit inland from this tidal zone on low, flat land kept moist to wet by the nearby water body. Black mangrove does best in full sunlight and is not at all drought or fire tolerant.

Mangroves are well known for their wildlife habitat values, being primary producers of food and energy in the form of detritus for a unique ecosystem that provides nursery areas for fish, shellfish, and crustaceans as well as habitat for a variety of birds and other wildlife (Hill 2009). In northern Florida, this is counterbalanced by the fact that the formation of new mangrove forest often occurs at the expense of salt marsh it replaces. Salt marsh is also very valuable wildlife habitat. Besides providing habitat for marine creatures and nesting and roosting habitat for birds, black mangrove produces flowers that are attractive to bees and other pollinators.

One valuable benefit provided by black mangrove is the storm protection mangrove forests offer to coastal developments. Similarly beneficial are the nursery areas it provides for game fish and other seafood resources. In addition, its dense wood can make good firewood, and it has also been used to make charcoal.

Sweetgum—*Liquidambar styraciflua*

Altingiaceae

Sweetgum is one of our largest trees. When young, it grows with a single, straight trunk and symmetrically conical crown with many nearly horizontal branches. When it reaches about 50 feet in height, there is an abrupt change in its growth habit. It ceases to have a single terminal bud and begins branching out with fewer, more upright, and more massive branches to form a spreading crown. In the forest, it produces a tall, straight, cylindrical trunk clear of branches for most of its height. Open-grown sweetgums develop large crowns nearer the ground. The 4- to 6-inch-wide, star-shaped, deciduous leaves are sometimes confused with maple leaves, but they are arranged alternately on the twigs, whereas maple leaves grow opposite each other. Also, when crushed, the green leaves have a distinctive spicy or piney odor.

When the tree is young, the twigs often have corky ridges on them. These are normal growths, which probably help to protect the young tree's twigs from being eaten by browsing animals. The larger stems and the trunk have a roughly furrowed, corky bark. The foliage of healthy trees is dense, particularly on young trees in the open. In summer, once they attain the age of 20 to 30 years, sweetgum regularly produces an abundance of round, spiny, 1-inch-diameter fruiting structures (gumballs) (Bonner 1974), which produce tiny winged seeds beginning about October 1st, with seed availability tapering off through late November, December, and January, and ending in mid-February. Most of the gumballs fall from the tree during the month of February. Sweetgum is monoecious, producing distinctive male and female flowers on the same tree in early spring.

Sweetgum commonly becomes 100 feet tall with a trunk diameter of 2 to 3 feet at maturity. Located on the edge of the Apalachicola River floodplain in Torreya State Park, the largest sweetgum reported so far in Florida has a height of 135 feet, a trunk diameter of 4 feet 4 inches, and a crown spread of 88 feet (Florida Champion Trees Database 2020). Even this tree is small by comparison to some sweetgums of the past. When Jimmy Miller was a forestry professor at the University of Florida, he used to tell about one on Hogtown Creek in Gainesville in the early part of the twentieth century that was so big the loggers of the area could not handle it. They brought in a special logging crew from somewhere else just for this one tree. A previous national champion from North Carolina—136 feet tall with a 278-inch trunk circumference (over 7 feet in diameter) at 4½ feet above the ground—is more representative of the maximum size potentially attained by sweetgum (American Forests 1994).

Sweetgum ranges from Tampa Bay and Cape Canaveral on the Florida peninsula northward and westward throughout Florida and as far north and west as New Jersey, southern Indiana, and eastern Texas (Kormanik 1990). It also occurs in parts of southern Mexico and northern Central America, where it grows to great size in upland tropical rain forests along with such trees as mahogany, *Ceiba*, and *Podocarpus*.

In Florida, sweetgum is associated with other hardwood trees in upland hardwood forests, slope forests, floodplains, and pine-hardwood forests. It prefers moist soils with some clay content and can withstand about a month of flooding during the growing season or somewhat longer flooding in winter. Along with laurel oak, water oak, and pignut hickory, sweetgum is one of the most common trees in the upland hardwood forests and slope forests of northern Florida. It is even more abundant in many lowland and floodplain hardwood forests, where its most common associates are live oak, laurel oak, water oak, hornbeam, loblolly pine, cabbage palm, red maple, and Florida elm. It is also common on the shallow edges of swamps in association with red maple, swamp laurel oak, pumpkin ash, green ash, and bald-cypress. Because of its poor salt tolerance, sweetgum is absent from forests close enough to the Atlantic and Gulf Coasts to get flooded by brackish water during hurricanes.

Sweetgum reproduces prolifically from seed and from root sprouts following logging or other disturbances, and it aggressively invades old fields and other openings. It prefers full sunlight, where it will grow rapidly if its soil and moisture requirements are met. It is not very shade tolerant but is a strong competitor on sites particularly favorable for it, where it will grow under pine and sometimes even oak canopies. One surface root is often much larger and longer than the others, and, on wet sites, where it is more easily wind-thrown than most other kinds of trees, it has a very shallow root system. On well-drained soil, it has both shallow and deep roots and is moderately wind-firm, but it is not very drought or fire resistant and is sometimes plagued by sweetgum dieback that repeatedly kills the upper branches. It also loses branches more easily than most other hardwood trees because of its long, slender branching habit and rather weak wood and because of the bark-feeding of gray squirrels, which can introduce rot into a branch. Branches are particularly prone to breaking during the months of July and August, when the new green cones reach maximum size and weight and when summer thunderstorms are at their peak. In spite of these problems and its rapid growth rate, sweetgum is a tough tree that can live a long time, sometimes reaching ages of 200 to 300 years (Harlow and Harrar 1958).

Because sweetgums usually have straight, clean, rot free trunks and because its wood is easily worked and adaptable for many uses, it has long been valued as a timber tree. The wood is well suited for making "turned" objects on a lathe. It is currently used in making furniture, paneling, veneer, plywood, pulpwood, and packing crates. Sweetgum wood is not easy to split, and the sapwood rots rapidly. It makes poor firewood. Historically, styrax, the aromatic gum or resin produced by wounds on the trunk, was used in the making of medicines, incense, and perfumes (Fordham 1961).

Because it has been repeatedly logged from most forests, sweetgum is not often found today as a giant old tree, and yet, potentially, its greatest wildlife value is as a den tree when it is old and hollow. Even sweetgums of moderate size with the dieback disease can be useful as den trees for some animals: woodpeckers will hollow out nest cavities in the dead branches, and the ends of branches that have died and rotted away will often rot back into the living branch, creating a small, long-lasting den site ideal for chickadees, titmice, flying squirrels, and other cavity nesters. In forests where logging has occurred, sweetgum stumps often survive for many years because of sprouts they have produced and by virtue of their being connected to the root systems of other sweetgums. As these stumps rot inside, they sometimes hold water, providing animals of all sorts with a watering hole, birds with a birdbath, and mosquitos with a good breeding site (all benefits from a wildlife perspective). There is even a legend about the health benefits of drinking this "stump water" as a cure for stomach aches (Dana Griffin, retired botany professor at the University of Florida, personal communication).

Sweetgum seeds are quite nutritious and are a highly valuable part of the diet of gray squirrels and several kinds of birds. Squirrels begin lightly feeding on the green fruits in July and are earnestly feeding by September 1st, which is about one month before the first fruits mature and about two months before most of the fruits are mature. Squirrels continue to feed on the seeds through January. American goldfinch and purple finch consume large quantities of sweetgum seed in the fall and winter by picking them out of the ripe fruits in the treetops (Martin et al. 1951). Many times I have seen flocks of goldfinches extracting seeds from the fruits in the crowns of sweetgums in northern Florida in November and December. Pine siskins also feed on the seeds in this manner (Terres 1980). I observed a flock of about 50 rusty blackbirds feeding from fruits in a large sweetgum in Gainesville on December 9, 1994; a flock of about 100 common grackles doing the same thing on October 16, 1997; and a flock of about 200 red-winged blackbirds doing the same on January 2, 1995. Wildlife ecologist Katie Greenberg has observed the same behavior in common grackles (personal communication). This kind of feeding continues into early February. Mourning doves feed heavily on the seeds when they fall to the ground in suburban situations, and I have seen house finches doing this in Gainesville. Both mourning doves and ground doves, along with bobwhite quail and several kinds of sparrows, feed on the fallen seeds on the edges of fields and pastures, and mourning doves, quail, fox sparrows, and white-throated sparrows feed on them to some extent on the forest floor beneath upland hardwood forests.

Both squirrels and beavers eat sweetgum bark, which they seem to prefer over the bark of most other trees. Squirrels do little damage to sweetgum by doing this, although when squirrels eat bark from the top of a branch near the trunk, rot that weakens the wood can be introduced, causing some of the branches to break. On the other hand, on river floodplains, beavers kill many sweetgums, including mature trees, by eating enough bark to girdle the tree. Sweetgum is a likely candidate for the most valuable tree in the American beaver's diet in Florida. Rabbits, mice, and cotton rats eat the bark and twigs of seedlings, and squirrels feed to a limited extent on the large male flowers in early spring. Finally, sweetgum is one of the tree species most often used as a source of sap and insects by the yellow-bellied sapsucker while it is here in the winter.

All things considered, the wildlife value of sweetgum ranges from moderate to very high. In a forest where sweetgum is uncommon, the wildlife value of a single, mature, mast-producing sweetgum is quite high. In landscaped situations, sweetgum trees older than 30 years are often extensively used by gray squirrels, flying squirrels, mourning doves, goldfinches, blackbirds, and sapsuckers and can be of great value to wildlife if they develop cavities useful for nesting. However, these wildlife values only begin once the trees are old enough to begin abundant seed production, which takes at least 20 years (Bonner 1974).

Sweetgum has considerable potential as a landscape tree but also has some serious drawbacks. It is attractive as a young tree because of its rapid growth, its symmetrical, conical growth form for the first 20 years, its dense foliage, and its striking fall colors. These colors, ranging from yellow to red to deep purple, are quite variable from tree to tree and are brighter on acidic soil where it is not as well adapted. It also gets quite large and often makes a good shade tree. When open grown in a landscaped situation, it is usually firmly rooted and not particularly vulnerable to being wind-thrown. Drawbacks include sweetgum's shallow root system, which often rises above the ground surface (raising curbs, sidewalks, and foundations, and getting in the way of lawn mowing), its susceptibility to the sweetgum dieback disease, and its tardily deciduous leaves and gumballs, which litter the ground over a period of several months. It is not a good tree to plant over parking spaces or buildings given both the frequency with which it loses large branches during the summer and the sticky resin which sometimes drips from wounds in the crown. One of its best features as a landscape tree is its good wildlife value once it is old enough to produce seed.

Winged Sumac (Shining Sumac)—*Rhus copallinum*

Anacardiaceae

Winged sumac is usually a sparsely branched, small- to medium-sized shrub. It is often only 3 to 5 feet in height with trunks about half an inch in diameter. Occasionally it gets larger, growing to 20 or 30 feet in height with trunks of 2 to 6 inches in diameter. The twigs are about a quarter of an inch thick and slightly roughened with hairs. The deciduous, pinnately compound leaves have a "wing" along the rachis (leaf stalk) between the leaflets. Male and female flowers, which bloom mostly in August in northern Florida, occur on separate plants (dioecious). The leaves often turn bright red before falling. The clusters of fruit at the ends of the main twigs on the female plants turn red as they ripen in September and then fade to brown and often remain on the plant into winter.

Winged sumac is a common shrub in high pine, sandhill, and pine flatwoods forests, whether they occur in native pine forests or pine plantations. It also occurs in old fields and to some extent in upland hardwood forests, especially along the edges and in openings. It is so common and widespread, occurring throughout Florida (Wunderlin et al. 2020) and most of the rest of the eastern United States (Godfrey 1988), that it almost becomes invisible as an expected and ignored part of the landscape. Only in the fall, when the foliage turns bright orange and red, does it attract attention.

Winged sumac grows best in the open in full sun. It is tolerant of moderate shade, but not the full shade of a closed-canopy hardwood forest. It tolerates a wide range of soil conditions from sand to clay, low pH to neutral, and from low fertility to high. It prefers well-drained situations but can do well in the flatwoods where drainage is rather poor. It does not survive flooding and is therefore not found in swamps and marshes.

Winged sumac is short lived in most situations, the individual stems probably not surviving more than 10 to 20 years. In a landscaped situation, they may survive somewhat longer. Sumac spreads by suckering from its widespread surface root system, forming clonal patches consisting of either dense thickets or, more commonly, widely spaced individual plants. Some of these clonal patches have been estimated to be very old, in some cases in the northeastern United States for other sumac species, perhaps thousands of years old. The vast majority of winged sumac plants in the pine forests of Florida are of root-sprout origin. Winged sumac can spread rapidly in this way, with new root suckers occurring as far as 50 feet from the parent plant. However, reproduction also comes from the abundant annual seed crop, which is widely disseminated by the birds that eat the fruit.

The wildlife value of winged sumac is moderate to good. It is a reliable supplier of fruit that is low in food value but usable for sustenance by many species of mammals and birds in fall and winter. Birds that make use of this food resource are wild turkey, bobwhite quail, bluebird, cardinal, catbird, mockingbird, brown thrasher, robin, and starling (Martin et al. 1951). The twigs, bark, berries, and leaves are also eaten for winter sustenance by deer, rabbits, and cotton rats. The seeds are eaten by golden mice (Peles et al. 1995) and perhaps cotton mice and oldfield mice. The flowers are attractive to honey bees and a wide assortment of native bees, flies, and wasps. Finally, some insects eat the leaves and become, in turn, a food source for birds, lizards, toads, frogs, and spiders.

Winged sumac has no commercial value, but the fruit clusters can be picked in autumn and boiled to make a tea that is similar to lemonade. This tasty beverage is a rich source of vitamin C. The plant has also been used for dyeing cloth and tanning leather (Nelson 1996).

Winged sumac is rarely used in landscaping, but it has some potential. It is more luxuriant and grows larger in a landscaped situation than in most places in the wild, and the autumn foliage is beautiful. One drawback is that it spreads rapidly and widely by root suckers.

Poison Sumac—*Toxicodendron vernix*

Anacardiaceae

Poison sumac is an uncommon wetland shrub or small tree that has toxic properties similar to poison ivy and, like it, belongs to the cashew family. Its maximum height is 20 feet and maximum diameter 4 inches, but most plants are half that size. It has alternate, pinnately compound leaves with 7 to 15 lanceolate leaflets per leaf. The leaf stem (petiole) and the new growth of twigs is reddish. Clusters of small, yellow-green flowers bloom in spring, followed by whitish drupes in summer that hang on until winter if not eaten. Male and female flowers occur on separate plants (dioecious). The bark is smooth and gray in color.

Poison sumac grows in sunny spots in seepage wetlands and bayheads, often on acidic soil within the pine flatwoods complex of plant communities that include cypress domes, wet prairies, creek bottoms, and bayheads. It occurs in wetlands in Morningside Nature Center in Gainesville. I have also seen it growing beside streams such as Mormon Branch in the Ocala National Forest and Deep Creek south of Interlachen. Poison sumac is found in widely scattered populations throughout most of the eastern United States, extending onto the Florida peninsula in the interior counties as far south as Highlands County (Kartesz 2015).

Poison sumac seems to require a constant water supply in a situation that does not flood excessively. Most plants specializing in this sort of habitat are not at all drought tolerant and also may not be especially flood tolerant even though they are obviously very tolerant of continuously wet soils. Poison sumac stems are not fire tolerant, but the plant will sprout back from the base after being killed to the ground by fire.

The wildlife value of poison sumac is probably good, as its fruits are similar to those of poison ivy. These fruits, which hang down in a loose cluster, are whitish in color, ripen in the late summer, and hang on through the winter, thus providing birds with a food source when most other fruits are in short supply. The spring flowers are attractive to bees and other insects.

Poison sumac is rather attractive in its native habitat and produces good fall color but is not used as an ornamental for obvious reasons.

Poison Ivy (Poison-Ivy, Eastern Poison Ivy)—*Toxicodendron radicans*

Anacardiaceae

Poison ivy is a common vine throughout Florida and the rest of eastern North America, occurring both as a ground-cover plant and as a high-climbing vine on tree trunks and other surfaces. It will also, on occasion, grow vertically on its own, producing a small, slender, unbranched shrub or, rarely, a moderate-sized, branching shrub. It is a well-known plant because of its widespread abundance combined with the well-known contact dermatitis it can cause on the skin of susceptible people (about 85% of people are susceptible). It is primarily identified by the fact that its deciduous leaves have three leaflets. The vine most similar to poison ivy that also occurs both as a ground-cover plant and a high-climbing vine on tree trunks and other surfaces is Virginia creeper, which occasionally produces a leaf with three leaflets though the majority of its leaves usually have five leaflets.

The other vine occurring in Florida that is similar to poison ivy is Atlantic poison oak, which has slightly fuzzy (pubescent) leaves, is restricted to fire-adapted upland pine areas, and does not climb on vertical surfaces. Poison ivy produces clusters of small, greenish-yellow flowers as the new leaves emerge in March here in Florida, followed, on female plants, by clusters of small fruits that remain on the plant through autumn and winter. It is dioecious, with male and female flowers occurring on separate plants, and it is in the cashew family.

The vine trunk of poison ivy adheres to surfaces with many tough little rootlets that attach it tightly to whatever surface it climbs. The vine trunk is literally covered with these little roots, giving it a distinctive woolly appearance that can help to identify the plant in winter when the leaves have dropped or when it has climbed so high into a tree that the leaves are out of sight in the canopy.

Poison ivy is well adapted to survive as a vine on tree trunks. It easily climbs any type of tree bark, and it is shade tolerant enough to be healthy in moderately shady forests as it makes its way up a tree trunk. In the dense shade of mesic upland hardwood forests, it is often restricted to gaps and edges. As it grows higher, poison ivy begins forming lateral branches that reach out several feet in all directions, creating its own little crown of foliage. It does this below, or even in, the crown of its host tree but always below the tree's leaf surfaces, thus avoiding competing with the tree for sunlight. In this way, poison ivy avoids harming the host tree.

Poison ivy occurs throughout most types of hardwood forest, being especially common along stream banks and other edge situations where there is additional light. It is also found in some pine forests. The limiting factor in pine forests seems to be fire, which will kill back any vines attempting to grow on the tree trunks. Poison ivy is uncommon in upland longleaf pine forests, where it is replaced by the closely related and very similar ground-cover-dwelling Atlantic poison oak. Poison ivy prefers rich, well-drained soils but is adaptable, being able to grow in swamps, floodplains, upland hardwood forests, and forests on poor, sandy soil. It can grow in pine forests that burn, and it can grow in open, nonforested land by remaining, in both situations, a ground-cover plant with stems below the ground surface that sprout vertical branches 1 to 3 feet tall.

The wildlife value of poison ivy is high. Its clusters of small whitish fruits, which hang on throughout the winter, are an important winter food source for a wide variety of birds including bobwhite quail, catbird, mockingbird, Carolina chickadee, northern flicker, yellow-bellied sapsucker, downy woodpecker, red-bellied woodpecker, ruby-crowned kinglet, white-throated sparrow, hermit thrush, yellow-rumped warbler, and Carolina wren (Martin et al. 1951). I have seen ruby-crowned kinglets feeding extensively on these fruits both in northern Florida and in eastern Texas. Poison ivy is also an important browse plant for white-tailed deer and swamp rabbit (Miller and Miller 1999) and is a source of pollen and nectar for wild bees, honey bees, and other insects when in bloom in early spring.

Of course, the main interest we humans have in this plant is how to avoid contact with it and how to keep it from growing in our own yards. Cutting a high-climbing poison ivy vine can be counterproductive, causing the plant to resprout and send up foliage and perhaps ground runners that will now be within our reach, increasing the threat of skin contact. Vines that are creeping underground as part of a lawn or planting bed can be removed by spraying them with a contact herbicide. Avoiding contact with the plant should include avoiding contact with smoke produced by burning debris containing poison ivy fragments.

Atlantic Poison Oak (Eastern Poison Oak)—*Toxicodendron pubescens*

Anacardiaceae

Here in the southeastern United States, the term poison oak (or more properly poison-oak, as this is clearly not an oak species) refers to the relative of poison ivy that is a common ground-cover component of upland longleaf pine forests on well-drained soils. It occurs from Marion County, Florida, northward into Virginia and westward into eastern Texas and Oklahoma (Kartesz 2015). It is similar to poison ivy in that it has three leaflets on each leaf. It differs in that the leaves and stems have a fine woolly coating of hairs (pubescence). It also differs by rarely, if ever, climbing tree trunks as a vine. Its stems are all underground, spreading widely and producing vertical, aboveground shoots 1 foot tall with one to three compound leaves, and, in season, either small flowers or a cluster of small whitish fruits. Male and female flowers occur on separate plants (dioecious).

Atlantic poison oak is a fire-adapted species that is often quite common in association with wire grass, sandhill dropseed, bracken fern, summer-farewell, blazing star, partridge pea, gopher apple, and other sandhill and high pine ground cover species. It is similar to poison ivy in its ability to induce a skin rash on susceptible people. Although fairly immune to this toxin, I have gotten a rash when hand planting pines by accidently cutting the underground stems of poison oak while preparing the planting hole and then getting the sap on my hands. Poison oak is less common overall than poison ivy because of its more specialized habitat, rarely being found outside of its pineland niche.

The wildlife value of Atlantic poison oak is fairly high, its fruit supplying a winter food source for birds such as bobwhite quail, wild turkey, catbird, and northern flicker (Martin et al. 1951) and its flowers a source of pollen and nectar for bees and other insects in spring.

Pawpaw—*Asimina* spp.

Annonaceae

Common Pawpaw—*Asimina triloba*

Common pawpaw is a shade-tolerant shrub or small tree found scattered about in hardwood forests in most of the eastern half of the United States. It reaches the western panhandle of Florida in a few spots of moist, fertile hardwood forest. Like the other species of pawpaw, its deciduous, alternate leaves are smooth margined with a distinctive green-pepper or petroleum-like odor when crushed, and it has naked buds covered with rusty-colored hairs. It produces maroon flowers 1 to 1½ inches wide on one-year-old stems. The oblong fruit is packed with somewhat flattened seeds and gets soft and mushy when ripe, at which time it is edible with a pleasantly fruity odor and taste. It is the only pawpaw species to commonly spread by root suckers, forming clonal patches.

Most of the pawpaw plants in Florida belong to a set of related species that are restricted to the coastal plain of the southeastern United States or exclusively to the Florida peninsula. These other species are smaller, and they tend to have fruit that is not as pleasant to eat as that of the common pawpaw, although all are edible.

Hammock Pawpaw—*Asimina parviflora*

Hammock pawpaw is very similar in appearance and habitat preference to common pawpaw, having thin, oblong leaves 3 to 7 inches long and 2 to 4 inches wide and maroon-colored flowers less than three quarters of an inch wide borne on one-year-old twigs. It is smaller, however, rarely more than 6 to 10 feet tall, with a smooth, gray stem no more than a couple of inches in diameter. It is a shade-tolerant understory shrub found mostly in upland hardwood forests, where it is uncommon. It occurs throughout the Florida Panhandle, the northern half of the Florida peninsula, and most of the rest of the southeastern coastal plain of the United States (Kartesz 2015).

The rest of the pawpaw species have much thicker, more leathery leaves and are much less shade tolerant, being adapted to living in full sun or partial light shade.

Narrow-Leaf Pawpaw—*Asimina angustifolia*

Narrow-leaf pawpaw, also known as *A. longifolia*, is usually found in pinelands or former pinelands. It occurs in pine flatwoods forests, upland longleaf pine forests, and in sand pine scrub but is most abundant in areas that are intermediate between these communities. It is especially common in scrubby flatwoods. Narrow-leaf pawpaw is usually a multistemmed shrub not more than 3 feet tall, and, as the common name implies, the 6- to 8-inch-long leaves are quite narrow. Its white flowers, with purple central patches on the inner petals, hang down below the current year's stem growth and are 2 to 3½ inches long. Narrow-leaf pawpaw has a rather small area of distribution in North Florida and southeastern Georgia, with a couple of isolated populations in the Florida Panhandle (Kartesz 2015).

Woolly Pawpaw—*Asimina incana* (*A. incarna, A. speciosa*)

Woolly pawpaw has many local common names, including "showy pawpaw," "flag pawpaw," "sandhill pawpaw," and "polecat bush." It is normally found as a fairly common small shrub in longleaf pine sandhills and also the more fertile types of upland longleaf pine forests or in pastures that were formerly longleaf pine forests. One reason it is common in pastures is that it is difficult to kill. It has a very deep root system, and if an area where it grows is cleared of vegetation mechanically, with a bulldozer or a harrow, for example, these little pawpaw bushes will sprout back and keep on growing. They have roundish leaves that are covered top and bottom with a thin layer of whitish hairs (pubescence). The hanging, 2- to 3-inch-long white blooms appear on the previous year's twig growth before the new leaves are fully developed. Woolly pawpaw occurs in southeastern Georgia and on the northern and central Florida peninsula (Godfrey 1988), with an isolate population in Gadsden County (Wunderlin et al. 2020).

Scrub Pawpaw—*Asimina obovata*

Scrub pawpaw, also called "big flower pawpaw" and "flag pawpaw," is mostly found in Florida scrub communities such as sand pine scrub. It is similar to woolly pawpaw but, instead of a whitish pubescence, has a rusty- or coppery-colored pubescence on its new stems. Scrub pawpaw also has somewhat larger leaves than woolly pawpaw, and they are shinier on top. It also tends to grow taller than woolly pawpaw. Like woolly pawpaw, scrub pawpaw has large, white flowers, but instead of appearing on last year's twig growth, they appear on the twig growth of the current year. Scrub pawpaw will sometimes occur in upland longleaf pine forests located near scrub communities, in which case, it sometimes crosses with woolly pawpaw, producing hybrid individuals intermediate in appearance. Scrub pawpaw is endemic to the central Florida peninsula, occurring primarily from Clay and Alachua Counties southward into St. Lucie and Glades Counties (Wunderlin et al. 2020).

Netted Pawpaw—*Asimina reticulata*

Netted pawpaw or flatwoods pawpaw is a pine flatwoods specialist, occurring most commonly in the driest types of pine flatwoods forests. Its leaves are thick, leathery, and shiny on top when mature, with the vein pattern showing on the undersides. The 1- to 3-inch-long white flowers are produced on the previous year's twig growth. This pawpaw is usually not more than 2 feet tall but is often a good fruit producer, and the fruit is sometimes quite good to eat. It primarily occurs on the peninsula of Florida from Bradford County southward into Dade County (Wunderlin et al. 2020).

Dwarf Pawpaw—*Asimina pygmea*

Dwarf pawpaw or pygmy pawpaw is the smallest species of pawpaw, rarely growing more than 1 foot tall. Its small flowers, produced on new growth, are at least partly maroon in color. It produces fruit sparingly. It grows in either the driest of the pine flatwoods forests or in forests that are intermediate between dry pine flatwoods forests and upland longleaf pine forests. Pygmy pawpaw occurs in northern Florida east of the Aucilla River from the Georgia border southward on the Florida peninsula into Brevard and Manatee Counties (Wunderlin et al. 2020).

As noted earlier, all of the pawpaw species emit a strong, distinctive odor when the leaves are crushed. They all have deep root systems and are very drought tolerant. They all produce fruit that contains hard, somewhat flattened seeds. They reproduce by seed and, with the exception of common pawpaw, do not spread by runners. None of them are weedy. Indeed, the individual plants seem to be long lived, and new plants seem to appear infrequently. They are not easy to grow from seed and are nearly impossible to transplant from the wild because of their deep root systems.

They all are adapted to fire, being able to sprout back from the base if killed to the ground by fire.

Pawpaw fruits ripen beginning in early July and ending in late August or early September. Some of this variation in ripening time depends on the time of flowering, which may be altered considerably by fires and other disturbances. Some of the variation is also the result of species differences, with netted (flatwoods) pawpaw ripening first and hammock pawpaw and scrub pawpaw ripening last.

The wildlife value of pawpaw bushes is fairly high, especially in relation to the small amount of space they occupy. The leaves are often eaten by insects including the larvae of the zebra swallowtail butterfly, which specializes on this genus (Glassberg et al. 2000). Pawpaw fruits are eaten by a wide range of animals including raccoon, opossum, gray squirrel, gray fox, and wild turkey (Miller and Miller 1999). The ripe fruit is quite aromatic, inviting animals to come and eat it, and it is no doubt eaten by many species in addition to those just mentioned. I'm sure neither a black bear nor a mouse would pass up a ripe, richly aromatic pawpaw fruit. Pawpaw flowers are primarily pollinated by beetles and flies.

Pawpaw plants are not important economically. They are rarely used for landscaping and are neither a benefit nor a hindrance to forestry or range management, although some pasture owners do attempt to eliminate them for reasons that have nothing to do with economic need. They are also eliminated incidentally by some forestry site preparation activities such as bedding or the use of herbicides.

Leafless Swallowwort (Marsh Milkvine)—*Orthosia scoparia* (*Cynanchum scoparium*)

Apocynaceae

This is a small, delicate, semievergreen, twining, milkweed vine with smooth-edged, hairless leaves 1 to 2 inches long and less than a quarter of an inch wide. It has milky sap and inconspicuous little white flowers in summer followed by twin seed pods. It is fairly common on the Florida peninsula and occurs also in Franklin and Gulf Counties in the Florida Panhandle and in a few spots along the Georgia and South Carolina coasts (Wunderlin 1998; Kartesz 2015).

This climbing milkweed vine rarely reaches higher than 5 feet, with very slender stems that sometimes twist tightly around each other. It usually forms a fairly dense covering on top of a low shrub or young tree in the shady understory of a hardwood forest. It mostly occurs in mesic to xeric hardwood forests but is also reported to occur on the borders of marshes and in scrub (Lance 2004).

The wildlife value of leafless swallowwort is mostly the result of its being a food plant for various insects including the caterpillars of the monarch and queen butterflies. It is reportedly a food plant for the caterpillars of the strikingly beautiful faithful beauty moth, which occurs at the northern edge of its range in South Florida, and it is also reported to be the only food plant for the giant milkweed bug (Rogers 2011).

Leafless swallowwort is not used commercially and is very rare as a landscape plant although a few native plant nurseries do grow it. It could be an interesting addition to a butterfly garden.

American Holly—*Ilex opaca* (including a note on Scrub Holly [*I. opaca* var. *arenicola*])

Aquifoliaceae

American holly is a well-known, medium-sized, evergreen tree with smooth, light gray bark and stiff, dark green leaves with five to ten prominent, sharp spines on the leaf margins. Because the plants are dioecious, the red berries (actually drupes) that decorate holly trees in winter occur only on female trees, which comprise about half the wild trees. Red berries appear on almost all planted holy trees, as people usually plant female hollies. The native range of American holly extends from New Jersey southward to the central Florida peninsula and westward into eastern Texas, reaching as far north in the interior as West Virginia and the southeastern tip of Missouri (Grelen 1990, "*Ilex opaca*").

Most mature American holly trees are 40 to 50 feet tall and 6 inches to 1 foot in trunk diameter. However, they sometimes grow much larger. The largest one reported for the United States, from Maryland, is 64 feet tall and about 4 feet in trunk diameter (American Forests 2015). The largest one I've seen in Gainesville, Florida, was in Bivens Arm Nature Park. In 1993, it was 65 feet tall with a trunk diameter of 22 inches (Ward and Ing 1997). In 1994, an American holly growing in Green Cove Springs, Florida, was 69 feet tall with a trunk diameter of 30 inches (Ward and Ing 1997).

American holly is a moderately adaptable tree. It is usually found growing in hardwood forests as an uncommon understory or midstory tree. However, when growing in live oak forests in places that flood on rare occasions, it is sometimes quite common. It is also common as an invader in some pine plantations on well-drained, sandy uplands.

A variety of American holly called "scrub holly" (*Ilex opaca* var. *arenicola*) grows in the deep sands of Florida scrub communities such as on the east side of the scrub in the Ocala National Forest and in some of the scrubs on the Lake Wales Ridge. It is smaller than American holly and has yellowish-green leaves but is otherwise similar in appearance to *Ilex opaca*.

American holly is not fire tolerant (although it will resprout vigorously from the root collar after a fire), is shade tolerant, is moderately tolerant of salt spray near the ocean, and is slightly flood tolerant, being able to withstand occasional floods of short duration. It does not do well in dense shade but can grow well in light to medium shade and does very well in full sun. It grows slowly and seems to be able to live a long time.

The wildlife value of holly is high. American holly trees provide nesting habitat for birds, and, perhaps more important, the red berries persist into winter, providing mammals such as raccoons and birds such as wild turkey, bobwhite quail, cedar waxwing, bluebird, mockingbird, brown thrasher, red-bellied woodpecker, northern flicker, pileated woodpecker, catbird, hermit thrush, eastern towhee, and blue jay with a winter food source (Miller and Miller 1999; Martin et al. 1951). Gray squirrels in Gainesville, Florida, begin eating the fruit on September 1st, and the fruit remains available, if not all eaten by then, until April 1st. When there is an American holly full of ripe fruit in the territory of a pair of mockingbirds, the mockingbirds will try to defend this food resource against other birds, chasing away any birds they can. What usually happens, however, is that at some point during the winter a flock of cedar waxwings or robins will come and strip the tree of its fruit, easily overwhelming the mockingbirds' defenses. American holly flowers, which bloom the last half of March in Gainesville, Florida, are a good source of nectar for honey bees, native bees, and butterflies.

American holly has been harvested in the past for its hard, uniform, white wood to create such specialty items as piano keys and wood carvings, and the foliage has been harvested for making Christmas decorations to such an extent that large numbers of trees have been severely damaged (Grelen 1990, "*Ilex opaca*"), causing holly in its native habitat to disappear from the vicinity of large cities (West and Arnold 1950).

The main economic value of American holly derives from its use as an ornamental tree. It is adaptable and trouble free and makes a beautiful evergreen tree (often with a conical, Christmas-tree shape) that has dark green foliage and bright red berries in winter. Several other holly cultivars such as East Palatka holly and Savannah holly have become much more popular as ornamentals. This is unfortunate, as American holly is a much longer lived, healthier, and more dependable tree. It is also a more beautiful tree, with denser and darker green foliage and larger berries, and the fruit of American holly seems to be preferred by birds. (See the following species account on dahoon for more details on East Palatka holly and Savannah holly.)

Dahoon—*Ilex cassine* and Myrtle Dahoon—*Ilex myrtifolia* (including notes on East Palatka Holly [*I. opaca* x *I. cassine*] and Savannah Holly [*I. opaca* x *I. cassine*])

Aquifoliaceae

These small- to medium-sized hollies are all similar to American holly in that they have stiff, evergreen leaves, red fruit, and separate, dioecious male and female plants. They also have smooth, light tan-gray bark, with some exceptions. In highly fire-prone areas, some large dahoon and myrtle dahoon trees have bark that is thickened with a corky outer layer on the lower part of the trunk. They all differ from American holly by having very few to no spines on the leaf margins. All of these forms are capable of getting 50 feet tall with 1 foot in trunk diameter at maturity, but they are typically much smaller. Myrtle dahoon is typically smaller than dahoon or East Palatka holly with much smaller and narrower leaves. Some hybridization has been reported between these forms (Godfrey 1988), and they are being treated together here because of their similarities.

Dahoon is the most widespread and common of these trees in the wild, occurring in bald-cypress hardwood swamps, river floodplains, and the cypress domes of pine flatwoods forests throughout most of Florida and reaching to some extent into nearby areas of adjacent states from South Carolina to Louisiana (Kartesz 2015). It is quite showy in October and November with its bright red fruit. Though evergreen, it often loses some leaves in late winter, and its berries (actually drupes) are a bit smaller and do not seem to last as long as those of American holly. The leaves are variable in size and shape, being mostly twice as long as wide and about 2 inches long on average.

Dahoon is almost always found in association with water and is fairly tolerant of flooding. It occurs in fire-adapted areas, such as pond-cypress domes and wet pine flatwoods but usually in wet areas that are somewhat protected from fire by the water. It is not tolerant of fire, drought, or salt spray. It can tolerate some shade, often growing in the understory under pines or cypress, but prefers full sun. Where it grows under hardwood forest in river floodplains, it is often situated on the edge, next to the water.

Myrtle dahoon, which has very narrow leaves about 1 inch long, is mostly restricted to pond-cypress domes in pine flatwoods forest areas. It seems to be quite flood tolerant and survives the occasional fires in these forests both by being located in the wetter parts of the swamps and being able to sprout back following a fire. It is not drought tolerant, is only somewhat shade tolerant, and is not known to be salt tolerant.

East Palatka holly is a commercial cultivar of holly that is popular in the nursery trade along with various other cultivars such as Savannah holly. These cultivars are derived by vegetative reproduction from hybrids between dahoon and American holly. They are similar in appearance to dahoon but have broader leaves and are more drought tolerant. Since they are genetically uniform cultivars with no variability to help them overcome disease, they are unreliable as landscape plants. American holly is a vastly superior species to select for planting as an ornamental.

The wildlife value of these hollies is very high, nearly as high as American holly in this regard. For details, see the discussion under American holly.

The ornamental value of the plants in this holly complex is somewhat limited, even though East Palatka holly and similar cultivars have been widely used for this purpose. Dahoon and myrtle dahoon are not widely adaptable to upland situations, and the various commercial cultivars such as East Palatka holly and Savannah holly have been plagued by a *Sphaeropsis* fungus disease that causes witches brooms and a serious decline in health that often leads to tree death. This problem is so severe on the central Florida peninsula that people there are being advised not to plant these cultivars.

Gallberry—*Ilex glabra* and Large Gallberry—*Ilex coriacea*

Aquifoliaceae

Gallberry, also called "inkberry," is an abundant and characteristic shrub in the pine flatwoods forests of the southeastern United States, occurring naturally in coastal plain pinelands and wetlands from New Jersey to Louisiana and throughout most of Florida with scattered populations as far north as Nova Scotia (Kartesz 2015). It has small, oblong, evergreen leaves that have a small tooth or two at the tips. It is dioecious, producing male and female flowers on separate plants, with the female plants producing black berries in the fall that hang on through the winter. It reproduces prolifically both by seed and underground runners, and it forms large thickets of numerous slender, dark, upright stems, which are killed back to the ground by fires but then quickly resprout. Though similar to gallberry, large gallberry is much less common, grows a bit taller, is less prone to form extensive clonal patches, and more commonly establishes in seepage bogs and along the margins of other wetlands than out in the drier areas of pine flatwoods forests.

Gallberry is most common in moist to wet pine flatwoods forests but also occurs in the wetlands that are imbedded within these forests. Both species of gallberry prefer acidic soil and can withstand moderate flooding. Gallberry is also drought tolerant and very fire tolerant, able to vigorously resprout following fires that kill its stems to the ground. If left unburned for several years, gallberry normally grows about 5 feet tall, but it can sometimes reach heights of 10 feet, with slender stems usually less than a half inch in diameter. By comparison, large gallberry can reach 15 feet tall with a trunk 1 or 2 inches in diameter.

The two most common shrub associates of these little black-fruited hollies are saw palmetto in the pine forests and fetterbush in the shrub bogs, but there are many other associates including various species in the following genera, to name a few: *Vaccinium*, *Gaylussacia*, *Smilax*, *Lyonia*, *Leucothoe*, *Eubotrys*, and *Morella* (*Myrica*). Longleaf, slash, and pond pine are common tree associates in the pine forests, as are slash pine, pond pine, pond-cypress, sweetbay, loblolly bay, and swamp tupelo in the wetlands.

Gallberry and, to a lesser extent, large gallberry, are important for wildlife. White-tailed deer browse on the foliage, and in some areas of the Florida pine flatwoods more than 20 percent of the winter diet of deer may be comprised of gallberry (Miller and Miller 1999). An assortment of mammals and birds eat the berries including raccoon, wild turkey, bobwhite quail, eastern towhee, American robin, and northern mockingbird (Miller and Miller 1999). Wild turkeys begin eating the berries green in July but not in great quantity (Hutto 1995). The flowers, which bloom in late spring, are an important nectar source for bees and other pollinators.

The importance of gallberry to humans is both negative and positive. Gallberry honey is a local favorite in southeastern Georgia and northeastern Florida. On the other hand, the timber industry often goes to considerable effort to reduce or eliminate gallberry when establishing pine plantations. The results of site preparation efforts by the timber industry are often mixed, with saw palmetto usually greatly reduced but with gallberry often able to sprout back and recover much of its former abundance.

Yaupon—*Ilex vomitoria*

Aquifoliaceae

Yaupon is a shrubby holly species with small, alternate, evergreen leaves, the edges of which lack spines and are wavy (crenate toothed). Sometimes a small tree, but more often a small to large shrub, it can occur individually or in thickets that result from spreading by seed or by underground runners. An unusually large yaupon in Texas was 45 feet tall with a trunk diameter slightly over 1 foot (Godfrey 1988). Yaupon is native from the Carolinas across the southern states into eastern Texas, the Florida Panhandle, and as far south on the Florida peninsula as Highlands County (Kartesz 2015). Male and female flowers are borne on separate, dioecious plants, and the red fruits produced by female plants ripen in the fall and hang on through winter. Its branches are stiff, and the light gray bark is thin and smooth.

Yaupon grows on uplands, usually under hardwood forest trees growing on fertile soil, but it will also invade upland pine forests that lack frequent fire. Although it is sometimes found in pine flatwoods forests, yaupon does not do well on the highly acidic soil preferred by gallberry, growing more commonly on soil with a pH that is closer to neutral. It is a normal understory component of hardwood forests growing on limestone outcrop areas. Yaupon can be abundant, especially from the Florida Panhandle west into eastern Texas on sandy or clay soils or on sand dunes along the coast.

Yaupon is highly adaptable. It is not fire tolerant but can persist in fire-adapted forests by continually resprouting after fires and spreading both by seed and rhizomes. It is both moderately flood tolerant and very drought tolerant. It can grow in full sun but can also withstand the dense shade of closed-canopy hardwood forests. It is tolerant of salt spray and of trimming by humans and browsing by deer. It is not bothered noticeably by insects or diseases.

The wildlife value of yaupon is high. It is an important browse plant for white-tailed deer, being more palatable to them than gallberry (Miller and Miller 1999). The berries (drupes) are an important winter food for raccoons and several species of birds including wild turkey, bobwhite quail, cedar waxwing, northern mockingbird, brown thrasher, American robin, eastern bluebird, hermit thrush, red-bellied woodpecker, pileated woodpecker, eastern towhee, and blue jay (Miller and Miller 1999; Martin et al. 1951). Yaupon thickets can also provide cover for deer and other species.

Yaupon is used by humans for landscaping and, increasingly, for producing tea. There are numerous cultivated varieties of yaupon sold commercially as ornamentals, ranging from dwarf hedge plants to a tall weeping variety. As for tea, Native Americans made considerable use of the leaves to make "black drink," which was consumed both as a ceremonial and as an everyday beverage. Early colonial Americans also used yaupon tea regularly, and it has now become popular again, marketed by high-end tea companies and whole foods stores and offered in fine restaurants where its historical novelty, nonbitter taste (it contains less tannin than regular tea), and high levels of both caffeine and antioxidants make it healthful and appealing (Burnett 2016).

Deciduous Hollies: Possumhaw—*Ilex decidua* and Carolina Holly—*Ilex ambigua*

Aquifoliaceae

These two small holly species have deciduous leaves and, on female plants, bright red, somewhat translucent berries (drupes). Like all holly species, they are dioecious, with male and female flowers borne on separate plants. Both species have upright growth habits and, though often multitrunked and shrubby, they will sometimes attain small tree size. The leaves of possumhaw are elongated, whereas those of Carolina holly are only slightly longer than they are wide. Neither of these hollies is particularly common, but both occur on the northern and central parts of the Florida peninsula, in the Florida Panhandle, and in much of the rest of the southeastern coastal plain.

The main difference between the two is that possumhaw is primarily a river floodplain species, whereas Carolina holly is an upland species. Possumhaw typically occurs along the major rivers of North Florida from the Suwannee and Santa Fe Rivers west throughout the panhandle and on into East Texas. It also occurs to some extent on limestone outcrop uplands. It is shade tolerant, usually growing in the shade of hardwood forest trees. The variety of possumhaw that grows along the Suwannee and Santa Fe Rivers is Curtiss possumhaw, *Ilex decidua* var. *curtissii*, which has smaller and narrower leaves than typical *Ilex decidua* plants growing elsewhere. Its fruits begin to ripen as early as August 1st.

Carolina holly can be found as a scattered shrub in well-drained sandy upland forests of either hardwood or pine trees. It is moderately shade tolerant but does best in partial to full sunlight.

The wildlife value of these hollies is good. Their fruit is eaten by the same mammals and birds that eat the fruits of other hollies. Possumhaw fruit remains on the plant through the winter, making it perhaps more valuable for wildlife than Carolina holly, which drops its fruit rather soon after ripening in September.

The only human use for these plants is their occasional use as ornamentals. Possumhaw is perhaps the better of the two for this purpose because of its longer fruit retention.

Devil's Walkingstick—*Aralia spinosa*

Araliaceae

A small understory tree with an unusual appearance, devil's walkingstick occurs naturally from New Jersey south onto the central Florida peninsula and west into eastern Texas (Godfrey 1988). Young individuals grow straight upward with a single, thick, unbranched, spiny stem that is topped with a round or umbrella-shaped head of very large, doubly or triply compound, deciduous leaves. Large, older devil's walkingsticks often have a few branches that form an open crown. In July and early August, some of the most vigorous tops of these trees produce a large (2 feet in diameter) cluster of small (1 inch in diameter) umbels of small white flowers. The thick twigs, from one quarter to 1 inch in diameter, and the large prickles, arranged in whorls on the stems and trunk, are helpful clues for identifying this species in summer and winter. However, on large, old trunks, the prickles may be absent. Devil's walkingstick is most often confused with Hercules club. While devil's walkingstick has doubly to triply compound leaves, very thick twigs, and large prickles attached directly to the trunk and arranged both in whorls and randomly on the stems, Hercules club has thinner twigs, singly compound leaves, and large prickles that are not arranged in whorls and grow on raised pyramids of cork on older trunks.

Devil's walkingstick requires moderately fertile, well-drained soil. It is found almost exclusively in mesic hardwood forests and the most fertile of xeric hammocks, where it is never more than a minor component of the total vegetation. The most common overstory trees associated with devil's walkingstick are laurel oak, sweetgum, pignut hickory, and magnolia. Associates in the understory include hophornbeam, hornbeam, dogwood, American holly, one-flower hawthorn, and waxmyrtle. Devil's walkingstick will survive in moderate shade but does best in openings such as those produced by logging, windfalls, and natural and man-made edges. It reproduces primarily by root suckers from its very shallow, wide-ranging root system, and it can be weedy and abundant in spots. It often grows in large clones. The individual stems do not live very long: the oldest ones observed by me were about 20 years old.

San Felasco Hammock has a thriving population of devil's walkingstick trees and has provided three national champions. The largest of these was 51 feet tall, 24 inches in trunk circumference (7½ inches in diameter) at breast height, and had an average crown spread of 23 feet (American Forests 1986). Large specimens also occur around the upper margins of steephead stream valleys in the panhandle at Eglin Air Force Base and in the Apalachicola Bluffs and Ravines Preserve. Devil's walkingstick is abundant in the upland hardwood forests of O'Leno State Park and the adjacent River Rise State Preserve and is found in most other inland northern

Florida upland hardwood forests. However, it is uncommon in the Gulf coastal hammocks, presumably because it cannot withstand the occasional flooding that occurs there.

Devil's walkingstick has considerable wildlife value in relation to the minor extent of space it occupies. The flowers attract an amazing swarm of bees and wasps when in bloom in late July and early August. I've never seen any other plant attract such a variety and abundance of these insects. The fruits (drupes), which begin to ripen in late August and remain on the tree into the winter, are particularly attractive to thrushes (John H. Hintermister V, expert birder and naturalist, personal communication) and are eaten by other birds as well. I have seen mockingbirds, catbirds, and rose-breasted grosbeaks feed on them in my yard in Gainesville and several species of thrushes feeding on them at Ichetucknee Springs State Park. Miller and Miller (1999) note the consumption of devil's walkingstick drupes by northern cardinal, northern mockingbird, brown thrasher, blue jay, eastern bluebird, and white-throated sparrow and the browsing of the foliage by white-tailed deer.

Because of its unusual appearance and its wildlife value, this little tree has some value for landscape planting. It is very easy to transplant and responds well to watering and fertilizer. The drawbacks to such use are the sharp prickles on its trunk and its ability to spread prolifically and widely by root suckering.

Saltbush—*Baccharis halimifolia, B. angustifolia,* and *B. glomeruliflora*

Asteraceae

These three very similar species are treated here as one group, collectively referred to as "saltbush." Other names for them include "baccharis," "groundsel," and "sea-myrtle." The appearance, size, habitat preferences, wildlife value, and horticultural value of the three are similar. They are semievergreen to evergreen, weedy, short-lived, sun-loving bushes that grow up to about 12 feet tall. The leaves are grayish-green to green, 1 to 3 inches long, 1 to 2 inches wide, variable in shape, but often coarsely toothed toward the tip. The stems are green until a year or two old, when they turn tan. The stems and main branches grow upright. The inconspicuous, whitish male and female flowers are borne on different plants, with the male plants producing a much more noticeable show of flowers. The female plants put on their best show in the fall when white, fluffy fruiting heads cover the plant tops with what looks from a distance like a mass of small white flowers.

Baccharis halimifolia is native to the southeastern coastal plain of the United States, ranges along the coast into New England, and grows throughout Florida; *Baccharis angustifolia* occurs in the southeastern coastal plain and is more narrowly restricted to coastal habitats; *Baccharis glomeruliflora* occurs mostly throughout Florida and on the coast of South Carolina (Kartesz 2015).

Saltbush is salt tolerant, drought tolerant, fast growing, short lived, and weedy, but not shade tolerant. It does not spread by root suckers but reseeds prolifically on bare soil. Saltbush can grow in very wet or very dry soil, inland or along the coast, but it usually either invades a disturbed site or grows on the edge of a habitat such as salt marsh or freshwater marsh. During dry periods, saltbush will invade marshes only to be killed back when the area burns or is flooded again. An example of this occurred on Payne's Prairie where saltbush expanded out from the edge during the very dry years of 1999, 2000, 2001, and 2002 to cover hundreds of acres of formerly open marsh. With the return of wetter conditions around 2013, the saltbush invasion ended, and the marsh returned to its former, open, wet, herbaceous condition. Saltbush will also invade pine plantations after the disturbances of logging, site preparation, and replanting create an ideal seed bed. Once the new crop of pine trees reach crown closure or when prescribed burning is conducted, the saltbush invaders die.

Saltbush is poor to moderate deer browse (Miller and Miller 1999), produces seeds that are sometimes eaten by sparrows (Martin et al. 1951), and can provide a nesting structure for birds in a marsh, but it is not considered especially valuable for wildlife in the southeastern United States. Similar species in desert habitats in the western United States are considered more important for wildlife (Martin et al. 1951).

Saltbush is used sparingly as an ornamental, especially along the coast, and several cultivars have been developed for this purpose.

Woody Goldenrod—*Chrysoma pauciflosculosa*

Asteraceae

Woody goldenrod is a weak, low, sprawling shrub, 1 to 3 feet tall with gray twigs and dull gray-green leaves about 2 inches long and one quarter of an inch wide. The leaves are rather thick and have no noticeable veins other than the central one (Godfrey 1988). Its dull gray-green appearance is distinctive in the rather open ground cover of pinelands, scrublands, and dunes. Spikes of bright yellow flowers appear above the crown of the low shrub in late summer that closely resemble the flower spikes of goldenrods of the genus *Solidago*. It is common in parts of the western panhandle of Florida and occurs in scattered locations along the fall line in the Carolinas and in scattered spots in Georgia, Alabama, and Mississippi (Christman 2008).

Woody goldenrod grows on deep, well-drained sands as a ground-cover plant beneath the widely scattered longleaf pines and turkey oaks of very open, dry, infertile, sandhill forests. It also occurs in sand pine scrub and oak scrub on former coastal sand dunes and on secondary coastal sand dunes located behind the primary dunes. Woody goldenrod is very drought tolerant and fairly salt tolerant. It prefers to be in full sun and does not tolerate shade. It is not fire tolerant or flood tolerant but can sprout from seed following fires.

The main wildlife benefit of this plant is to the insect pollinators that visit the flowers in abundance in late summer and autumn.

Woody goldenrod does not transplant well from the wild. It is sometimes grown from seed by native plant nurseries and sold as an ornamental.

Garberia—*Garberia heterophylla* (*G. fruticosa*)

Asteraceae

This is a 2- to 5-foot-tall, long-lived, slow-growing shrub with dull grayish-green, sticky-feeling (tacky), roundish leaves about 1 to 2 inches long with entire margins. It is most noticeable in the fall, when it produces showy clusters of pinkish-lavender flowers on spikes that rise above the crown of the bush.

Garberia grows in scrub habitats and on the coastal dunes of the central Florida peninsula from Clay County to Highlands County (Wunderlin et al. 2020) with a large population in the Ocala National Forest in Marion County and much smaller populations in various isolated scrub habitat locations from there south to the end of the Lake Wales Ridge. It grows on deep, excessively well-drained, sandy soils and will invade longleaf pine sandhill forests if fire is excluded. It will grow in partial shade but does best in full sun. It is very drought tolerant and salt tolerant but not at all flood tolerant.

Garberia is a Florida endemic species (occurring only in Florida). It is frequent in the scrub in the Ocala National Forest but infrequent to rare elsewhere. It is listed by the state of Florida as a threatened species because of its rarity and because many existing populations of the species, other than those in the Ocala National Forest, are threatened by real estate development.

The wildlife value of garberia is related primarily to its flowers, which are attractive to a wide variety of insect pollinators.

A beautiful plant when in bloom, garberia has considerable potential as an ornamental for well-drained sandy soils, both inland and along the coast. Impossible to transplant successfully from the wild, it is sometimes grown from seed by native plant nurseries, which then sell to the public. It requires full sunlight and well-drained sandy soil to thrive.

Marsh-Elder—*Iva frutescens*

Asteraceae

Marsh-elder is a semiwoody shrub that grows along the Atlantic coast southward from New England to the southern end of the Florida peninsula and westward along the Gulf Coast from South Florida to Texas (Kartesz 2015). It has dull green, narrow, opposite, tardily deciduous leaves about 2 inches long with serrated edges. It grows multiple branches, often forming a dense shrub. The small flowers produced on terminal spikes are green and are wind pollinated. The stems are green at first, becoming tan on older stems.

Marsh-elder grows along the coast, often just above the high tide line, and on the edges of salt marshes and salt flats. It is sun loving, very salt tolerant, and not at all shade tolerant. It grows rapidly and has a short life span, probably not more than six or seven years. It commonly grows from 3 to 6 feet tall.

Marsh-elder probably benefits wildlife by providing nest sites for birds and food for insects.

Marsh-elder is sold primarily in native plant nurseries as a hedge or foundation plant for use near the coast in places too salty for most other plants.

Saltwort—*Batis maritima*

Bataceae

Saltwort grows in sunny spots exposed to occasional flooding by salt water such as the upper edges of salt marshes, salt flats, mud flats, and mangrove swamps along the coast from North Carolina southward into the Florida Keys and, following the Gulf Coast, into Texas (Godfrey 1988). It is a low shrub, usually not more than 2 to 3 feet tall, with swollen, succulent leaves. Its leaves are edible but very salty.

Saltwort is highly salt tolerant, withstanding periodic inundation by salt water and even growing into salt pans where salt concentrations are higher than in the ocean. Saltwort is not shade tolerant, almost always growing in full sun.

Saltwort grows in dense patches that slowly expand as the stems take root at the nodes. These patches tend to accumulate additional sand over time, thereby slightly elevating the surface. Saltwort can serve as a pioneer species, invading the edges of open flats and disturbed sites such as former mangrove swamps, potentially aiding the recruitment of mangrove trees onto mud flats and areas of former mangrove swamp that were previously destroyed (Milbrandt and Tinsley 2006).

Saltwort flowers are visited by butterflies, and its foliage is used as a larval food by the great southern white butterfly and perhaps the eastern pygmy-blue butterfly (Glassberg et al. 2000).

Saltwort leaves, roots, and seeds can also provide food for people. It is sometimes planted as an ornamental in coastal areas.

Hazel Alder—*Alnus serrulata*

Betulaceae

Hazel alder is a deciduous woody shrub in the birch family with alternately arranged dark green leaves, 2 to 3 inches long, that are slightly toothed on the margins. Its twigs are initially covered with shaggy brown hairs that wear off as the twigs become woody. Hazel alder is monoecious, with male and female flowers occurring in different structures on the same plant. Male flowers occur as hanging catkins about 2 inches long, while female flowers emerge in cone-like structures that bear small, winged nutlets (Godfrey 1988). The flowers are wind pollinated.

The alder genus is widespread, occurring in dense thickets from the Arctic tundra of Siberia, Alaska, and Northern Canada southward across much of Europe, Asia, and North America. Hazel alder ranges from Nova Scotia to Missouri and eastern Texas, with populations in the Florida Panhandle and scattered, isolated populations on the northeastern Florida peninsula from Marion County northward (Kartesz 2015). I have seen it in a few spots in the Ocala National Forest's forested wetlands and also on the north edge of Newnan's Lake in the delta area of Hatchet Creek, where a good stand of it occurs in a very wet and boggy stream delta.

Alders generally occur in wetlands and on stream banks and stream floodplains. They are not shade tolerant and are usually found in sunny situations. They are unusual in that they have nodules on their root systems containing bacteria that convert atmospheric nitrogen to organic nitrogen molecules usable by vascular plants (Bond 1956). This enables alders to grow well in nitrogen-deficient soils, and it also benefits the surrounding plant community by increasing the available nitrogen in these soils.

The wildlife value of alders in general is fairly low (Martin et al. 1951). Here in Florida, where alders are uncommon, the wildlife value is even lower although the fruits may be eaten to some extent by birds and the foliage browsed to some extent by deer.

Alders have no commercial value and minimal landscape use this far south, but, where they are more common, they are used for stream bank stabilization and wetland restoration.

River Birch—*Betula nigra*

Betulaceae

This southernmost species of birch is a medium-sized deciduous tree with simple, alternate, doubly serrated, arrowhead-shaped leaves 2 to 3 inches in length, and interesting bark. The leaves often have a soft, white- to cinnamon-colored fuzz on their undersides. The monoecious river birch produces separate male and female catkins on the same tree in the spring. The bark on saplings often appears as flat, peeling sheets that are light cinnamon to reddish-brown in color. On twigs, the bark is often silvery and, on old trunks, it is broken up into small scales that are dark gray.

River birch is a pioneer tree of riverbanks and floodplains from the Santa Fe and Suwannee Rivers west throughout the Florida Panhandle (although not within several miles of the coastline) and from there north and westward throughout much of the eastern United States (Grelen 1990, "*Betula Nigra L.*"). In Florida, it grows in scattered locations in association with black and coastal plain willows, bald-cypress, water-elm, Florida elm, water hickory, water locust, red maple, sweetgum, overcup oak, water oak, and swamp laurel oak. Farther north and west, sycamore and cottonwood are also associates. It grows fast and has a rather short life span. It is of moderate size, growing as tall as 90 feet with a trunk diameter of 2 feet, though most individuals are much smaller. It prefers soils containing clay or silt and will grow on very acidic soils as well as soils of neutral pH (Grelen 1990, "*Betula Nigra L.*"). It is not fire tolerant, drought tolerant, shade tolerant, salt tolerant, or tolerant of prolonged flooding. It can withstand the short-term flooding of the river floodplains where it grows but does not grow in swamps or marshes.

River birch has little recognized value for wildlife (Martin et al. 1951) although insects feed on the foliage, deer make some use of the seedlings and sprouts for browse, and beavers sometimes eat the twigs, small branches, and bark. It is one of the favorite trees of yellow-bellied sapsuckers, who, in wintertime Florida, drill lines of holes in the bark and then come back repeatedly to feed on both the sap and the insects attracted to the sap.

River birch is used occasionally for landscaping because of its unique appearance, the fact that it is a birch (name recognition), and because it is easy for nurseries to grow. However, it is short lived and usually not well adapted to the urban and suburban landscapes where it is planted: it often does not stay healthy or last very long. Nonetheless, with sufficient care, good soil, and a good irrigation system, it can be quite attractive.

American Hornbeam (Ironwood, Blue-Beech)—*Carpinus caroliniana*

Betulaceae

This small understory tree in the birch family is native to the eastern half of the United States, growing naturally as far south in Florida as Manatee County on the south-central Florida peninsula (Wunderlin et al. 2020). Its oldest name, of European origin, is hornbeam. However, in this country, it is often known as "ironwood," "musclewood," "blue-beech," or "water-beech."

American hornbeam is an attractive small tree with smooth, thin, light gray bark often mottled with lichens, and small, bright green, shiny, deciduous leaves with serrated edges. Male and female wind-pollinated flowers are borne on separate catkins on the same monoecious tree. The trunk is often sinuous in shape rather than being evenly rounded, thus giving rise to the name "musclewood." The crown is usually spreading and open. In Florida, this tree rarely gets over 40 feet tall or has a trunk over a foot in diameter, and the average mature tree is probably 30 feet tall with a 6-inch trunk diameter.

Hornbeam is an understory tree in hardwood and mixed hardwood and pine forests on moist to wet soils. Some of its most common associates are sweetgum, swamp laurel oak, water oak, and red maple. It is common to abundant along and on the floodplains of creeks and rivers and on the edges of swamps throughout Florida north of Orlando except for the western panhandle. Its range includes most of the eastern United States and a small area of southeastern Canada, plus parts of central and southern Mexico, Guatemala, and western Honduras (Metzger 1990). It is especially common in the low, moist to wet hardwood forests of the Big Bend region of Florida such as Gulf Hammock (although not near the coast because of its salt intolerance). It is also common in wet places in interior hardwood forests such as San Felasco Hammock and O'Leno State Park. Its growth rate is slow to intermediate, and it rarely lives more than 60 years. It is very shade tolerant, moderately flood tolerant, but not at all drought, salt, or fire tolerant. Hornbeam begins leafing out rather early each year and is sometimes damaged by late freezes.

The wildlife value of hornbeam is moderate. From my observations, its foliage provides good forage for white-tailed deer and for insects, and it produces nuts that are eaten to some extent by gray squirrel, wood duck, bobwhite quail, and yellow-rumped warbler (Martin et al. 1951). Wild turkeys also sometimes eat the nuts (Eaton 1992). Gray squirrels and beavers occasionally feed on the bark.

The strong, heavy, hard wood of hornbeam is not commercially valuable, but it is sometimes used locally for tool handles and other objects requiring strength and hardness. The wood is not at all rot resistant, but it makes good firewood if used before rot begins.

Hornbeam is potentially a good landscape tree. It grows well in fertilized and watered landscapes of all sorts and is quite attractive. Its leaves are small enough to filter down into grass, thus eliminating any need for raking. This is an ideal tree to plant under tall utility lines or in situations too wet or too shady for dogwood. Because of its thin, easily damaged bark, the base of the trunk needs to be protected from weed whackers and mowers in landscapes maintained by landscape contractors.

Hophornbeam (American Hophornbeam, Ironwood)—*Ostrya virginiana*

Betulaceae

Also known as "ironwood," hophornbeam is a small- to medium-sized understory tree in upland hardwood forests. It is slow growing and rarely lives more than 60 years. Taller growing and somewhat larger on average in Florida than hornbeam, the largest one currently reported, from Virginia, is 67 feet tall with a 20-inch trunk diameter (American Forests 2020). The average mature size in Florida is 20 to 40 feet tall with a trunk diameter of 4 to 8 inches. Hophornbeam is in the birch family, but it looks more like a small elm tree. Its deciduous leaves are shaped rather like elm leaves, have serrated edges, are somewhat soft and hairy, and are not shiny. The tree is monoecious, with male and female flowers borne on separate catkins on the same tree, and the flowers are wind pollinated. The bark is tan-brown and has a shredded appearance. The most distinctive attribute of hophornbeam is its fruit, which resembles hops hanging from the ends of the branches of sexually mature trees from August until December.

Hophornbeam occurs throughout the eastern half of the United States, ranging as far south on the Florida peninsula as Marion and Hernando Counties (Kartesz 2015). It is the most common understory tree in the upland hardwood forests of San Felasco Hammock in Alachua County. It only occurs on well drained, moderately fertile soil, usually in association with pignut hickory, white ash, winged elm, southern magnolia, laurel oak, water oak, swamp chestnut oak, basswood, black cherry, and sweetgum in Florida. It is very shade tolerant, only moderately drought tolerant, and not at all tolerant of flooding, fire, or salt.

Hophornbeam does not appear to have a high wildlife value. However, some mammals and birds eat the nuts it produces. Wildlife ecologist Katie Greenberg has seen pine siskins feeding on them (personal communication). Martin et al. (1951) add downy woodpecker, rose-breasted grosbeak, and purple finch to this list and note that deer browse the twigs to some extent. Eaton (1992) reports that wild turkeys occasionally eat the nuts.

Because of its small size and high shade tolerance, hophornbeam has some potential for landscape use. It is often well shaped, produces dense shade, and has an interesting appearance in autumn when in fruit. However, it has the drawbacks of not having pretty flowers or particularly pretty foliage and of having irritating hairs on the twigs, leaves, and fruit that can easily dislodge onto a person's skin causing severe itching.

Southern Catalpa—*Catalpa bignonioides*

Bignoniaceae

Southern catalpa is a medium-sized tree native to the banks and floodplains of major rivers in the Florida Panhandle and adjacent areas of southwestern Georgia, southern Alabama, and southeastern Mississippi (Godfrey 1988). The large, heart-shaped, deciduous leaves are arranged in either opposite or whorled fashion on the rather thick twigs. The showy clusters of April flowers are white with numerous flecks of dark purple and bright yellow color toward the center. The fruit is a long, flattened, bean-like capsule containing flattened, winged seeds.

Elsewhere in Florida and the rest of the southeastern United States, catalpa is planted as an ornamental. It escapes into the wild occasionally but is never problematic.

Catalpa is rather slow growing on dry upland sites, where it is often a small- to medium-sized tree, but it can grow to be 60 to 80 feet tall with a 3- to 4-foot trunk diameter at maturity on fertile soil. It does especially well on floodplains associated with streams and rivers. Catalpa appears to be able to live a long time, probably at least 100 years.

Catalpa is fairly tolerant of flooding, though not as tolerant as trees that commonly grow in swamps. It is moderately drought resistant and is not seriously bothered by insects and diseases except for the catalpa "worms" (actually caterpillars) that occasionally feed on the foliage. Even these do little harm. It will grow in light shade but prefers full sun, and while catalpa prefers rich soil with good moisture-holding capacity, it will grow on a wide variety of soils.

The wildlife value of catalpa is low to moderate. I have seen ruby-throated hummingbirds and bees visiting the flowers in April. The catalpa worms that sometimes feed in large numbers on the foliage may provide some bird species with a food resource. These so-called worms are, in fact, the caterpillars of the catalpa sphinx hawk moth. In addition to feeding these caterpillars, young, early spring catalpa leaves are sometimes fed upon by grasshoppers, which are also good food for birds. Old catalpa trees can live a long time after becoming hollow, providing long-lasting cavity sites useful for denning and nesting.

Catalpa is occasionally used in landscaping, mostly at homesites and beside boat ramps in rural areas by people who value catalpa worms as fish bait. The heartwood is rot resistant and has been used to make fence posts.

Crossvine—*Bignonia capreolata*

Bignoniaceae

Crossvine is a high-climbing vine with evergreen leaves that are arranged opposite each other, each with two leaflets also opposite each other, often with a tendril used for attaching to surfaces located between the two leaflets. The leaflets are much longer than wide, are heart shaped at the base, usually hang downward, and are evenly spaced along the length of a young, growing vine, giving the vine a unique appearance. The showy, 2-inch-long, stout blossoms, which bloom mid-March to mid-April, are trumpet-shaped red-orange flowers with yellow-orange color inside. They are shorter and wider than trumpet creeper flowers. The flowers are borne in clusters of two to five. The fruit is an elongated capsule containing winged seeds. The stem of a mature vine, which does not get as long or nearly as thick as that of a trumpet creeper or grape vine stem, has a covering of ridged, gray-brown bark.

Crossvine is an adaptable vine found commonly in both hardwood and mixed hardwood pine forests on both uplands and floodplains. It is a common invader of fencerows and subdivisions. When climbing trees, it remains on the trunks and branches rather than growing out over the foliage and therefore does no harm to the tree. It is not adapted to fire, although it will resprout following a fire. It is drought tolerant and somewhat flood tolerant. It grows well in partial shade, but it does best in full sun. It occurs throughout the southeastern United States, reaching as far south on the Florida peninsula as Polk County (Wunderlin et al. 2020).

While crossvine is blooming in March and April, it is an important nectar source for ruby-throated hummingbirds, and the foliage is browsed by swamp rabbits and white-tailed deer (Miller and Miller 1999). The flowers are also visited by native bees, butterflies, moths, flies, and ants.

Crossvine is occasionally used as an ornamental, with several cultivars for sale at nurseries. It climbs walls easily and is more manageable than trumpet creeper given its less wild and aggressive growth habit. It can be quite showy, producing a denser display of blooms than trumpet creeper, but it has a shorter blooming season.

Trumpet Creeper—*Campsis radicans*

Bignoniaceae

Trumpet creeper is a high-climbing vine with deciduous, pinnately compound, opposite leaves and red-orange, trumpet-shaped, 2½- to 3-inch-long flowers that bloom in clusters from May through October. It occurs throughout most of the eastern United States, growing as far south on the Florida peninsula as Collier County (Kartesz 2015). This vine can creep along the ground as a ground cover but more commonly climbs tree trunks and other surfaces by means of clinging rootlets at the nodes that stick tightly to any surface. When growing as a ground cover, or as a combination ground cover and high-climbing vine, trumpet creeper sometimes occurs as a large patch that is one large clone. Old vines have a light tan bark that is shredded into loose, vertical strips. The vine can climb high into the crown of tall trees, and the trunk of the vine can be from 1 to 4 inches in diameter. Because this vine climbs by clinging to surfaces, it generally remains below the upper crown, doing little or no damage to the tree.

The largest trumpet creeper vine I ever saw was hosted by a large loblolly pine tree in Gainesville. The vine had a trunk 6 inches in diameter and was growing on the branches of the pine throughout its 100-foot canopy without doing any harm to the pine tree because all of the vine's foliage remained below the pine's foliage. In spite of this and in spite of the beautiful flowers the trumpet vine produced each summer, a landscape worker eventually cut its trunk, killing it. (The old, naked-looking pine is still there, without its beautiful companion.)

Trumpet creeper is one of the kinds of vines that is being cut and eliminated from many public lands in Alachua County by persons unknown for unknown reasons, thereby greatly reducing the wildlife and ecological value of these sup-posedly "protected" natural areas.

Trumpet creeper is a widely adaptable vine, growing in both upland and floodplain hardwood forests and sometimes also in upland pine forests. It also invades fencerows and subdivisions, climbing on all sorts of surfaces. It is most common in hardwood forests, growing on river and creek floodplains and in the ecotones between swamp and lowland or upland hardwood forest. Moder-ately flood and shade tolerant, trumpet creeper stems are not fire tolerant, but it sprouts back after fires and will sometimes grow as a ground cover in places where fire is frequent and then climb tree trunks vigorously between fires. It prefers fertile soil with good moisture but is adaptable and moderately drought tolerant.

Trumpet creeper foliage is browsed by white-tailed deer, mostly in spring and summer, and the flowers are a preferred source of nectar for ruby-throated hum-mingbirds (Miller and Miller 1999). Trumpet creeper appears to be primarily adapted to pollination by the ruby-throated hummingbird (Weidensaul et al. 2019). Orchard orioles, sphinx moths, honey bees, bumblebees, and various other native bees (including some tiny species that are less than one centimeter long), and various native flies also visit the flowers, which bloom throughout the summer.

Trumpet creeper is sometimes planted as an ornamental for its showy flowers. It will cling to a pole or wall and make a dense cluster of foliage. Out in the sun, it will bloom prolifically all summer and into the fall. Perhaps one of the best ways of using it in a landscaped situation is to plant it next to a tree that is out in the open, so that it will grow up the trunk, out of reach of people, and produce flowers without hurting the tree. Another suggestion is to plant it in a large pot (to keep it from spreading) or in the ground next to a trellis or a 5- to 10-foot pole so that its flowers and the hummingbirds visiting them can be easily seen. One drawback to using this vine on surfaces within reach of people is that some people get a rash from touching the foliage. Perhaps for this reason, "cow-itch vine" is one of its common local names. Another drawback is that trumpet creeper can be very weedy, spreading both by runners and by seed.

Cat's Claw Vine—*Dolichandra unguis-cati (Macfadyena unguis-cati)*

Bignoniaceae

This extremely invasive, high-climbing, exotic vine from Central and South America is somewhat similar in appearance to crossvine when not flowering or fruiting. The leaves of cat's claw vine are opposite and nearly evergreen or tardily deciduous. The leaves on upwardly growing vines of this species are narrow and pointed and about 1 inch long and one quarter of an inch wide. The leaves on stems running along the ground as a ground cover are much broader, being about 1 inch wide and nearly 2 inches long. The climbing foliage has small claws in groups of three that cling very effectively to almost anything, including human skin. As the climbing vine grows larger, it grows many small roots that adhere tightly to tree trunks or other surfaces. The large, trumpet-shaped flowers are bright yellow. The seed pods are long, very flattened, and woody and contain flattened seeds with wings on both ends.

This vine has become quite common in the Gainesville urban area and many other cities and counties throughout Florida, including suburbs and rural areas containing old homesites. It seems to do best in fertile soil and can grow either in sun or shade, including dense shade. It can grow both as a ground cover and a high-climbing vine. It forms an oblong tuber on the taproot when young, which enables it to resprout repeatedly if the stem is cut or broken. As the vine spreads along the ground as a ground cover, it produces new taproots and tubers at many places along the way, thus continually increasing its hold on new ground. Cat's claw vine is extremely tenacious and can eventually produce such dense and abundant foliage that it smothers and kills its host tree, even if the tree is 100 feet tall. It spreads both by wind-distributed seed and by runners that can grow 30 feet or more per year.

Cat's claw vine is a serious threat to the ecology, flora, and fauna of native hardwood forests throughout Florida and the rest of the southeastern United States. It has no natural enemies here, and it is capable of outcompeting and smothering all of the native plants in these forests. The only reason it has not been a large-scale disaster to date is that it spreads rather slowly, mainly because the seeds are not spread by birds or other wildlife. However, once established in an area, it persistently increases its abundance and dominance. Its foliage and seeds are not eaten by insects or any other wild animal. One of the greatest invasive exotic plant threats to the upland hardwood forests of Florida, cat's claw vine should be exterminated whenever possible. Unfortunately, this is nearly impossible to do because its numerous tubers make both systemic herbicides and mechanical removal only marginally effective. The only real hope for controlling or eradicating it is for a biological control agent to be found and released, similar to what has been done with air-potato. In the meantime, it would be helpful to outlaw the sale of this plant, which is still widely promoted and advertised for sale on the Internet.

Pricklypear Cactus—*Opuntia* spp.

Cactaceae

Pricklypear cacti (genus *Opuntia*) grow "pads" that are actually broadly flattened stems. Single long spines and clusters of tiny short spines grow in evenly scattered patches on the pads along with tiny leaves that are often gone by the time the pad is fully formed. Both kinds of spines have backward pointing barbs on their tips that cause the spines to hold fast when imbedded in materials such as skin or leather. There are several forms of pricklypear cactus, with the taxonomy in flux to such a degree that my interpretation here may be incorrect. All of the species in our area have 2-inch-diameter yellow flowers in spring and red-purple fruits that ripen in late summer or fall. (There are additional species of *Opuntia* and additional genera of cacti in South Florida.)

Eastern Pricklypear—*Opuntia mesacantha lata* (*O. humifusa*)

Eastern pricklypear, the common pricklypear of Florida, is a common small shrub of sandy uplands throughout the state, except for the Florida Keys

(Wunderlin et al. 2020). Another variety, *Opuntia mesacantha* var. *mesacantha*, occupies the Florida Panhandle from Bay County westward, with an isolated population in Gadsden County (Wunderlin et al. 2020). Both varieties have long spines that are round in cross section and gray in color. They sometimes grow prostrate along the ground but often form clusters of semi-upright pads. They survive fires by sprouting back from the root collar and are very drought tolerant but not shade tolerant.

Beach Pricklypear—*Opuntia austrina* (*O. ammophila*)

Beach pricklypear is an upright form of pricklypear occurring in the Ocala National Forest and along some areas of the Atlantic coast of Florida. It grows to 3 or 4 feet tall with a woody trunk often 2 or 3 inches in diameter near the ground and numerous pads in the crown of the tree-shaped shrub. The spines are similar to those of eastern pricklypear. It is salt tolerant and very drought tolerant but not shade tolerant.

Erect Pricklypear—*Opuntia stricta*

Erect pricklypear (upright pricklypear, shell-mound pricklypear) occurs in scattered locations on sunny, well-drained uplands in scrub, along coastal dunes, and elsewhere in the coastal counties of Florida and in the other coastal southeastern states from North Carolina into Texas (Kartesz 2015). It has long spines that are yellow in color and somewhat flattened in cross section at the base. It is larger and more upright than eastern pricklypear but not as strongly upright and single-trunked as beach pricklypear. Erect pricklypear is salt tolerant and very drought tolerant but not at all shade tolerant.

The cochineal insect is often common to abundant on pricklypear pads, forming white, fuzzy patches wherein the insect hides. Cochineal is used to produce a scarlet-red dye for cloth and is also used as food coloring. The cactus moth, *Cactoblastis cactorum*, an invasive exotic moth that arrived in the Florida Keys in 1989, is currently a threat to cacti throughout Florida (Zimmermann et al. 2001). The adult moths lay eggs on the cacti. When the eggs hatch, the larvae burrow into the pads and eat the insides.

Pricklypear cacti support bees and other insect pollinators with their flowers and provide food for gopher tortoises and other wildlife such as birds, deer, and rodents with their pads, their fruit, and the seeds within their fruit. Martin et al. (1951) rate pricklypear cacti an outstandingly important wildlife food plant in areas where they are abundant.

Pricklypear cacti are sometimes used as ornamental plants because of their showy flowers and interesting growth forms, and sometimes the pads and fruit are used as human food. They are also sometimes considered a pest when found in pastures or lawns.

Sweetshrub—*Calycanthus floridus*

Calycanthaceae

Sweetshrub is a deciduous, thicket-forming shrub 5 to 10 feet tall with multiple, thin, stiff, upright stems and simple, pointed leaves about 1 inch wide and 3 inches long attached opposite each other on the stem. It produces fragrant flowers in the spring that are usually deep purplish-red to brownish-red. It is most common in the southern Appalachians but reaches the panhandle of Florida in scattered locations along streams on fertile soil.

As an understory shrub, sweetshrub is moderately shade tolerant but does best in full sun. It spreads by rhizomes and, also, much less frequently, by seed. It grows naturally on moist soil with good access to water but is moderately drought tolerant. It is free of noticeable insect or disease problems.

Sweetshrub is not often used for food by insects or browsing animals, but it is sometimes browsed by white-tailed deer (Miller and Miller 1999) and cottontail rabbits (Dorothy Lanning, landscape gardener, personal communication). As the flowers produce very little nectar, they are seldom visited by bees, wasps, flies, butterflies, or moths. Instead, they are regularly visited by several kinds of beetles that are attracted to their aroma, burrow into the flowers, and provide for pollen transfer (McCormack 1975).

Sweetshrub has been used as an ornamental for centuries and several cultivars are available in nurseries. It is attractive for its unusual flowers, which have a strong, pleasant, fruity odor. Because it forms thickets of upright stems from underground runners, it must be actively managed to prevent its spread into areas where it is not wanted. Historically, its dried flowers were used by early settlers in clothing storage drawers and cabinets to provide a sweet odor in an era when mildew and mold were common (David Hall, professional botanist, personal communication).

Sugarberry (Southern Hackberry)—*Celtis laevigata* (including notes on Hackberry [*C. occidentalis*] and Dwarf [Georgia] Hackberry [*C. tenuifolia*])

Cannabaceae

Sugarberry is a medium- to large-sized, very fast-growing, deciduous tree found on calcareous soil throughout Florida and the rest of the southeastern United States from Virginia into Texas (Kennedy 1990, "*Celtis laevigata*"). The closely related sister species hackberry (*Celtis occidentalis*) grows in the eastern United States mostly north of sugarberry's range, and a number of botanists consider some, or even all, of the sugarberry trees in Florida to be of this species. I am sticking with Godfrey's (1988) and Kennedy's (1990, "*Celtis laevigata*") interpretations of the distinctions between and ranges of these species. In any case, the ecological attributes of these two species are nearly identical.

Sugarberry's smooth, light gray bark supports a variable density of corky knobs and ridges. Some trunks are perfectly smooth while a very few have no smooth bark at all. Sugarberry typically produces cork in response to injuries, so, for instance, the dotted bands of old sapsucker holes are often represented by a band of corky protuberances. The leaves are asymmetrical, long-pointed, thin, and light green, which, along with the light color of the trunk and branches, gives the crown a pleasingly light and airy appearance. The tree is finely branched but without the excessive number of live branches and abundance of dead twigs and branches commonly occurring in some other kinds of trees. This is because of sugarberry's unusual ability to shed small twigs by forming an abscission layer at the base of a twig. On the other hand, when sugarberry is young and growing in full sun, it sometimes produces an abundance of stout, thorn-like twigs.

Sugarberry grows in several different habitats, preferring calcareous soils with some clay content. It is often one of the dominant trees in calcareous (limestone outcrop) upland hardwood forests on well-drained soil. Common associates here are red bay, sweetgum, winged elm, white ash, live oak, swamp chestnut oak, bluff oak, and pignut hickory. It also grows well in calcareous floodplain forests that inundate for no longer than one month at a time. Common associates here include sweetgum, live oak, swamp laurel oak, water hickory, overcup oak, red maple, and Florida elm. In low elevation hardwood forests along the Gulf Coast, sugarberry grows in areas occasionally flooded by brackish water and is moderately salt tolerant. Finally, sugarberry is a common invader under power lines, along fencerows, and in residential and urban areas because of its ability to reproduce prolifically from seeds distributed widely as they are carried in the guts of the birds that eat its red berries. Sugarberry will also reproduce by root sprouts. Both seedlings and root sprouts are moderately tolerant of shade but grow poorly in shady situations. Sugarberry requires full sun to thrive, and mature trees are not shade tolerant.

Sugarberries are particularly common on the hillsides around the rim of Payne's Prairie in Alachua County. Sugarberry was once abundant in Sugarfoot Hammock on the west side of Gainesville and is common on the upper edge of the floodplains of the Ocklawaha, Silver, Santa Fe, and Suwannee Rivers. Several mature sugarberry trees can be seen from the trail south of Millhopper Road in San Felasco Hammock Preserve State Park. In 2014, the largest sugarberry in Florida, from Hamilton County, was 92 feet tall with a trunk circumference of 259 inches (nearly 7 feet in diameter) and a crown spread of 76 feet (Florida Champion Trees Database 2020). A seemingly distinctive form of this tree is moderately common in what is left of Gulf Hammock, where it rarely gets over 80 feet tall or 18 inches in trunk diameter. The bark of the Gulf Hammock sugarberry rarely has corky knobs but sometimes has a rough, flaky texture.

Sugarberry rarely lives more than 100 years. Its weak wood eventually contributes to the breaking of a large branch somewhere, and this quickly leads to the rotting away of the inside of the trunk. Once this process begins, the tree is doomed because the wood is not at all rot resistant and because sugarberry is very poor at healing its wounds. It is often severely damaged by the actions of tree surgeons for this reason. Even if a large branch is carefully cut and treated with "tree paint" of some sort, the resulting wound will heal very slowly and internal rotting will commence. (Incidentally, tree paint is now generally regarded as unhelpful as a wound protectant for most trees.)

Sugarberry is quite beneficial for many species of wildlife. Along the Suwannee and lower Santa Fe Rivers, sugarberry is one of the tree species most favored as food by beavers. Many of the large trees near the river have been killed by having all the bark eaten from their lower trunks. Squirrels also eat the bark of sugarberry, sometimes damaging the branches to such an extent that they break after beginning to rot inside. This is particularly damaging to sugarberry, with its inability to effectively heal wounds and its ready susceptibility to decay. Deer, cattle, and rabbits eat the seedlings, sprouts, and twigs of this tree in preference to most others, and the fruits, which ripen in September and then remain available on the tree throughout the winter until March, are eaten by a wide variety of birds and mammals. Martin et al. (1951) record significant consumption of sugarberry fruit by Florida birds, including wild turkey, bluebird, cardinal, fish crow, flicker, yellow-bellied sapsucker, mockingbird, brown thrasher, robin, and cedar waxwing. Terres (1980) lists sugarberries as a food item for the rusty blackbird. Other species that eat some quantity of sugarberry fruit include gray fox, opossum, raccoon, striped skunk, squirrels, and bobwhite quail (Martin et al. 1951).

Sugarberry is host to a number of gall-forming and leaf-eating insects which, in turn, provide food for insect-eating birds and other creatures. The caterpillars of the American snout, question mark, hackberry emperor, and tawny emperor butterflies feed on the foliage. Two species of exotic insects have recently become established in this country that suck sap from the foliage and do the tree some damage (Dr. James Meeker, USDA Forest Service entomologist, personal communication). One of these tiny insects, the Asian woolly hackberry aphid, is now an important part of the diet of small migrating songbirds such as kinglets, warblers, and vireos as they pass through North Florida. Overall, sugarberry has a relatively high wildlife habitat value.

Native North Americans used sugarberry fruits to make a sweet drink. The fruits are sweet and edible but are dry and contain very little flesh. The soft, weak wood is used commercially for making packing crates, plywood, and pulpwood.

Sugarberry has a mix of good and bad qualities for landscape use. On the plus side are sugarberry's rapid growth under good conditions, its unique and beautiful appearance as a large shade tree, its being easy to grow and transplant, its adaptability to calcareous conditions which makes it well suited for growing where there is a lot of pavement, and its being able to survive the air pollution of inner cities. It is also quite drought resistant. On the negative side, its weak, rot-susceptible wood and poor ability to heal wounds make it a liability near buildings, utility lines, streets, and parking areas unless its trunk and larger branches are in excellent condition. It is also one of the trees most vulnerable to uprooting or breakage by high winds. In areas where gray squirrels are common, sugarberry is likely to be damaged by bark feeding unless it is located well out in the open where squirrels rarely visit. Because of its undesirable qualities, sugarberry is rarely planted or grown in nurseries. Most of the sugarberry trees currently located in landscaped situations came up on their own.

One other species of *Celtis* occurs in northern Florida—dwarf hackberry or Georgia hackberry (*Celtis tenuifolia*)—which is an uncommon small tree of the panhandle. There are at least two populations of this tree in the vicinity of Gainesville, one on the University of Florida main campus and the other east of Gainesville just north of Payne's Prairie. Dwarf hackberry is fairly common at Wakulla Springs State Park and in the bluff and ravine country on the east side of the Apalachicola River floodplain. Dwarf hackberry is a shrub or small tree occurring on dry uplands and slopes in places that are, or were, longleaf pine, southern red oak, and mockernut hickory forest (perhaps typically where limestone is at or near the surface) (Godfrey 1988). Its leaves are rougher on the surface and more rounded in shape than sugarberry leaves and the berries, which begin ripening around May 15th, turn black instead of red.

Coral Honeysuckle (Trumpet Honeysuckle)—*Lonicera sempervirens*

Caprifoliaceae

Coral honeysuckle is a small native vine with semievergreen opposite leaves and bright red, 2-inch-long, tubular bisexual flowers in small clusters at the ends of twigs. It produces small, succulent red berries containing a few hard seeds. The best identification features are the flowers and the leaves immediately behind the flowers or fruits, which are fused together to form a disc behind each flower cluster. Coral honeysuckle occurs throughout the southeastern United States, ranging as far south as the southcentral Florida peninsula (Wunderlin et al. 2020).

This vine never gets large or climbs higher than about 10 feet. It is most often seen along fencerows, on forest edges, or in shrubby situations on well-drained soil where it can take advantage of the sunlight. It is neither shade tolerant nor flood tolerant. It is not an aggressive vine and is not especially common, being scattered about here and there in sunny locations.

Coral honeysuckle's wildlife value is good. Its flowers are a favorite of ruby-throated hummingbirds and are attractive to butterflies and bees. The berries are eaten occasionally by wild turkeys, bobwhite quail, and songbirds, and the foliage is browsed by white-tailed deer and cottontail rabbits. Since this vine is not nearly as common as the exotic Japanese honeysuckle, it is not considered as valuable to wildlife by some authors (Martin et al. 1951; Miller and Miller 1999), but, in my experience, where it is present, it is more valuable for hummingbirds and butterflies and at least as valuable for the other wildlife species with the exception of rabbits, which like to both eat and hide in the thick tangles of Japanese honeysuckle. The golden mouse makes nests in the vine tangles and eats the berries and/or seeds of both honeysuckle species.

This beautiful little vine is available from plant nurseries and is often used as an ornamental. If provided with a sunny place to climb on a fence or trellis, it will produce flowers all summer long, attracting butterflies and hummingbirds.

Japanese Honeysuckle—*Lonicera japonica*

Caprifoliaceae

Brought in from Asia and widely naturalized in the eastern United States and elsewhere, this very aggressive invasive exotic vine can damage native ecosystems. It is a semievergreen vine that has 1- to 2-inch-long, soft, dull green leaves and produces 1½-inch-long white bisexual flowers in abundance in May and black berries in the fall. The flowers turn yellow as they age and have abundant, sweet nectar.

Japanese honeysuckle is abundant and highly aggressive in the Piedmont and upper coastal plain of the southeastern United States but is less of a problem in lower coastal plain areas such as Florida. In places where it has been planted near Gainesville, it grows well as long as it remains in full sunlight, but as other plants grow up and begin producing shade, the honeysuckle is crowded out. Like other honeysuckle species, it does best where it is in full sun for at least part of the day. One reason it has become a threat to native ecosystems is that it was widely planted in the past and frequently escaped cultivation.

Japanese honeysuckle reproduces by seed, spread widely by birds that eat the fruit, by underground runners, and by the aboveground growth of vines, which can creep along the surface and attach to the soil by roots that form at the nodes.

The wildlife value of this vine is a mix of good and bad attributes. On the benefits side, deer and rabbits like to eat the foliage, which can be a significant part of their diet (Miller and Miller 1999). Rabbits like to hide and nest in the dense vine tangles that sometimes form at ground level, the flowers are valuable nectar sources for bees and butterflies, and the black berries are eaten by birds. The bad news is that a thick growth of this vine can smother other plants, including entire native ecosystems. In many cases, these smothered ecosystems were essential habitat for many species of wildlife.

Japanese honeysuckle has been used in the past both as an ornamental plant for landscaping and as a plant used for stabilizing roadside banks. Such use should now be discouraged, as it has resulted in damage to native ecosystems and great expense in eradication efforts.

Strawberrybush—*Euonymus americanus* (including a note on Eastern Wahoo [*E. atropurpureus*])

Celastraceae

Strawberrybush is a small, multistemmed shrub 3 to 6 feet tall, with tardily deciduous, opposite, shiny green leaves and slender, somewhat four-sided, green stems. The 1- to 3-inch-long leaves have serrated edges and pointed tips. It produces small, greenish-white bisexual flowers in spring and oddly shaped fruits in the fall: capsules with a red husk that somewhat resembles a strawberry. As the fruit ripens, the thick and roughly textured husk splits into five pointed partitions that lift up to expose up to five orange, fleshy seeds that hang down from the husk's edges.

This species occurs throughout most of the southeastern United States including the Florida Panhandle and as far south on the peninsula as Highlands County (Kartesz 2015). There is a related species called "eastern wahoo," *Euonymus atropurpureus*, that grows farther north with a widely isolated population along the Apalachicola River in the Florida Panhandle. Both of these species are uncommon, growing on moist, fertile soil in hardwood forests, often along small streams. Although small, a strawberrybush can live at least 50 years as evidenced by one in my yard. They are very shade tolerant but not drought tolerant and will not withstand browsing by large numbers of deer. Where deer are abundant, these plants disappear from the forest (Miller and Miller 1999).

The wildlife value of these shrubs is modest. They are preferred deer and rabbit food but are not abundant enough to be greatly important in the overall diet of deer. The seeds are eaten and distributed to some extent by wild turkeys and other birds (Miller and Miller 1999).

Both strawberrybush and eastern wahoo are used as ornamental shrubs to a limited extent, mostly well north of the state of Florida.

Gopher-Apple—*Geobalanus oblongifolius* (*Licania michauxii*)

Chrysobalanaceae

Gopher-apple is a low-growing, deciduous, clonal shrub with stout underground stems from which slender stems rise 3 to 9 inches above the surface. Its oblong leaves, 2 to 3 inches long by half an inch wide, may be shiny green on both sides or have a white pubescence underneath. Terminal clusters of small pale yellow to greenish-white bisexual flowers appear in May and June, followed in late summer and fall by an olive-sized and shaped fruit (a drupe) with a single hard seed inside. The fruit starts out green, becoming white as it ripens and often developing some red or purple color at one end.

Gopher-apple occurs from South Carolina southwestward throughout the coastal plain into Louisiana, including all of the high, sandy parts of Florida. Its range closely mirrors the range of the gopher tortoise. Gopher-apple grows clonal patches that can be extensive. It is very drought tolerant, fire tolerant, and salt spray tolerant but not flood tolerant. It requires at least partial sunlight and does best in full sun, inhabiting sunny habitats such as longleaf pine–turkey oak sandhills, sand pine scrub, oak scrub, rosemary scrub, and coastal sand dunes.

The wildlife value of gopher-apple is high. Its clonal patches provide cover for small ground-dwelling species such as lizards, snakes, and mice. The flowers are a nectar source for bees, wasps, flies, and butterflies, and the fruit is highly sought-after by many species, including gopher tortoises.

Gopher-apple has some ornamental potential as a ground cover on well-drained, sunny, sandy sites, including sites next to the ocean, but it is nearly impossible to transplant from the wild or grow from either stem or root cuttings. It is best to buy plants from a nursery or to plant seeds an inch or two deep in the ground. (The seeds should not be allowed to dry.)

Sweet Pepperbush (Summersweet)—*Clethra alnifolia*

Clethraceae

Sweet pepperbush grows on the Atlantic coastal plain from Maine southward into northernmost Florida and then, in scattered locations, south to Polk County and west into eastern Texas (Kartesz 2015). In Florida, it is most common in the central and western panhandle. It is a deciduous shrub with alternate, 2- to 5-inch-long leaves similar in appearance to alder leaves and with sweet smelling, showy white flower spikes (racemes) in midsummer. It is normally 3 to 6 feet tall, spreading by root suckers to form clonal thickets.

This shrub usually grows on the edges of streams and wetlands and in seepage bogs, bayheads, and very wet pine flatwoods forests. It prefers constantly moist to wet, acidic soil and does best with ample sunlight but also tolerates moderate shade. It is moderately tolerant of salt spray. Sweet pepperbush is fire tolerant in that it will sprout back vigorously following a fire. It is not tolerant of lime, non-acidic soils, drought, or prolonged deep flooding.

Sweet pepperbush is a preferred browse plant for white-tailed deer, and the flowers are very attractive to pollinating insects (Miller and Miller 1999).

Sweet pepperbush, usually called "summersweet" in the nursery trade, is used as an ornamental plant with several named cultivars. It is easy to grow if provided with acidic soil and ample moisture, much as one would provide for a blueberry bush or a wild azalea.

St. John's Wort—*Hypericum* spp.

Clustiaceae

According to Godfrey (1988), 18 woody species in the genus *Hypericum* occur in northern Florida; of these he notes that "Identification using only small pieces of branches is virtually impossible" (346). Wunderlin (1998) lists 29 species for Florida as a whole. Although individual species sometimes have interesting individual names, most plants in this genus are commonly referred to as St. John's wort. Internationally, the best known species, common St. John's wort, *Hypericum perforatum*, native to Eurasia, is a problem invasive exotic weed on the high plains of North America and in many other places around the world but is not known from Florida. The native Florida species of *Hypericum* are small shrubs or herbs with small, opposite, evergreen leaves and yellow, bisexual flowers. Most species occur in regions of pine flatwoods forest on wet, acidic soil, but some are upland plants that occur either on fertile soil or on deep, infertile sandy soil. Some species are broadly adapted.

The most broadly adapted species is St. Andrew's cross (*Hypericum hypericoides*), which is probably the most common species in Florida, occurring in almost any habitat and sometimes coming up in suburban landscapes. Four other common species are St. Peter's wort (*Hypericum crux-andreae*), which inhabits acidic pinelands ranging from very wet to very dry; fourpetal St. John's wort (*Hypericum tetrapetalum*), which inhabits pine flatwoods forests; coastalplain St. John's wort (*Hypericum brachyphyllum*), which inhabits pine flatwoods forests; and flatwoods St. John's wort (*Hypericum microsepalum*), which inhabits wet pine flatwoods in the Florida Panhandle from Dixie County to Walton County (Godfrey 1988; Wunderlin et al. 2020).

The woody species of *Hypericum*, typically small shrubs with weak, slender stems, are usually a minor component of their surrounding vegetation. Two of the species, sandweed (*Hypericum fasciculatum*) and Apalachicola St. John's wort (*Hypericum chapmanii*), are exceptions in that they can be abundant in open, sunny, shallow wetlands, and they can also be fairly tall. Sandweed, which grows in scattered spots in acidic wetlands throughout most of Florida, is often 3 to 6 feet tall while Apalachicola St. John's wort, which is restricted to similar habitats in the Florida Panhandle, is often 6 to 9 feet tall (Godfrey 1988).

St. John's wort species are poor deer browse but supply some seed for birds such as bobwhite quail (Miller and Miller 1999). They also supply food for some insects, and the flowers are visited by insects.

Human use of the Florida species is limited, but some species are planted occasionally as ornamentals. The common St. John's wort of Eurasia is one of the most ancient, commonly used, and thoroughly studied of herbal medicinal plants. Its most common use is for treating mild to moderate depression.

Flowering Dogwood—*Cornus florida*

Cornaceae

This beautiful little tree is easily identified in early spring when it produces blossoms consisting of central clusters of small greenish-yellow, bisexual flowers surrounded by four white petal-like bracts, the whole bloom being about 2 inches across. The opposite leaves, and thus the opposite branching pattern, help identify the tree the rest of the year. The distinctive appearance of the twigs, the newest of which are green, combined with the distinctive rough brown bark that forms a pattern of small squares on the main trunk, are additional reliable clues. One trick to identifying dogwood, if you have a green leaf, is to tear the leaf in half, being careful not to pull the halves more than one eighth of an inch apart. If there are fine threads stretched between the two halves, you know you've found a dogwood. Only the dogwood genus (*Cornus*) has these threads. Flowering dogwood is a native tree in most of the eastern United States, occurring as a native tree as far south on the Florida peninsula as Orlando and occurring as a planted tree in many other areas.

Flowering dogwood rarely grows over 40 feet tall or achieves a trunk diameter of more than 1 foot, and its growth rate is slow to moderate. The largest one reported from Florida is 43 feet tall with a trunk diameter of 29 inches (Florida Champion Trees Database 2020). Dogwood does best in light shade but can grow in full sun if its crown is able to spread widely enough to shade the trunk from sunlight. It can also grow in moderately dense shade but is less healthy and does not bloom as well. It is not as shade tolerant as hophornbeam or American hornbeam.

Although best known as an ornamental, flowering dogwood is also native, growing in openings in well-drained upland hardwood forests but being most common and growing best in transition areas between hardwood forests and upland pine forests. For such a small tree, it is quite fire tolerant and well adapted to living beneath a pine overstory on well-drained soils of moderate fertility. It does not survive flooding, poorly drained soils, or salty soils at all well. Dogwood twigs are often killed by stem borers, which do not severely threaten the overall health of the tree. The main trunk is also sometimes attacked by borers, especially if it is exposed to direct sunlight.

A much greater threat is dogwood anthracnose (dogwood blight), *Discula destructiva*. This disease is spreading south since its introduction in 1978 in the northeastern United States, and so far it has reached central Alabama and central Georgia (Anderson et al. 1993). Hopefully, it will either not spread to Florida or will not be too serious here, as the disease prefers moist, cool climates. From 1994 to the present, the dogwoods in Gainesville, Florida, have been severely damaged by another fungal disease, a white powdery mildew caused by fungi in the genera *Microsphaera* and *Phyllactinia* that grow on the leaves. Many dogwood trees have died of the disease. This particular fungal disease was not a problem prior to 1994 and is much worse some years than others—perhaps years with higher than average humidity in spring and summer.

The wildlife value of dogwood is high. Its bright red fruits (drupes) ripen in early September and stay on the tree into winter, providing a wide range of birds and mammals with a nutritious food high in calcium and fat at a time when animals need to build up fat reserves in preparation for winter. One bird that occasionally visits as far south as Gainesville, the rose-breasted grosbeak, concentrates on dogwood fruits while it is here, at least until the mulberries are ripe. I saw more of these birds in the winter of 1992–1993 than ever before. As dogwoods farther north die from the blight, this bird, and perhaps other migratory fruit eaters such as robins, thrushes, and cedar waxwings, may be forced to visit Florida in greater numbers in their search for food.

Resident birds that feed on dogwood fruit include wild turkey, bobwhite quail, wood duck, cardinal, bluebird, catbird, mockingbird, wood thrush, red-eyed vireo, pine warbler, flicker, and most of the other woodpeckers (Martin et al. 1951). The red-headed woodpecker is especially fond of them, and they are also eaten by the rusty blackbird (Terres 1980). Other birds observed feeding on them in the Gainesville area include the summer tanager, scarlet tanager, Swainson's thrush, and veery (Rex Rowan, expert birder and naturalist, personal communication). Mammals that feed on dogwood fruits (and also stems in the case of deer) include white-tailed deer, gray squirrel, and fox squirrel (Martin et al. 1951). The spring flowers are visited by various native bees and other insects.

The wood of dogwood is very hard and strong and was once used for making shuttles for textile mills. The wood is also beautiful and highly suitable for wooden carvings, tool handles, and other specialty items. The tree has become so uncommon in the wild, however, that the use of its wood now usually occurs only when one in a landscaped situation dies or needs to be removed for some other reason. Then, if it is used at all, it is usually cut for firewood.

The landscape value of dogwood is high. It was one of the first native trees to be recognized for its beauty and has remained the most popular small, native ornamental tree ever since. It has a pleasing shape, beautiful and abundant blooms in March, pretty fall leaf color, and pretty red fruit in fall and winter. Planting dogwood in greater abundance than it formerly grew in the wild is justified by its high wildlife value and the fact that it is rapidly disappearing farther north as it falls victim to dogwood anthracnose. Dogwood is increasingly less common here as well because of other disease problems to which it is susceptible. Planting it in full sun and away from other trees to avoid crowding seems to be the best way to minimize its current disease vulnerabilities. Also, some trees are more disease resistant than others, so growing seedlings from hardy trees should increase its ability to combat infection. When planted in full sun, a dogwood tree should not be pruned at all, and the lower branches especially should not be removed. If pruned, the exposed main trunk may attract stem borers that will weaken and potentially kill the tree.

Swamp Dogwood—*Cornus foemina*

Cornaceae

This small tree has leaves similar to those of flowering dogwood except that they are thinner and narrower. It differs from both flowering dogwood and roughleaf dogwood in having leaves that are smooth to the touch. The trunk of swamp dogwood is often multistemmed, and the bark is smooth except on unusually large stems. The flowers and fruits occur in clusters. The small bisexual flowers, in full bloom on April 15th, are cream colored, and the small roundish fruits (drupes), which ripen in mid-August, are blue when ripe.

On average, swamp dogwood is much smaller than flowering dogwood, rarely achieving more than 1 or 2 inches in trunk diameter or growing more than 10 feet tall. An unusually large one in Gad's Bay in Levy County was 36 feet tall and had a trunk diameter of 4 inches.

Swamp dogwood grows throughout the southeastern coastal plain from Virginia into Texas and as far south on the Florida peninsula as the Big Cypress Swamp in Monroe County. It typically grows in wet hammocks, bottomland hardwood forests, river floodplains, and the edges of swamps as an understory tree or shrub under a canopy of hardwood trees. It will also grow into the edges of freshwater marshes. While swamp dogwood usually grows in the shade and is quite shade tolerant, it does best in full sun, especially in hardwood forest areas that are opened to additional sunlight by cattle grazing. It is tolerant of flooding but not at all drought tolerant. Its life span is short, probably not more than about 20 years, although resprouting from the root collar may extend its life somewhat.

The wildlife value of this little tree is pretty high. Its flowers are attractive to insects, its foliage is browsed by deer (and cattle), and its quarter-inch-diameter fruits, when they ripen in mid-August, are consumed by many of the same wildlife species that consume flowering dogwood fruits.

This little dogwood is not currently used in landscaping, but it could be purposefully installed along the edges of retention areas and small streams and ditches as wildlife plantings.

Roughleaf Dogwood (Dwarf Dogwood)—*Cornus asperifolia*

Cornaceae

This is a very small dogwood. Its leaves are small, narrow, and rough to the touch. Like those of flowering dogwood, its leaves are opposite each other on the stem, and consequently the stems have an opposite branching pattern as well. The small, cream-colored, bisexual flowers and small blue fruits (drupes) are borne in clusters. The bark is not as smooth as that of swamp dogwood, being roughened on stems as small as 1 inch in diameter. Although very small in stature—getting no more than about 10 feet tall (and usually half that height) with a trunk 1 or 2 inches in diameter—this species is normally single trunked and tree shaped, often with a flat-topped crown.

Roughleaf dogwood grows as an understory tree in coastal plain upland hardwood forests from southeastern North Carolina south into the central Florida peninsula and west through Georgia and the Florida Panhandle into Alabama and Mississippi (Godfrey 1988; Kartesz 2015). It prefers well-drained, fertile soil and is more common where there is limerock near the surface. It is very shade tolerant, but it grows best in partial shade or full sunlight. It reproduces both by seed and by underground runners. Although browsed to some extent by both deer and cattle, this little tree does well in hardwood forests that are opened to increased sunlight by cattle grazing.

The wildlife value of this little tree is high relative to the small amount of space it occupies. Its flowers are attractive to insects, the foliage is browsed by deer, and, once they ripen in mid-August, the quarter-inch-diameter fruits are consumed by many birds. Mockingbirds, brown thrashers, Carolina wrens, and cardinals eat them in my yard. Roughleaf dogwood fruit is the only fruit I have seen wrens eating other than poison ivy fruit. Brown thrashers seem to be especially fond of it.

This little dogwood is not ordinarily used in landscaping, although I enjoy it in my yard. Its small size, high wildlife value, and dwarf tree form provide a unique combination that might be desirable in situations where space is limited, such as beneath low power lines. Its high shade tolerance makes it a good choice for dense shade conditions most other trees and shrubs cannot tolerate.

Black Titi—*Cliftonia monophylla*

Cyrillaceae

A small- to medium-sized wetland tree of the Florida Panhandle from Jefferson County westward, black titi also occurs across southern Georgia and in the southern tip of Alabama and Mississippi. There is a small, isolated black titi population in Clay County, Florida, and two even smaller isolated populations in the Ocala National Forest in Marion County. In 1993, the largest reported tree was 58 feet tall with a 2-foot trunk diameter and a 30-foot crown spread (Ward and Ing 1997), but trunks are usually less than 1 foot in diameter.

Black titi's growth habit is much straighter and taller than that of white titi. Its dark green, somewhat leathery, evergreen leaves are typically 1 to 4 inches long and three quarters of an inch to 1 inch wide. They are simple, alternate, entire margined, without a noticeable leaf stalk, and with veins, other than the midvein, barely, if at all, visible. The leaves are shiny on top with a somewhat bluish-white waxy coating underneath. Conspicuous clusters of white, bisexual flowers are borne in early spring, followed quickly by small, dry drupes possessing two to five small wings. The bark is dark gray-brown to nearly black.

Black titi is usually either entirely absent or extremely abundant, forming dense thickets. It can be the dominant plant on wet acid soil on stream banks and in large shrub bog and seepage bog areas, often occurring in nearly pure stands. If fire is excluded or prescribed burning is not frequent enough, black titi invades uphill from its original site into surrounding pine flatwoods and downhill into herb and pitcher plant bogs, spreading prolifically by root suckers and by seed. Common associates include Atlantic white cedar, slash pine, pond pine, pond-cypress, white titi, sweet pepperbush, gallberry, large gallberry, waxmyrtle, swamp candleberry, odorless waxmyrtle, fetterbush, sweetbay, and swamp red bay.

Black titi is not really a swamp tree. Swamps that dry out and then flood deeply for long periods are usually devoid of this plant, as it does not tolerate this kind of flooding. It also does not tolerate the dense shade of swamps with closed canopies of tall trees, and it is not fire tolerant. What it does tolerate better than most other plants is constantly wet, extremely acidic soil of low fertility.

The wildlife value of black titi is quite variable, depending on the situation. The value of tree-sized thickets is largely restricted to the value of its flowers for bees and other pollinating insects. It is an excellent honey plant (Miller and Miller 1999). Shorter plants, especially those repeatedly pruned back by fire, are valuable browse for white-tailed deer, especially in winter (Miller and Miller 1999). The leaves are rarely eaten by insects, and the fruits do not seem to have much value for wildlife. In fact, it could be argued that black titi sometimes has a negative wildlife impact. Areas where this species has established large, dense, tall monocultures due to fire suppression provide less varied and valuable habitat for wildlife in general than areas kept open and titi-free with frequent fires.

Black titi is rarely used in landscaping, but its dark, shiny, evergreen foliage and pretty, early spring flower clusters could make it an attractive small tree in wet seepage areas in full sun. Caution is advised, however, because of its ability to spread aggressively.

White Titi (Titi)—*Cyrilla racemiflora*

Cyrillaceae

White titi is normally a wetland shrub or small tree. The simple, alternate, smoothly glabrous, slightly leathery leaves are from a half inch to 4 inches long, deciduous to somewhat persistent, and have smooth, unlobed margins. They are generally more than twice as long as wide, almost without a noticeable leaf stalk, and with veins on either side of the midvein visible. The leaves can produce some red color in the late fall. The first-year twigs are tan and slightly angled. The bark is medium reddish-brown and usually thin and slightly roughened. However, in Mallory Swamp in Lafayette and Dixie Counties, there are several large individuals with thick, blocky, corky bark on their lower trunks. The small, white, bisexual flowers and dry fruits are borne on long, narrow, cylindrical spikes (racemes) that radiate out from the tip of the previous year's growth. The flowers bloom in June.

White titi grows on the coastal plain from southeastern Virginia southwestward into southeastern Texas (Kartesz 2015): the wetland form extends as far south in Florida as Marion and Levy Counties while the scrub form reaches south through the central Florida peninsula to Highlands County. This species is quite variable in overall size, leaf size, bark form, and habitat. The largest one, a national champion from Washington County, Florida, measured 16 inches in trunk diameter, 44 feet in height, and 30 feet in crown spread (Ward and Ing 1997). White titi's trunk is usually short and crooked but may be tall and straight. The leaves of white titi plants on the Lake Wales Ridge are usually quite small, and this "scrub form" is considered a distinct species by some botanists. However, there are many intermediate forms elsewhere that confuse the picture. Even in scrub habitat, the scrub form is always associated with at least some moisture (David Hall, professional botanist, personal communication).

White titi is usually found in acid bogs and bayheads, seepage wetlands, wet depressions in the pine flatwoods, and in floodplains along streams. It often extends uphill from these habitats to some extent. Though somewhat erratic in its distribution, it can be abundant where it grows. In some bogs, often called "titi bogs," the plants are so abundant that walking through them becomes a difficult task, especially as, in such bogs, their most abundant associate is usually laurel greenbriar. Other common associates include slash pine, pond pine, pond-cypress, bald-cypress, red maple, swamp tupelo, sweetbay, loblolly bay, swamp red bay, waxmyrtle, swamp dogwood, elderberry, tall blackberry, gallberry, large gallberry, fetterbush, Virginia-willow, Sebastian-bush, sweet pepperbush, and, in the panhandle, black titi.

White titi reproduces both by seed and by root sprouts. The root suckering sometimes produces dense thickets. It is very tolerant of acidic soils and constantly wet conditions but is less tolerant of prolonged, deepwater flooding than bald-cypress, swamp tupelo, and buttonbush. It is moderately shade tolerant. It tolerates fire by sprouting back vigorously from the root collar and roots after the stems are killed. Stems with the kind of thick corky bark that occurs, for example, on white titi in Mallory Swamp, would apparently be somewhat resistant to stem damage from ground fires.

Although the hard fruits have not been observed as useful for wildlife, the young shoots of white titi are a preferred browse for white-tailed deer (Miller and Miller 1999), and the flowers attract insects, including honey bees. Along streams, it is often browsed by white-tailed deer, and it provides a habitat structure in shrub bogs for birds such as the common yellowthroat, the hooded warbler, and Swainson's warbler. Titi bogs provide important escape cover for deer and, especially, for black bears (where they still exist), particularly when they are being hunted.

White titi has no commercial value. The stems could be used for firewood and to make walking sticks. It could possibly also be used in landscaping as its dark green, lustrous foliage and delicate, graceful bloom spikes are attractive. It is hardy, requiring no special care, and appears to be long lived. The one planted in my yard is healthy and vigorous at age 50.

Persimmon—*Diospyros virginiana*

Ebenaceae

Persimmon is usually a small- to medium-sized tree with a single, upright, moderately slender, and usually rather straight trunk. Its dark green leaves are often spotted with black, and the black bark on old trunks and branches is divided into distinctive small squares. Although the simple, smooth, shiny, alternate leaves with entire margins look similar to those of blackgum and swamp tupelo (genus *Nyssa*), these trees can be distinguished from persimmon by the bud at their twigs' ends. All the tupelos have twigs that end in a terminal bud. Persimmon does not have a terminal bud. Its branches end with the small dead remnant of the end of the twig, with the last lateral bud beside it serving as a terminal bud.

Blackgum, swamp tupelo, and persimmon leaves all display fall colors, but blackgum and swamp tupelo leaves become bright scarlet, whereas persimmon leaves turn purple. The black bark of old persimmon trees is probably the easiest way to spot them. The small, yellow, bisexual flowers are not particularly noticeable, but, when in fruit, persimmon is unlike any other Florida tree. The roundish, thin-skinned, yellow-orange to red-orange fruit is 1 to 2 inches in diameter and has a distinctive woody calyx at its base. When ripe, it is soft inside and contains several hard, flattened, shiny chocolate-brown seeds. Only about half of persimmon trees ever produce fruit, however, because the other half are male trees, persimmon being a dioecious species. Persimmon occurs throughout Florida and the rest of the southeastern United States (Halls 1990).

Persimmon is moderately fast growing at first, especially as most reproduction is from root suckers. However, growth beyond the sapling stage is rather slow, and large trees can be fairly old. Three mature trees I cored in Gulf Hammock were between 100 and 160 years old. The maximum age for a persimmon tree may be about 200.

On well-drained soil, persimmon develops a strong root system with a deep taproot and an extensive system of lateral roots. It is quite drought tolerant and wind resistant. Unless they are rotten or hollow, persimmon trunks, although slender, are robust because of the tree's very strong wood.

One oddity about persimmon trees is that many are very late to leaf out in the spring. Some of the large trees in Alachua and Levy Counties do not leaf out until the first of May, which is easily a month later than most other trees in this area and two months later than some early leafing trees. Individual persimmons in residential and urban settings are sometimes thought to be dead because of this late leafing and are cut down, even though they are actually alive and healthy.

Persimmon is a very adaptable plant. It grows best and gets largest in hardwood forests on fertile upland soils (mesic hammocks) or on moist to rather wet soils (hydric hammocks, slope forests, and swamp edges). In these situations, persimmon occurs as a widely scattered, uncommon tree, and it sometimes grows to 100 feet tall and 18 inches in trunk diameter. The largest one recorded in Florida, in 1993 in Torreya State Park, was 18 inches in diameter at breast height and 118 feet tall. On the edges of prairies and in some old fields, persimmon trees are sometimes abundant, often occurring in large groups that originate from root suckers. It also occurs as a scattered small tree in upland pine forests on a variety of soils ranging from sand to clay. In such habitats, however, it doesn't grow well or get large and individual stems don't live long, although the root systems that support these upland clones may live a very long time.

One reason persimmon is able to grow in so many different situations is that, in addition to germinating from seed, it spreads by sprouting from very long lateral roots. In this way, it is able to establish itself in locations where a seedling would have a hard time getting started. Persimmon is also able to withstand a moderate amount of flooding and a bit more fire than most hardwood trees. While it grows in flood zones, it typically remains near their edges and is perhaps able to survive there because its long lateral roots can reach dry soil.

One thing persimmon cannot survive is dense shade. Although several publications list persimmon as having moderate to high shade tolerance (Fowells 1965; Halls 1990), this is clearly not true. In the uniform environment of dense pine plantations on old fields in Suwannee County, persimmon has a shade tolerance roughly equal to that of bluejack oak. It is less shade tolerant under these conditions than live oak, sand live oak, laurel oak, southern red-cedar, and black cherry. It is somewhat difficult to judge shade tolerance of individual persimmon trees in hardwood forest situations because some of the young trees are vigorous root sprouts that derive sustenance from a parent tree that may be some distance away. This is probably the reason for the confusion on this matter. However, even here, all the medium-sized to large persimmons have at least the top of the crown in full sunlight, and young persimmons in hardwood forests are clearly less shade tolerant than sweetgum, which is considered shade intolerant (Kormanik 1990).

Persimmon has a high wildlife value because of the nutritious fruit it produces. Available from early September to late February, persimmon fruit is most abundantly available in November and December with considerable variation in timing from tree to tree. Mockingbirds are particularly fond of persimmons. Mammals very fond of them are opossum, raccoon, fox, and deer, and persimmons are also consumed by skunks, wild turkeys, bobwhite quail, crows, rabbits, squirrels, hogs, and cattle (Glasgow 1977). Other birds and mammals that eat persimmons are catbird, robin, yellow-rumped warbler, cedar waxwing, and black bear (Martin et al. 1951). I have seen ruby-crowned kinglets and Baltimore orioles feeding on the fruit and have also observed butterflies feeding on the fruit juices where birds have pecked holes in ripe fruit. I found persimmon seeds in the stomach remains of a coyote that was shot in Levy County. Wildlife biologist James Perrin Ross reported to me that he has found persimmon seeds in the scats of coyotes and that deer eat most of the persimmons that fall to the ground on his property next to Payne's Prairie (personal communication). Similarly, experienced birder and naturalist Michael Meisenburg has watched gopher tortoises eating ripe persimmons that have fallen to the ground (personal communication). In the wild, persimmon fruit rarely goes to waste.

Another wildlife benefit of persimmon trees is that their leaves are sometimes eaten by fall webworms, which, in turn, provide food for insectivorous birds, particularly cuckoos and orioles. Longhorn beetles often girdle vigorous twigs that their larvae then feed on, often after the twigs have broken from the tree. Finally, both honey bees and native bee species are abundantly attracted to the flowers in late May, thus making persimmon useful in the production of honey (Halls 1990).

Historically prominent in what are now largely bygone manufactures, persimmon wood was once the traditional wood for crafting the wooden heads of golf clubs used as drivers. Most driver heads today, of course, are made of metal. Of similar historic interest was the use of persimmon wood for making the shuttles in textile machinery. The reason persimmon wood has been preferred for such demanding applications is that it is very strong, shock resistant, split resistant in all directions, and remains smooth under wear (Brown et al. 1949).

Even now, these same qualities make it a good specialty wood for the crafting of highly detailed wooden carvings and objects turned on a lathe. Persimmon wood is too hard to be easily carved with a knife, but it can be carved very well with sharp chisels. Although persimmon is in the ebony family, with dark brown to jet-black heartwood, only the sapwood is normally used. This is because most persimmon trees are made up almost entirely of sapwood and because the heartwood checks (splits) excessively during drying. (The three trees I cored in Levy County, all over 100 years old, were all sapwood throughout.) Uniform, stable, and fine grained, persimmon sapwood is yellow when fresh and wet, soon becoming light tan-gray in color with abundant small ray flecks.

Although some wild persimmon trees remain in subdivisions, the native persimmon is very rarely planted as an ornamental or shade tree. This is unfortunate, because it has considerable wildlife value, and the fruit of many wild trees is excellent for human consumption as well. It is a slender, attractive tree of modest size, and it produces fall color here in Florida (usually a deep purple). Persimmon is difficult to transplant but is easy to grow from seed. For very patient people, the best and most enjoyable way to obtain a really fine persimmon tree of one's own is to sample a lot of wild persimmon fruits until an exceptionally fine one is found and then to grow several seedlings from its seed. Several trees need to be grown because some will be nonfruiting male trees, and there may be considerable variation in the quality of fruit on the females. An alternate method is to graft or bud a scion of an exceptional tree onto a seedling or sapling wild persimmon.

Most persimmons that are planted are Japanese persimmon varieties grafted onto native persimmon rootstock. Although the Japanese persimmon fruits are several times larger than the native ones, are less astringent when not fully ripe, and are usually seedless, their flavor is not as good, in my opinion, as that of a good native wild persimmon. Persimmon fruits are best when fully ripe (mushy). When less than fully ripe, they are usually exceedingly astringent. They do not need a frost before ripening, but freezing temperatures do cause unripe fruit to ripen. Storing nearly ripe fruit in the freezer will do the same thing. Fruit that ripens after a freeze seems to be sweeter.

Persimmon fruits are very good when eaten fresh, are excellent dried, and are excellent frozen and eaten like popsicles. Persimmon pudding is an old, traditional treat still enjoyed throughout the southeastern United States, and homemade persimmon ice cream is also good. Another use of persimmon is to make a tea from the young leaves in the late spring (retired University of Florida botany professor Dana Griffin, personal communication). The dried, roasted, ground seeds have historically been used as a substitute for coffee (Halls 1990), although when I tried this, the aroma and taste were not at all similar to coffee, being more like a mild green tea.

Florida Rosemary—*Ceratiola ericoides*

Ericaceae

Florida rosemary is a dense, evergreen shrub with half-inch-long, needle-like leaves. It usually grows in a symmetric dome or ball-like clump. It can grow as large as 6 feet tall and 6 feet wide but is often half that size. Its maximum life span is about 50 years (Menges 2014). It is dioecious, with only the female plants producing small, eighth-inch-diameter, fleshy yellow fruits (drupes) containing two seeds each (Godfrey 1988). This fruit becomes available in the fall and persists into winter.

Florida rosemary occurs on excessively well-drained sandy soils from Duval, Alachua, and Gilchrist Counties south throughout the Florida peninsula, in Taylor County, and in the western panhandle (Wunderlin et al. 2020). It also occurs westward into southern Mississippi, on sand ridges in South Carolina, and on sand dunes adjacent to rivers in southeastern Georgia such as the Ohoopee Dunes.

Highly drought tolerant, Florida rosemary has a shallow root system adapted to the deep, droughty sands where it typically grows. While it can survive in the partial shade of sand pine forests, it is a sun-loving plant that grows best in full sunlight. Besides growing in association with some of the other Florida scrub plants such as sand pine and scrub palmetto, it also grows in nearly pure stands on the very poorest, driest sands of some of the interior sand ridges of peninsular Florida such as the Lake Wales Ridge and the Brooksville Ridge. It occurs on dunes along Florida's Atlantic coast and is a common plant in the 200,000-acre sand pine scrub area of the Ocala National Forest.

Florida Rosemary has adapted to fire by producing an abundance of seed that remains dormant in the soil for up to eight years (Menges 2014). Fruit production begins when the plants are between 10 and 15 years of age, peaks at ages 20 to 40, and declines after age 40, which indicates that Florida rosemary is adapted to a fire recurrence cycle of 20 to 40 years (Johnson 1982). Florida rosemary grows on diverse sites with a wide range of fire return cycles, which Menges (2014) estimates to be between 15 and 100 years. When the scrub burns, all of the Florida rosemary plants within the area of the fire will often be killed, but many of the dormant seeds will then sprout to create a new population of plants.

Florida rosemary produces the chemical ceratiolin which inhibits the germination of its own seeds plus the seeds of many other plants (Menges 2014). This chemical makes Florida rosemary allelopathic to most other plants, a feature that results in areas of bare sand around and among the bushes. This chemical warfare with other plants benefits Florida rosemary by reducing competition that might otherwise deprive it of sunlight and soil moisture.

Florida rosemary is capable of invading and thriving in longleaf pine–turkey oak–wire grass sandhills but is kept from doing so if these areas are maintained by regular ground fires. A large-scale example of this occurred on the Brooksville Ridge southwest of Archer, Florida, along Highway 24, where Florida rosemary had massively invaded an area of poor, dry longleaf pine–turkey oak sandhill, almost totally eliminating the other ground-cover vegetation over a large area. Ecologist, pioneering conservationist, and zoologist Dr. Archie Carr watched the progression of this invasion from the 1930s through the 1960s, when no prescribed burning was done (personal communication). Much of this area has since been converted to a longleaf pine plantation, eliminating most of the Florida rosemary and resulting in a return of herbaceous and other sandhill ground-cover plants.

Florida rosemary is one of the least flood tolerant of plants. Even the temporary soil saturation of a few days' hard rain will kill these plants on soils other than the high, deep, very well-drained sands where they naturally grow.

The wildlife value of Florida rosemary is moderate. Its fruits are eaten by birds, especially Florida scrub jays and rufous-sided towhees, which in turn redistribute the seeds; the seeds are also eaten by mice and Florida harvester ants, which do not aid in seed distribution (Johnson 1982; Menges 2014). Areas of scrub strongly dominated by Florida rosemary provide habitat for Northern cardinal, Florida scrub jay, common yellowthroat, mourning dove, yellow-rumped warbler, and gray catbird (Woolfenden 1973). The Florida scrub habitats in which Florida rosemary often grows provide habitat for a number of rare and interesting Florida endemic species such as the sand skink, scrub lizard, and the Florida Scrub jay as well as for more common species like the oldfield mouse, coachwhip snake, and eastern towhee. I have observed yellow-rumped warblers feeding on the fruits in winter in both Levy and Marion Counties. Two uncommon insects, the rosemary grasshopper and a rare plant bug, *Keltonia balli*, are found exclusively on Florida rosemary (Menges 2014).

Florida rosemary has potential for use in a xeriscape landscape design, provided that the very well-drained sandy soil it requires can be provided. It is easy to grow from seed in a pot and to transplant when less than 3 inches tall provided an undisturbed ball of the sand where it was growing remains around the roots. It is nearly impossible to transplant from the ground after the plants are more than 6 inches tall. In spite of its name and in spite of the honey-like odor it emits on hot days, Florida rosemary is not useful as a cooking herb or spice. It is not related to the rosemary plant of commerce that is used for flavoring, and it is not edible.

Pipestem (Florida Leucothoe)—*Agarista populifolia* (*Leucothoe populifolia*)

Ericaceae

This tall, slender, upright-growing, multistemmed, evergreen shrub is largely restricted to moist or wet forests on seepage areas and floodplains, especially spring-run floodplains, in and near the St. Johns River drainage basin from Duval County south through Clay, Putnam, Marion, and Lake Counties into Polk County (Wunderlin et al. 2020). There are also small, isolated populations in extreme southern South Carolina (Godfrey 1988). The 2- to 4-inch-long shiny leaves have entire or finely toothed margins and come to a narrow point at the tip. New, actively growing foliage is coppery colored. The upper branches are usually long bending in a graceful arch. The stems are hollow with diaphragms (Walter Judd, University of Florida botanist, personal communication). Pipestem produces clusters of small, sweetly aromatic, creamy white, urn-shaped flowers in springtime.

This uncommon plant is located in isolated spots, often on acidic, organic soils, and usually occurs in the shade of overstory trees. It is shade tolerant but can withstand full sun. It appears to be free of trouble from insects and diseases.

The wildlife benefit of this plant is limited to the value of its flowers for insect pollinators.

Usually called "Florida leucothoe" in the nursery trade, with an occasional nursery still using the former Latin name *Leucothoe populifolia*, this plant has become somewhat popular as an ornamental and is sold in native plant nurseries. It responds well to pruning, even severe pruning, and is a graceful evergreen that can make a natural-looking hedge or provide erosion control along creeks and ditches. It is easy to propagate from cuttings and is cold hardy well north of its natural range. Because the foliage contains andromedotoxin (now more commonly called "grayanotoxin") it is highly toxic to livestock and humans if eaten (North Carolina Cooperative Extension, n.d.). As with other flowering shrubs in the heath family, honey bees foraging largely on the flowers of this species would produce honey known as "mad honey" that contains grayanotoxin.

Tarflower—*Bejaria racemosa* (*Befaria racemosa*)

Ericaceae

Tarflower is a slender, upright-growing, evergreen shrub of the pine flatwoods forests of southeastern Georgia and the Florida peninsula from Nassau and Taylor Counties south to Dade County (Wunderlin et al. 2020). It is usually about 3 to 8 feet tall, often being slightly taller than the rest of the shrub community surrounding it. It has stiff, 1- to 2-inch-long, leathery leaves that are shiny green on top and whitish below, twice as long as wide, and usually slightly bent or twisted into something of a boat shape. It is most noticeable when in flower in June, at which time it is quite showy, providing a splash of pinkish-white in the pine flatwoods landscape. The flowers are clustered at the top of the bush and are about 2 inches wide, consisting of seven spatula-shaped, widely spaced petals that are white except at the base, where they are pink. The flowers are pleasingly aromatic and are sticky with a resin that gives the plant its name.

Tarflower is narrowly adapted to the drier parts of the pine flatwoods forests, being found in mesic and scrubby flatwoods. It reproduces by seed, produced in capsules, and by runners, sometimes forming clonal patches. It is not shade tolerant, requiring full sun to do well. It tolerates acidic soil that is poor in nutrients and appears to be quite drought tolerant.

Tarflower has no known value for wildlife other than providing nectar and pollen for pollinating insects. Several species of bees and wasps have been observed visiting the flowers at the Archbold Biological Station in Highlands County,

Florida (Deyrup and Deyrup 2015). Even this value is a bit problematic, as the sticky flowers trap some of the insects.

Tarflower can have value as an ornamental plant because of its sweet-smelling, showy flowers. It is sometimes available from native plant nurseries.

Mayflower (Trailing Arbutus)—*Epigaea repens*

Ericaceae

This low-growing, evergreen shrub with prostrate stems has an extensive natural distribution in eastern North America from beside Hudson Bay in Canada all the way south into Liberty County and the westernmost panhandle of Florida, but its distribution in the southeastern United States is localized and spotty. The tallest leaf or flower is rarely more than 4 inches above the ground. The dull, 1- to 2-inch-long, roundish to oblong, evergreen leaves have prominent veins and a varying number of rusty colored hairs on both leaf tops and bottoms. Male and female flowers occur on separate, dioecious plants. Female plants produce clusters of showy, sweet-smelling, pink to white flowers in spring and fleshy capsules with white, interior flesh in the fall (Dolan 2004).

Mayflower is a forest ground-cover plant that prefers to be on well-drained, moist, acidic soil in shady places that do not get covered deeply with leaves in the fall. It is very slow growing and not well adapted to disturbance, whether that disturbance is mechanical or caused by fire, livestock grazing, deer browsing, or collecting by humans. In Florida it grows in shade on the slopes of forested bluffs and ravines (Godfrey 1988).

Mayflower has no reported significance for wildlife other than the bumblebees that pollinate its fragrant flowers (Dolan 2004).

In northeastern North America and in northern Europe, mayflower (trailing arbutus) is sometimes sold in nurseries for use as an ornamental plant. There are several named cultivars. However, it is not widely adaptable and is reported to be difficult to grow horticulturally and nearly impossible to transplant. Attempting to transplant this species from the wild amounts to killing wild plants for no good reason. In Florida, this species is listed as endangered.

Hairy Wicky (Hairy Laurel)—*Kalmia hirsuta* (including notes on Mountain Laurel [*K. latifolia*])

Ericaceae

Hairy wicky is a dwarf shrub about 1 foot tall, with thin, hairy stems, small, hairy leaves, and small, bisexual flowers that look like miniature mountain laurel flowers. It occurs across the southeastern coastal plain from the Carolinas into southeastern Louisiana, ranging south into the Florida Panhandle and into the Florida peninsula as far south as Lake and Citrus Counties in isolated populations (Godfrey 1988).

Mountain laurel (*Kalmia latifolia*) is a common plant in the Appalachian Mountains that occurs in a few steep-sided ravines in the Florida Panhandle, where it grows underneath the canopy of hardwood forests.

Hairy wicky is fairly common in its special habitat, a variant of pine flatwoods with highly diverse, low-growing ground cover containing grasses, wildflowers, bracken fern, and dwarf shrubs such as dwarf huckleberry, dwarf waxmyrtle, dwarf live oak, runner oak, and shiny blueberry. The pines may be longleaf pine, slash pine, or pond pine. For instance, it is common in the ground-cover vegetation of a large area of pond pine flatwoods at the north end of the Ocala National Forest. In such habitats, the areas of low, diverse ground cover often occur as patches interspersed with patches of much taller ground-cover shrubs such as saw palmetto and gallberry. The soils where hairy wicky grows are acidic and have moisture not far below the surface. Hairy wicky is a fire-adapted plant and will disappear in the heavy shade that results when fires are excluded long enough to allow larger shrubs and trees to invade and dominate the habitat. It reproduces both by seed and by the spreading of underground stems to form clonal patches. As with other shrubs that form clonal patches, hairy wicky patches may be quite old.

Both hairy wicky and mountain laurel produce summer flowers that are attractive to insect pollinators.

Although quite attractive when in bloom, hairy wicky is rarely used as an ornamental, even though it is available from some native plant nurseries. Mountain laurel, on the other hand, is a popular ornamental shrub in the eastern United States north of Florida.

Crookedwood (Rusty Lyonia)—*Lyonia ferruginea* (including notes on *L. fruticosa*)

Ericaceae

Crookedwood is a slow-growing, upright shrub or small tree with evergreen leaves that are a rusty-tan color when young and a rusty-gray color underneath when older. The leaves are variable in size but are typically about 2 inches long and slightly less than 1 inch wide. The stems and trunks are crooked and covered with a thin bark finely ridged with gray- to tan- to reddish-brown vertical scales. Clusters of small, white, urn-shaped flowers appear in April and May, followed by small capsules in the fall (Miller and Miller 1999). Crookedwood occurs from the southeastern tip of South Carolina throughout most of southeastern Georgia and all of Florida except for the southern end of the peninsula and the extreme western end of the panhandle (Wunderlin et al. 2020). It is usually 5 to 15 feet tall but can grow to be 40 feet tall with a trunk diameter of 9 inches (Godfrey 1988). *Lyonia fruticosa*, sometimes called "staggerbush," is similar to crookedwood though much smaller, rarely getting over 5 feet tall. Compared to those of crookedwood, staggerbush leaves are less elongate, flatter, thicker, and usually about an inch long.

These two shrubs both grow in poorly drained pine flatwoods forests, being most abundant in the drier parts that are sometimes called "scrubby flatwoods." Crookedwood is often common to abundant in the various forms of Florida scrub such as sand pine scrub and oak scrub. It is one of the Florida scrub shrubs that persists as the sand pines and oaks grow up and shade out many other plant species. It is also common in xeric hammocks. Staggerbush, by comparison, is less common in scrub and rare in xeric hammocks. Crookedwood growing in xeric hammocks and in scrub that has not been harvested, chopped, bulldozed, or burned for many decades often grows to small tree size, with a single, tall, crooked trunk.

Although somewhat shade tolerant, crookedwood does best in full sun. It is extremely drought tolerant but not very flood tolerant. For a shrub, crookedwood is long lived, perhaps living 100 years or more. It is a fire-adapted species that sprouts back from the root collar after being burned to the ground, although the scrub and pine flatwoods habitats where it occurs do not burn with a high frequency. It does not normally occur in longleaf pine sandhills because fires there occur too often.

Although perhaps browsed a bit by white-tailed deer and occasionally defoliated to varying degrees by caterpillars, the main wildlife value of these shrubs is that of their flowers for insect pollinators.

Crookedwood is harvested commercially to make the stems of artificial decorative plants, with the Ocala National Forest being one of the main sources of material for making this product. It is also sometimes used as an ornamental shrub or small tree and is available for sale from a few native plant nurseries. It does best in full sun on well-drained, acidic, sandy soils and is trouble free once established, never needing spraying, watering, or fertilizing, although caterpillars in the genus *Datana* will sometimes defoliate part or all of a plant.

Fetterbush—*Lyonia lucida* (including notes on Maleberry [*Lyonia ligustrina*]; Piedmont Staggerbush [*Lyonia mariana*]; Coastal Doghobble [*Leucothoe axillaris*]; and Swamp Doghobble [*Eubotrys racemosa (Leucothoe racemosa)*])

Ericaceae

Fetterbush is a common to abundant, thicket-forming shrub with shiny, leathery, evergreen leaves about 2 inches long, each having a vein that encircles the leaf just inward from the untoothed leaf margin. Fetterbush produces clusters of light pink to dark pink urn-shaped flowers in spring and dry capsules in fall. It is usually 3 to 6 feet tall, and it usually occurs in dense thickets that spring from rhizomes. It grows throughout Florida and the rest of the southeastern coastal plain (Kartesz 2015).

Although capable of growing on dry land, fetterbush is primarily a wetland species. It is common in pine flatwoods forest areas, being most abundant around the edges of wetlands or, in many cases, throughout wetlands such as bogs, bayheads, and cypress domes. It can form dense, nearly pure stands in extensive bogs such as Impassible Bay in the Osceola National Forest and the Santa Fe Swamp in Alachua County and in some areas of very wet pine flatwoods forest such as Tate's Hell in the Florida Panhandle. In dense stands in bogs, it is associated with laurel greenbriar, which binds the shrubs together and makes attempting to traverse such areas on foot very difficult. Other plants in these bogs include slash pine, loblolly bay, swamp tupelo, and hooded pitcher plant.

Fetterbush along with several other plants in the heath family (Ericaceae) that share similar ecological characteristics such as maleberry, Piedmont staggerbush, coastal doghobble, and swamp doghobble often form a dense shrub border or ecotone around cypress domes, bayheads, and bogs in the pine flatwoods landscape of the southeastern coastal plain. The common names "fetterbush," "staggerbush," and "doghobble," which are somewhat problematic because of inconsistent usage, are based on the dense, entangled shrub thickets these shrubs often form that impede the passage of people and animals through them. All of these species reproduce both by seed and rhizomes.

Fetterbush and the four other species mentioned above are very tolerant of wet, acidic soils and shallow, intermittent flooding. They are all moderately shade tolerant but do best in full sun. Occurring in fire-adapted communities such as wet pine flatwoods, they are able to sprout back vigorously after a fire.

Although browsed occasionally by deer (Miller and Miller 1999) and sometimes partially defoliated by caterpillars, the main wildlife value of fetterbush and the other four species seems to be that of the flowers for insect pollinators. Honey bees are apparently not among the pollinators so benefited, however, as their tongues are too short to access the nectar in fetterbush's bell-shaped flowers (Christman [1997] 2008). Bumblebees are their main diurnal pollinators with swallowtail butterflies also visiting the flowers, but most pollination occurs at night by an assortment of moth species (Benning 2015).

Fetterbush, as well as the other four species, have some potential for use as ornamentals because of their delicate, dangling, urn-shaped flowers in various hues of pink and white. They are available from some native plant nurseries but are infrequently planted.

Climbing Fetterbush (Vine Wicky)—*Pieris phillyreifolia*

Ericaceae

Climbing fetterbush is an unusual woody plant. It is evergreen, with elliptic leaves 1 or 2 inches long, and small, white, urn-shaped flowers in clusters of three to nine in early spring. The bracts and stems of the flowers may be white but are often salmon-pink, often giving the flower display a pinkish-white appearance. The most unusual aspect of this species is that it often takes the form of a vine and climbs up underneath the outer bark of a pond-cypress or Atlantic white cedar tree, sometimes to a height of 20 feet or more, producing widely scattered horizontal branches that appear to sprout from the trunk of the host tree. At other times, climbing fetterbush is a low shrub that creeps over hummocks, tree stumps, logs, and along the surface of the ground or sticks up out of shallow water. The red-brown stems are slender, rarely more than a half inch in diameter.

Climbing fetterbush occurs from coastal South Carolina southwestward into coastal Mississippi (Kartesz 2015) with most of the population in the Okefenokee Swamp in southeastern Georgia and in the Florida Panhandle. Smaller, isolated populations occur along a few spring runs such as Alexander Springs Run in Lake County and Mormon Branch and Salt Springs Run in the Ocala National Forest on the north central Florida peninsula. Climbing fetterbush is restricted to wetlands and seems to prefer those with some organic soil accumulation and with either pond-cypress or Atlantic white cedar in the overstory. It is somewhat shade tolerant but does best with ample sunlight. It is flood tolerant but not drought tolerant.

Climbing fetterbush spreads by seed and by the rooting of the creeping vines, producing clones that spread out over a considerable area.

The main wildlife benefit of climbing fetterbush would seem to be the value of its early spring flowers for bees, moths, and other insect pollinators.

Climbing fetterbush is grown and sold by some native plant nurseries. It is an attractive plant when in bloom.

Sourwood—*Oxydendrum arboreum*

Ericaceae

Sourwood is a tall-growing, slender, medium-sized tree in the heath family with deciduous leaves about 6 inches long and 1 or 2 inches wide with fine serrations on the edges. The leaves turn yellow-orange and then bright scarlet in autumn. Small, urn-shaped, white, bisexual flowers bloom on spray-like racemes that extend out from the ends of the current year's twigs in the summertime. The thick, deeply furrowed bark is gray tinged with red and is longitudinally furrowed with scaly ridges.

Sourwood occurs throughout the southeastern United States from West Virginia and southern Ohio into eastern Texas, reaching into the Florida Panhandle in scattered locations (Kartesz 2015). Its size is variable, with Florida trees typically not exceeding 50 feet in height and 1 foot in trunk diameter. The largest one reported nationally in 1972 from the Nantahala National Forest in the Appalachian Mountains was 118 feet tall with a trunk diameter of slightly over 2 feet (Godfrey 1988).

Sourwood is an upland species, growing on acidic soils in places where it can get ample sunlight. It often grows on the upper edge of hardwood forests on the slopes of ravines or bluffs adjacent to fire-adapted pine forests.

The main wildlife benefit of sourwood is the value of its flowers, which are very attractive to pollinating insects, especially honey bees and native bees. Sourwood foliage is browsed to some extent by white-tailed deer.

Sourwood is valued as a nectar source for honey bees that produce high-quality sourwood honey from it, and it is also valued as an ornamental tree, both because of its attractiveness to honey bees and because of the beauty of its flowers and fall foliage. Sourwood can be difficult to establish and grow, as it requires acidic soil and an undisturbed root zone. It does not transplant well, so if establishing one is an objective, a nursery-grown plant should be obtained.

Chapman's Rhododendron—*Rhododendron chapmanii* (*Rhododendron minus* var. *chapmanii*)

Ericaceae

Chapman's rhododendron is a very rare, low-growing, evergreen shrub with stiff leaves 1 to 2 inches long that are usually curled under on the sides, have a somewhat wrinkled surface, and often have reddish-brown spots on their undersides. An attractive show of pale to bright pink, 1- to 1½-inch-wide flowers bloom for a couple of weeks around April 1st. The flowers, borne in terminal clusters, bloom before the leaves begin to grow. The rust-colored seed pods, which also occur in clusters and hang on year round, are half an inch long. Chapman's rhododendron is usually 2 to 4 feet tall, with a maximum height of 9 feet. It usually has stiff, upright branches and a leggy growth form.

Chapman's rhododendron has been on the United States Fish and Wildlife Service's endangered species list since 1979 and is also on Florida's endangered species list.

The native population of Chapman's rhododendron is endemic to three small areas in North Florida which are divided into 18 known, isolated spots. The seemingly most secure and viable location is along the border between Liberty and Gadsden Counties near Hosford in the Florida Panhandle, where the plant is scattered over perhaps 3 or 4 square miles with a total area of occupied habitat of about 200 acres. There is one especially favorable 10-acre site in this locale with perhaps 1,000 to 2,000 plants (rough estimate done in 1982). The rest are scattered on eight or more additional sites with 10 to 200 plants per site. The second isolated population is located in Gulf County on either side of Port St. Joe in the Florida Panhandle. In 1982, it had a population of about 700 plants that occupied a total area of perhaps 200 acres with eight known sites, each with 20 to 300 plants per site. The last population, containing perhaps 20 plants, is on a one-acre, highly disturbed site on Camp Blanding in Clay County on the northeastern Florida peninsula. There is a small, introduced population of about ten plants on one tenth of an acre in a pine plantation in Suwannee County that has been slowly expanding since being grown from seed and planted in 1982. Numerous plants are in cultivation in various yards and gardens.

It is sometimes difficult to get accurate plant numbers in the wild because the plant groups expand as the outermost branches recline on the ground and take root to create dense patches that originate from one plant. As these patches grow together, the distinction between individual clones becomes impossible to see. A rough estimate of the total wild population back in 1984 was 3,000 plants (Simons 1984).

Chapman's rhododendron is a fire-adapted species that occurs in transition zones between black titi bogs and either longleaf pine–saw palmetto–gallberry flatwoods or longleaf pine–turkey oak–wire grass sandhill habitats, or, alternatively, on slightly elevated pine flatwoods habitat between linear black titi bogs. With the possible exception of the Camp Blanding site, Chapman's rhododendron always grows underneath longleaf pine (or where longleaf pine used to be), and it always grows adjacent to a bog dominated by black titi on sands with abundant organic matter that are well drained at the surface and permanently saturated just below the surface with soft, acidic water, yet never subject to flooding (Simons 1984). The areas that support Chapman's rhododendron invariably have nutrient-deficient soils, enabling the habitat to remain sunny with very low shrub growth and a light overstory of pine trees. Such habitat has been characterized as "scrubby flatwoods," but in reality, it is a very specific and rare subset of scrubby flatwoods where all the shrubs and hardwood trees are dwarfed and where flooding never occurs.

The dominant flora associated with Chapman's rhododendron includes long-leaf pine (often replaced by humans with slash pine), saw palmetto, sand live oak, crookedwood, shiny blueberry, gallberry, wire grass, and bracken fern (Simons 1983). There are usually additional plant species mixed in, such as wild olive, myrtle oak, large gallberry, slash pine, white titi, hairy laurel, flatwoods St. John's wort, dwarf waxmyrtle, fetterbush, swamp doghobble, huckleberry species, blackberry species, water oak, roundleaf greenbriar, earleaf greenbriar, and deertongue (Simons 1983).

Chapman's rhododendron is a slow-growing, long-lived shrub that spreads primarily by the rooting of branches. Some of the clonal patches thus produced may be very old. Abundant seed is produced every year, but the exacting conditions needed for a seedling to become established seem rarely to occur in the wild.

The wildlife value of Chapman's rhododendron is rather low. The flowers are visited by pollinators such as swallowtail, sulfur, skipper, and other kinds of butterflies and by honey bees, bumblebees, other native bees, and other insects. Hummingbirds may also visit the flowers. The low, dense thickets of Chapman's rhododendron can provide cover for rabbits, nesting wild turkeys, and other animals.

Chapman's rhododendron is grown and sold by native plant nurseries for ornamental use. Healthy individual wild and cultivated plants produce thousands of tiny seeds every year that can be grown by germinating them on milled sphagnum and then growing the small plants in containers on a porous mix of sand and peat moss. Plants should never be dug from the wild, as this is one of the main threats to the continued existence of this species.

Wild Azaleas

Florida Flame Azalea (Orange Azalea)—*Rhododendron austrinum*; Sweet Pinxter Azalea (Mountain Azalea)—*R. canescens*; Swamp Azalea— *R. viscosum*; Alabama Azalea—*R. alabamense*

Ericaceae

Wild azaleas are deciduous shrubs with multiple, slender, brittle, tan-colored stems and simple, slightly fuzzy, alternate, deciduous leaves. The flower buds, which occur on the tips of the more vigorous twigs, are much larger than the leaf buds. The bisexual flowers are somewhat tubular, resembling honeysuckle flowers. Wild azaleas are called "wild honeysuckle" in the Carolinas. The fruits, which ripen in the fall, are elongated capsules full of very tiny seeds.

There are several kinds of wild azalea in our area. The most widespread and common species are the white-flowered swamp azalea and the pink-flowered sweet pinxter azalea. There are several other kinds occurring locally in the vicinity of Tallahassee and the Apalachicola River. Two of these discussed here are the orange-flowered Florida flame azalea and the Alabama wild azalea, with its white flowers tinged with yellow on the upper petal.

The swamp azalea blooms in June. It occurs on moist to wet acidic soil that floods neither very often nor for very long. This situation occurs most often in bayheads and on the outer edges of bayheads, swamps, and stream floodplains in pine flatwoods forests. Swamp azalea can be abundant in such areas and sometimes occurs well out into the saw palmetto thickets of the adjacent pine flatwoods. Swamp azalea spreads by underground runners to form thickets. It is the wild azalea best adapted to withstand fires, which burn it to the ground but do not kill the underground stems that readily resprout. In nature, the swamp azalea only grows on soils that are quite acidic and kept moist to wet by seepage within the plant's shallow root zone. It occurs naturally throughout the Florida Panhandle and as far south on the Florida peninsula as Highlands County (Wunderlin et al. 2020). It grows as far north as New England and as far west as eastern Texas (Kartesz 2015).

Sweet pinxter azalea blooms in late February to early March. It is most commonly found in or next to the floodplains of streams and rivers on sandy soil in regions of pine flatwoods or high pine forest. The forests where sweet pinxter azalea grows are usually hardwood forests, but the watershed of the nearby stream or river is usually dominated by pine forest (or what used to be pine forest). This azalea is tolerant of a somewhat wider set of moisture and flooding conditions than swamp azalea, but it is still not drought tolerant and requires soils that are at least mildly acidic. Sweet pinxter occurs naturally throughout most of the southeastern states, with an outlier population in southeastern Pennsylvania (Kartesz 2015). In Florida, this azalea occurs in the Florida Panhandle and as far south on the peninsula as Marion County (Wunderlin et al. 2020).

The orange flowers of the Florida flame azalea normally bloom in early March, somewhat later than those of the sweet pinxter azalea. In the wild, it is largely restricted to stream and river banks, slopes, and bluffs in the Apalachicola River basin and elsewhere in the western Florida Panhandle and nearby areas in southern Alabama and southwest Georgia, where it grows in the shade of hardwood trees on a variety of soils. Florida flame azalea and, perhaps, the Alabama azalea can tolerate a wider range of soil conditions than the other azaleas and, consequently, both are easier to maintain in cultivation.

The Alabama azalea blooms in April with white flowers bearing a yellow streak on the upper petal. They are fragrant with a lemon scent. This species is restricted in Florida to the northern part of the central panhandle, occurring in upland hardwood forest, usually on slopes. It also occurs in Tennessee, Alabama, and the western edge of Georgia. It often spreads clonally by rhizomes, is rare, grows on drier soil than the other azaleas, and is listed as endangered by the state of Florida.

Wild azaleas do not tolerate prolonged flooding and are not drought tolerant. They also are not very tolerant of nonacidic soils. Consequently, in nature, they only occur in isolated spots that provide for their specialized needs.

The wildlife value of wild azaleas is fairly low. The flowers are visited by hummingbirds, bees (including the specialized azalea miner bee), and various other insects, but the leaves and fruits provide little wildlife benefit. I have observed an Io moth caterpillar feeding on the leaves of the sweet pinxter azalea, and Knopf et al. (1995) report that the caterpillars of the striped hairstreak butterfly sometimes feeds on the leaves. Red spider mites are sometimes a pest on the leaves of all four species. The azalea caterpillar is another potential pest although it seems to prefer exotic ornamental indica azalea bushes (Dekle and Fasulo 2000).

The main benefit provided by these plants is the beauty of their flowers. Wild azaleas are used sparingly in landscaping, often struggling in these situations because of occasional droughts or because the soil, fertilizer, or irrigation water is not acidic enough. The sweet pinxter, Florida flame, and Alabama wild azaleas are usually more successful and attractive than the swamp azalea, sometimes doing quite well if placed where they get water runoff from the roof of a building (water that does not raise the soil pH). Even here, additional watering is often needed during droughts.

Wild azaleas can be grown from seed by gathering the capsules in the late fall and sprinkling the tiny seeds on a medium of milled sphagnum moss. Once the seedlings are an inch or so tall, they can be transplanted to a soil mix of peat moss and sand.

Blue Huckleberry (Hairy Dangleberry, Hairytwig Huckleberry)—*Gaylussacia tomentosa* (*G. frondosa* var. *tomentosa*) and Confederate Huckleberry (Blue Huckleberry)—*Gaylussacia nana* (*G. frondosa* var. *nana*)

Ericaceae

The common names for these plants are a bit confusing, as both forms are, at one time or another, identified by the umbrella term "blue huckleberry." Together, these plants are the common, medium-sized, blue-fruited huckleberry bushes of the bogs and pine flatwoods forests of the southeastern coastal plain. There are at least two varieties or species in northern Florida.

One of the common forms is called "confederate huckleberry" or "blue huckleberry" with its Latin name being either *Gaylussicia nana* or *Gaylussacia frondosa* var. *nana*. Occurring in both the Florida peninsula south to Polk County and in the panhandle and parts of Georgia, Alabama, and South Carolina (Kartesz 2015), it is usually only 1 to 2 feet tall and has leaves about an inch long that are glaucous (powdery white) on the underside.

The other common form is also often called "blue huckleberry" but given the Latin name *Gaylussacia tomentosa* or *Gaylussacia frondosa* var. *tomentosa*. In scientific literature, *Gaylussacia tomentosa* is often identified by the common names "hairytwig huckleberry" or "hairy dangleberry." This species, occurring in the Florida Panhandle, on the Florida peninsula south to Highlands County, and in parts of Georgia, Alabama, and South Carolina (Kartesz 2015) is often 2 to 5 feet tall and somewhat hairy, but the undersides of its leaves are green rather than glaucous white. The leaves of both *G. nana* and *G. tomentosa* are dull green on top with a textured surface. Both forms produce greenish-white to white to pinkish-white flowers in clusters in early spring followed by bright blue drupes containing 10 pits.

Because the fruits of these plants are bright blue, they are sometimes mistakenly called "blueberry," whereas black-fruited blueberry plants are often mistakenly called "huckleberry." The most recognizable difference between huckleberries and blueberries is that huckleberries have much bigger seeds (actually pits). These are quite noticeable when one eats the berries (drupes), which are crunchy to chew.

For the rest of this discussion, both forms will be lumped together under the name "blue huckleberry." Blue huckleberries grow on acidic sandy soils. They can occur on well-drained upland sands but are far more common near or in seepage areas or wetlands, where acidic water is within easy reach of their roots. They are mostly found on the ecotones surrounding cypress domes and bogs in pine flatwoods areas. Accessible places where they are common include the Apalachicola National Forest in the Florida Panhandle, the Okefenokee Swamp National Wildlife Refuge in southeast Georgia, the Pinhook Swamp and the Osceola National Forest in northeast Florida, and the Green Swamp Wildlife Management Area on the central Florida peninsula

Blue huckleberries spread by seed and by underground runners, often forming low thickets. They are sun-loving plants but can grow well in light shade. They are not highly tolerant of either drought or flooding. Extremely fire tolerant, however, huckleberries sprout back quickly following a fire and are more commonly found in areas that burn frequently. This is because, without fire, some of the associated plant species, such as saw palmetto, gallberry, and fetterbush, if they are not regularly trimmed back by fire, will grow over and smother the huckleberry bushes.

Although not as prolific at producing fruit as blueberry bushes, blue huckleberry bushes have good wildlife value based on their flower and fruit production. A wide assortment of mammals and birds eat the fruit including gray fox, wild turkey, northern bobwhite quail, catbird, Florida scrub jay, and eastern towhee (Martin et al. 1951). White-tailed deer occasionally browse the foliage in Florida (Miller and Miller 1999). The flowers are visited by both native bees and honey bees.

Blue huckleberries are attractive plants. Although not commonly sold in nurseries, they have good potential as ornamentals provided their need for moist acidic soil is met. They provide good tasting, edible fruit, and they are good, early-flowering honey bee plants that produce a lot of nectar (Krochmal 2017).

Woolly Huckleberry—*Gaylussacia mosieri*

Ericaceae

Woolly huckleberry is a medium-sized, black-fruited huckleberry that mostly occurs in the Florida Panhandle from Suwannee County westward and into nearby areas of southern Georgia, Alabama, and Mississippi (Kartesz 2015). Silvery hairs line the young twigs and most other parts of the plant including the fruit (Godfrey 1988). It grows 3 to 6 feet tall and in most other aspects visibly resembles blue huckleberry.

Woolly huckleberry inhabits seasonally wet pine flatwoods and pine savannas and the edges of bogs and bayheads within these pinelands. It is more narrowly restricted to wetlands than the other Florida huckleberry species.

The wildlife value of woolly huckleberry is similar to that of blue huckleberry, although it is perhaps somewhat less valuable given its being less common and widespread. The same suite of wildlife species make use of the foliage, fruit, and flowers.

Because woolly huckleberry is less common and more restricted in its habitat preferences than the other Florida species of huckleberry its use as an ornamental is limited.

Dwarf Huckleberry—*Gaylussacia dumosa*

Ericaceae

This small shrub is usually less than 1 foot tall although it occasionally reaches 2 feet. It has obovate, tardily deciduous leaves that are about 1 inch long, and it produces clusters of whitish flowers around April 1st followed in June by round, shiny black fruit (drupes) about one third of an inch in diameter. Some fruit in the Osceola National Forest is mottled white and pink when ripe though most plants produce black fruit. Although occurring as far North as New Jersey, with a closely related species in Newfoundland and Nova Scotia, it is most common in the southeastern coastal plain of North America and occurs in appropriate habitat throughout most of Florida (Kartesz 2015).

Dwarf huckleberry spreads by underground stems to form a low, sparse, shrubby ground cover in pine flatwoods forests kept open with sufficiently frequent fire. In areas that have become densely vegetated with saw palmetto, gallberry, and/or fetterbush, dwarf huckleberry is rare or absent. As with many of the ground-cover plants of fire-adapted pine forests, dwarf huckleberry is dependent on fire to maintain the habitat conditions it requires.

The habitat of dwarf huckleberry is typically the drier parts of pine flatwoods forests, where it is associated with shiny blueberry, saw palmetto, gallberry, bracken fern, and a wide assortment of other plants. It also occurs in shrubby flatwoods and ranges into the edges of longleaf pine–turkey oak sandhills. It is found only on acidic, sandy soils in our area. Dwarf huckleberry is a sun-loving plant that tolerates the shade of a pine canopy but not the additional shade of competing shrubs or invading hardwoods.

The wildlife value of this small shrub is fairly high because its spring flowers provide both pollen and nectar for bees (Krochmal 2017) and its early summer fruit is eaten by a wide variety of mammals and birds including wild turkeys and bobwhite quail (Martin et al. 1951; Halls 1977). The fruit grows low enough to the ground for gopher tortoises to pluck it from the plant and eat it.

The fruit of dwarf huckleberry is edible with an agreeable taste, but it is not as sweet as blue huckleberry or shiny blueberry. Dwarf huckleberry is one of the early blooming wild shrubs considered to be valuable for honey bees and honey production (Krochmal 2017). Dwarf huckleberry has some potential for use as an ornamental landscape plant. It could make an interesting ground cover in areas not destined for foot traffic.

Deerberry—*Vaccinium stamineum*

Ericaceae

Deerberry is a common shrub of well-drained upland pinelands throughout most of the eastern United States, growing as far south as the central peninsula of Florida (Kartesz 2015). It is usually a knee-high, multistemmed shrub but can grow to 16 feet tall with a trunk 6 inches in diameter (Godfrey 1988). The largest and tallest I've seen have been in sand pine scrub in the Ocala National Forest. The thin to somewhat thickened, deciduous leaves are variable, averaging about

2 inches in length and 1 inch in width, with varying amounts of whitish color on the undersides. Some of the plants in fire-maintained upland longleaf pine forests and in pine plantations on former longleaf pineland often have thicker, dark blue-green leaves with brightly white undersides, whereas other plants in the same habitat, as well as all the plants in scrub habitats, have thin, dull green leaves that are dull white underneath. These two variants seem to be different genetically but are lumped into the same species. It is the thin-leaved plants that occasionally grow more than 10 feet tall.

The small, white to greenish-white to pinkish-white, bisexual flowers are produced in clusters. The fruit, which ripens in June and July, averages about a half inch in diameter and is somewhat variable in color, often becoming black at maturity. Deerberry often produces abundant fruit. The taste of deerberry fruit is more like apple than blueberry and, though somewhat bitter and sour, is easily edible. Reproduction is both by seed and by underground runners, and deerberry often occurs in small clonal patches.

Deerberry is most common in longleaf pine sandhills and other upland longleaf pine habitats but also occurs in xeric hammocks and scrub habitats. It prefers full sunlight but is moderately tolerant of shade, often occurring in the shade of pine and fire-adapted hardwood trees. It is very drought tolerant but not tolerant of flooding. It is exceptionally fire tolerant, resprouting quickly following fires of any intensity. Indeed, it does best in places that burn frequently.

Deerberry is very valuable for wildlife. Joe Hutto (1995) described how his flock of wild turkeys in the Florida Panhandle always sought out and ate deerberry fruits when they were available. The fruit is important for many species including black bear, wild turkey, bobwhite quail, brown thrasher, and eastern bluebird, (Miller and Miller 1999). Hill (2002) observed white-tailed deer eating the fruit, leaves, and stems; and he notes that white-tailed deer, black bear, raccoon, fox, wild turkey, and bobwhite quail eat the fruit and that native bees visit the flowers.

Deerberry can have considerable visual interest, especially those whose leaves are dark blue-green on the upper surfaces and bright white on the lower surfaces. The tiny, clustered flowers resemble delicate white bells with slightly protruding inner yellow tassels at the center. Because of its beauty, edible fruit, and high wildlife values, deerberry has potential as a landscape ornamental and is sold in some native plant nurseries.

Sparkleberry (Farkleberry)—*Vaccinium arboreum*

Ericaceae

Sparkleberry, also called "tree sparkleberry" and "farkleberry," is a common large shrub or small tree with small, 1- to 2-inch, roundish, thick, evergreen to tardily deciduous leaves that are shiny green on top and dull green on the bottom. The growth habit is usually rather crooked or twisted, with twigs and branches growing in zigzag patterns. It may have a single trunk 3 or 4 inches in diameter or have multiple stems. The grayish-brown to reddish- or purplish-brown bark on larger stems sloughs off to expose brighter, reddish-brown inner bark. Clusters of small, white, bisexual flowers are produced in late April or early May, and the small, quarter-inch, black, rather dry and bitter berries begin ripening around September 1st and then remain on the plant at least into January. Sparkleberry usually grows 5 to 15 feet tall, though an exceptionally large individual reportedly reached 47 feet tall with a trunk diameter of 16 inches (American Forests 2004). Sparkleberry occurs throughout the southeastern United States, growing as far south on the Florida peninsula as Hendry County (Kartesz 2015; Wunderlin et al. 2020).

Sparkleberry differs from other blueberry species in several ways. It is more adaptable, growing well on both acidic and nonacidic soils; it produces fruit unappealing as food for human consumption; and the fruit remains on the plant for a long time after ripening. It does best in full sun but will tolerate considerable shade. It is very drought tolerant but will also withstand some short-term flooding. Sparkleberry is common to abundant on well-drained sandy or clay soils, occurring in dry, open hardwood forests, upland pine forests, dry flatwoods pine forests, pine plantations, and sand pine scrub. It reproduces abundantly both by seed and by its many root suckers that allow the plant to spread widely and form thickets.

The wildlife value of sparkleberry is good. Although not as appealing to most species as other blueberry fruit, sparkleberry fruit is nonetheless consumed by many species and is a valuable food resource because it is still available long after the other blueberries are gone. For instance, I observed a yellow-bellied sapsucker eating the fruit in eastern Alachua County on December 15, 2019. The flowers are very attractive to native bees, wasps, and butterflies and are also visited by honey bees. The foliage is occasionally browsed by white-tailed deer. Thickets of sparkleberry bushes provide cover for white-tailed deer, and the often dense, twisted branches of the bushes make good places for songbirds to build nests.

Sparkleberry can be a desirable ornamental. Its flowers are attractive in late spring, and its nearly evergreen foliage and crooked, twisted growth habit, together with the exposed reddish-brown inner bark, make for an attractive addition to a landscape. It is not at all easy to transplant from the wild but is available from many native nurseries. It is long lived, durable, trouble free, and rather slow growing. Once established, it requires no watering, fertilizer, or spraying.

Shiny Blueberry—*Vaccinium myrsinites*

Ericaceae

Other common names for this plant include dwarf blueberry, low-bush blueberry, ground blueberry, and Florida evergreen blueberry. It occurs throughout most of Florida as well as parts of eastern South Carolina, southern Georgia, and southern Alabama (Kartesz 2015). Shiny blueberry is a small shrub, usually only 1 or 2 feet tall, with very small, nearly evergreen, nearly sessile (stalkless) leaves only about a half inch in length that have some gland-headed hairs on the underside when young. The white to pink, bisexual flowers are borne in clusters in late March followed by quarter-inch-diameter black to blue-black berries from mid-May to mid-June. This common to abundant little shrub sometimes forms extensive clonal patches that can be quite old. Clonal patches over a half-mile in diameter and at least 1,000 years old have been reported (Tirmenstein 1990). It is the most common wild blueberry species in Florida.

Shiny blueberry is very adaptable, occurring in upland longleaf pine forests, longleaf pine–turkey oak sandhills, pine flatwoods forests, scrubby flatwoods, the saw palmetto prairies of the central Florida peninsula, sand pine scrub, oak scrub, and rosemary scrub. It requires acidic soil and sufficient sunlight and, while it is not very tolerant of flooding or salt spray, it is quite drought tolerant. Indeed, very few plant species are sufficiently drought tolerant to exist in rosemary scrub. It reproduces both by seed and by underground stems (rhizomes). Shiny blueberry is very tolerant of fire and is benefited by frequent fires that prune back competing vegetation such as saw palmetto, gallberry, sparkleberry, waxmyrtle, sand blackberry, and the sprouts of trees and large shrubs.

The wildlife value of shiny blueberry is high. It is an extremely important food resource for black bears and is also important for white-tailed deer, native mice, red fox, gray fox, skunks, eastern towhee, eastern bluebird, bobwhite quail, and wild turkey (Tirmenstein 1990). The fruit is produced low enough to the ground for gopher tortoises to make some use of it. The flowers are an important nectar source for bees, especially native bee species. Deyrup et al. (2002) reported five species of native bees visiting shiny blueberry flowers at Archbold Biological Station on the south central Florida peninsula. Perhaps the most important pollinator of native, wild blueberry plants in our area is the southeastern blueberry bee (Legare 2014; Cane and Payne 1988).

Shiny blueberry provides sweet, tasty blueberries for people to gather and eat. It also provides honey bees with an important early nectar source. The ornamental value of this little shrub is potentially good. It is variable in appearance, with some plants being quite attractive little bushes. It is available for sale from several native plant nurseries.

Darrow's Blueberry (Scrub Blueberry, Glaucous Blueberry)— *Vaccinium darrowii*

Ericaceae

Darrow's blueberry ranges from southeastern Louisiana eastward throughout the southern parts of the Gulf states and the Florida Panhandle and throughout the Florida peninsula southward into Collier County, but it is absent from the northeastern corner of Florida and southeastern Georgia (Kartesz 2015; Wunderlin et al. 2020). Bearing strong resemblance to shiny blueberry, Darrow's blueberry has similarly sized small leaves and flowers (perhaps slightly larger) and berries that range in color from powder-blue to dark blue to purple to black. The undersides of the leaves lack the gland-headed hairs that are present on newly emerged shiny blueberry leaves (Godfrey 1988). The size of Darrow's blueberry bushes tends to be, on average, about twice that of shiny blueberry. Darrow's blueberry typically grows 2 to 3 feet tall and, compared to shiny blueberry, has a wider, denser crown of foliage with a silvery tint created by the glaucous coating on its leaves. The new growth usually has a silvery-pink cast. Less common than shiny blueberry, Darrow's blueberry usually occurs in the higher, drier habitats. It is especially common in scrub, such as the sand pine scrub parts of the Ocala National Forest (hence the name "scrub blueberry"). It is a slow-growing, long-lived plant that sprouts back rapidly if pruned by fire or mechanical injury.

Darrow's blueberry occurs in the driest parts of pine flatwoods forests and scrubby flatwoods, longleaf pine sandhills, and various scrub communities such as sand pine scrub, oak scrub, and rosemary scrub. It reproduces both by seed and by underground stems and is very drought and fire tolerant. It prefers full sunlight but can endure partial shade. It is not salt tolerant and requires acidic soil.

The high wildlife habitat value of this species is similar to that of shiny blueberry and benefits the same species: black bear, white-tailed deer, raccoon, gray fox, native mice, wild turkey, bobwhite quail, scrub jay, bluebird, eastern towhee, northern mockingbird, and so on. The berries are produced low enough to the ground for gopher tortoises to eat them (Huegel 2010). Being larger than shiny blueberry and having denser foliage, it is perhaps a bit better at providing cover for some of the larger species of wildlife. The flowers support the same assortment of native and domestic bees.

Darrow's blueberry is one of the blueberry species used to produce, by way of hybridization, the southern highbush blueberry varieties that are grown in commercial blueberry farms. It is also sold as a native plant, along with several selected varieties, in native nurseries. It is an attractive, small, evergreen shrub with pretty foliage, pretty flowers, and sweet, tasty fruit.

Highbush Blueberry—*Vaccinium corymbosum* (Black Highbush Blueberry—*V. fuscatum*)

Ericaceae

Highbush blueberry occurs throughout most of the eastern United States and most of Florida north of Lake Okeechobee (Kartesz 2015). It is a semievergreen shrub with an open, irregular growth habit that can potentially reach heights of 10 feet, though it is usually no more than 6 feet tall. Highbush blueberry's variable leaves are 1 or 2 inches long. Highbush blueberry produces white, bisexual flowers in spring and sweet, tasty, powdery blue to shiny black fruit in early summer. Most Florida bushes produce black fruit and are classified as black highbush blueberry (*V. fuscatum*) by some botanists. It is fairly common in pine forest uplands in the Florida Panhandle but is rather uncommon elsewhere in Florida.

Highbush blueberry is quite variable in its size, appearance, and fruit characteristics and has been treated as either one highly variable species or as a group of closely related species. Godfrey (1988), who lumps seven entities into this one species, provides what seems to me to be the most practical way of approaching this issue. The plants in the western panhandle of Florida, where this species (or group of species) is most common, have larger, better developed crowns than those growing farther east and south and produce more fruit. They prefer to grow on upland soils that are acidic and have a clay subsoil. These areas were all originally longleaf pine forest that burned frequently. In southeastern Georgia, northeastern Florida, and on the Florida peninsula, highbush blueberry is most often found crowded in among the saw palmettos and gallberry bushes in pine flatwoods forests. It also occurs on the ecotones between pine flatwoods forests and swamps, bayheads, and cypress domes.

The wildlife value of highbush blueberry is high, in large part because of the desirability of the fruit for species such as wild turkey, bobwhite quail, black bear, eastern bluebird, eastern towhee, gray catbird, yellow-breasted chat, orchard oriole, brown thrasher, and others (Martin et al. 1951). The flowers are important for wild and domestic bees, and the foliage is browsed to some extent by white-tailed deer.

Commercially, highbush blueberry is very important, being the most frequently used species in the commercial blueberry production business in North America. Some of the commercial plants are pure highbush blueberry of several selected varieties, and some were produced through hybridization with other species such as Darrow's blueberry and Elliott blueberry. Highbush blueberry is also offered for sale as an ornamental plant. It needs acidic soil and abundant sunlight and moisture for good growth and prolific berry production.

Elliott Blueberry (Mayberry)—*Vaccinium elliottii*

Ericaceae

Elliott blueberry is a small- to medium-sized shrub with slender green twigs and small, usually narrow leaves that may persist into winter but eventually turn red and drop before the new leaves start growing in spring. The older twigs may have both gray and green stripes, and older main stems are gray with slightly flaky bark. Elliott blueberry's overall appearance is one of a finely branched, slender shrub, which is usually 2 to 3 feet tall but may occasionally grow 6 to 10 feet tall. The small white to light pink flowers are produced in February, followed by sweet, tasty, black berries about one third of an inch in diameter in early May (occasionally as early as April 10th in Alachua County). Elliott blueberry occurs in the Piedmont and coastal plain from southeastern Virginia to Arkansas and Louisiana, growing as far south into Florida as Alachua and Levy Counties with an outlier population in Orange County (Kartesz 2015).

Elliott blueberry requires acidic soil but is not as picky about soil conditions as some blueberry species, being able to grow in the floodplains of the Santa Fe and Suwannee Rivers and on the basin of Sanchez Prairie in San Felasco Hammock, where it grows abundantly on soils that are not particularly acidic. It seems to do best on the moist, fertile soil of floodplains, where it withstands occasional deep flooding and moderate shade. However, it also grows in sunnier places on poorer, well-drained soils under longleaf pines on upland sites north and west of peninsular Florida, although there is the possibility that these populations are somewhat different genetically.

Highbush blueberry species, of which Elliott blueberry is one, are notoriously difficult for botanists to figure out taxonomically, with a number of different theories on how to classify them being employed. One thing that separates Elliott blueberry from some of the other highbush blueberry species is that it is a diploid species, that is, a species having one set of chromosomes. Another difference is that Elliott blueberry does not normally spread by underground runners (rhizomes).

The wildlife value of Elliott blueberry is high. Its berries, which ripen earlier than other blueberries, are highly sought after by many species of birds and mammals. I have seen flocks of cedar waxwings, which normally prefer to forage in the crowns of trees, feasting within a foot of the forest floor on Elliott blueberries. The same species of wildlife that eat other blueberries also eat these, and the same species of insects visit the flowers.

Elliott blueberry is one of the species of blueberry used to produce, through hybridization, some of the commercial blueberry varieties, partly because of its early flowering and fruiting and partly because of its greater tolerance of less acidic, better-drained soils. Elliott blueberry can also be used as an ornamental and is available from some plant nurseries.

Chinese Tallow Tree (Popcorn-Tree)—*Triadica sebifera* (*Sapium sebiferum*)

Euphorbiaceae

This introduced species from Asia is now a common tree in northern Florida, Texas, and elsewhere in the southeastern United States. It is a medium-sized deciduous tree with small, broadly ovate or triangular leaves that are sharply pointed on each end. (Were there an indentation where the leaf stem meets the leaf, it would be heart shaped.) Chinese tallow tree has milky sap and thin, tan-gray bark. Its three-partitioned fruits, once they open, reveal a cluster of white seeds that somewhat resemble popcorn, hence the common name "popcorn-tree."

This tree is a state-listed invasive exotic pest that has been shown to damage native ecosystems (FLEPPC 2013). In some cases, Chinese tallow will form an almost pure monoculture, overgrowing and crowding out all native species of plants (David Hall, professional botanist, personal communication). According to the Florida Fish and Wildlife Conservation Commission, possession of Chinese tallow with the intent to sell, transport, or plant it is illegal in Florida. Extremely invasive and hard to eliminate once established, Chinese tallow spreads quickly by seed and by root sprouts and grows rapidly. It is now established in the wild from the Carolinas across the southern United States into Texas, including the northern and central Florida peninsula and the Florida Panhandle. It is also established in California. It was first introduced to the United States in the eighteenth century, and, unfortunately, is still available for purchase in many places, including on the Internet. It has long been a problem on Payne's Prairie, is a minor problem in San Felasco Hammock, and is a major problem in the coastal prairies of Louisiana and Texas.

Commonly seen as a yard and street tree, Chinese tallow is widely adaptable, growing well in sandy soil, clay soil, and very wet soil, and doing well in highly disturbed inner city soils. It withstands some flooding and is at least moderately salt tolerant. It is not fire tolerant, but resprouts prolifically from the base and roots whether it is burned down or cut down.

Chinese tallow reproduces by root suckers and by seeds, which it produces in abundance. Birds distribute the seeds, allowing the tree to spread widely and prolifically. It is particularly invasive on wetland sites.

This tree should be eliminated wherever possible. New seedlings are easily pulled, and older plants can be killed with herbicide. Cutting the tree down without first killing it is not an effective way to get rid of it, because it will resprout vigorously from both the stump and the root system.

The wildlife value of this tree is mostly negative. As an invasive exotic plant, it vigorously displaces plants of much greater wildlife value in many wildlife habitat areas. Also poisonous, its foliage is not eaten by insects or browsing mammals. The fruits are occasionally eaten by birds, especially pileated woodpeckers.

In China and Japan this tree was grown for the production of the solidified oil (tallow) that was obtained from the seeds and used in the making of candles and soap.

Unfortunately, Chinese tallow tree is still used as a landscape ornamental. Because it is easy to grow and plant, widely adaptable, has a fast growth rate and leaves that turn pretty colors in the fall, it has become popular over time. However, it is a serious threat to wildlife and wild places and should not be planted. Indeed, it would be a blessing if it could be eradicated from the western hemisphere.

A note of warning: the leaves, stems, and sap of this tree are poisonous. It would be unwise to use the stems of this plant for roasting hotdogs or marshmallows or to attempt to eat the fruit, which resembles popcorn. Indeed, the use of the name "popcorn-tree" should be discouraged lest it encourage consumption of the fruit.

Gulf Sebastian-Bush—*Ditrysinia fruticosa* (*Sebastiania ligustrina*)

Euphorbiaceae

Gulf Sebastian-bush is an uncommon, slender, loosely branching shrub in the spurge family (Euphorbiaceae). It has a number of former Latin names, including *Sebastiania ligustrina*, *Sebastiania fruticosa*, *Stillingia ligustrina*, and *Stillingia fruticosa*. Its native range extends from North Carolina into eastern Texas (Godfrey 1988), ranging into Florida as far south as Levy, Alachua, and Putnam Counties (Wunderlin et al. 2020). A small shrub at 2 to 6 feet tall, it has shiny, alternate, semievergreen leaves with entire margins 1 to 3 inches long and half as wide. It has very small, yellow, unisexual flowers on small spikes in the spring, with the male flowers on the bottom part of the spike and the female flowers on the top part (Godfrey 1988). Three-partitioned, one third–inch-diameter capsules mature in late summer. Like most members of the spurge family, Gulf Sebastian-bush has milky sap. It is the only member of its genus.

This inconspicuous little shrub grows on the moist, fertile soil of river floodplains, stream banks, and adjacent slopes in the shade of hardwood forests (Godfrey 1988). It can also occur in mesic flatwoods as part of the ground-cover vegetation in association with wire grass, shiny blueberry, dwarf huckleberry, saw palmetto, coastalplain staggerbush, pigmy pawpaw, and bracken fern. Such habitat use can be observed, for example, in the mesic longleaf pine flatwoods forest in Austin Cary Memorial Forest in northeastern Alachua County. Gulf Sebastian-bush is also fairly common in various locales in the floodplain of the Suwannee and Santa Fe Rivers and along Hatchet Creek in Alachua County. Similarly, it can be seen along the trails in the Santa Fe River Preserve in northern Alachua County south of Worthington Springs and in O'Leno State Park and River Rise State Preserve in western Alachua County.

To date, Gulf Sebastian-bush has not been well studied in the wild. In the Florida locales where it can be observed, it has shown itself to be both shade tolerant and somewhat flood tolerant. In the Austin Cary Memorial Forest where it currently grows in pine flatwoods habitat, it has also proven fire tolerant, sprouting back vigorously following the periodic fires that occur there every three or four years. It seems to be spreading there by underground runners as well as by seed.

The wildlife habitat value of Gulf Sebastian-bush seems good. I have observed abundant evidence of white-tailed deer browsing the plant both along the Santa Fe River and in the Austin Cary Memorial Forest, and the Iowa State University's BugGuide database photographically documents the beetle *Orsilochides guttata* (one of the shield-backed bugs in the family Scutelleridae) living on and going through various life stages on this plant (Iowa State University 2020).

To date, there appears to be no documentation of this plant's use in commercial, artisanal, or culinary enterprises or as a landscape ornamental.

Water Toothleaf (Corkwood)—*Stillingia aquatica*

Euphorbiaceae

Water Toothleaf is a small, upright, often single-stemmed plant with exceptionally lightweight wood. It occurs in shallow water in the southern half of the Florida peninsula and in southwestern Alabama and the Florida Panhandle east to Dixie County but is absent from northeastern Florida and the northern peninsula of Florida (Wunderlin et al. 2020).

Water Toothleaf can be found in wetlands in the Everglades and the Big Cypress Swamp in South Florida, in the Green Swamp in Central Florida, and in Tate's Hell and St. Marks National Wildlife Refuge in the Florida Panhandle. It is semievergreen, with 6- to 12-inch-long, narrow, alternate leaves with toothed edges that are sometimes tinged in red. The leaf petioles and newest twigs are also often tinged in red. The older stems are grayish brown. The plant is usually single stemmed and sparingly branched, about 2 to 4 feet tall, with an upright, slender trunk where all the foliage clusters at the top. The tiny yellow flowers are borne on short spikes. The three-partitioned fruit is a capsule that opens explosively when ripe and dry to forcefully scatter the seeds. Water toothleaf is in the spurge family (Euphorbiaceae) and therefore has milky sap.

This plant always occurs in nearly permanent shallow water, often in marshes or in sunny spots on the edges of swamps or in ditches. It is clearly very flood tolerant. It usually grows in full sun but sometimes occurs in partial shade. It has a minimal root system consisting primarily of a short taproot (Godfrey 1988).

To my knowledge, no wildlife habitat value has been documented for this plant to date, although the seeds of the two species of *Stillingia* native to Florida (the other one being "queen's delight" [*Stillingia sylvatica*]) have a fleshy, nutritious aril called a "caruncle" at one end of each seed to attract ants, which aids in the dispersal of the seeds (Rogers 2015).

This plant gets the name "corkwood" because, like the other plant with the name "corkwood," *Leitneria floridana*, it has very lightweight wood that historically may have been used by fishermen for making bobbers or fishnet floats (Godfrey 1988).

Chinquapin—*Castanea pumila* (*C. alnifolia*, *C. ashei*, and *C. floridana*)

Fagaceae

Chinquapin in its various forms occurs from New Jersey and West Virginia south into the central Florida peninsula and west across the Gulf coastal states into Arkansas and eastern Texas (Kartesz 2015). It is usually a shrub. Its deciduous leaves and the burrs that contain the nuts look similar to the related American, Japanese, and Chinese chestnut trees, only smaller. Small stems have smooth bark, but large stems have rough bark with scales or vertical, flat-topped ridges.

There are several forms of chinquapin in Florida, which earlier botanists designated as different species and most modern botanists consider to be variations of one species. In any event, these forms have always been difficult to differentiate with the exception of one form, which spreads widely by runners (rhizomes), easily separating it from the others that do not. By far the most common form is the dwarfed form that spreads widely by underground stems. This dwarfed form was identified as *Castanea alnifolia* by Small ([1933] 1972), who recognized six species of chinquapin, only one of which spread by underground stems. All of the forms of chinquapin occurring in Florida are currently identified as *Castanea pumila* by botanists who lump them all together (Godfrey 1988; Wunderlin 1998; Wunderlin et al. 2020).

In places that burn regularly, the dwarfed, running chinquapin forms large clonal patches that grow about 1 foot tall the first year after a fire. The bushes will continue to grow about 1 foot taller each year until some reach heights of 4 to 8 feet. There is a lot of variation from clone to clone, with some clones never growing over 1 foot tall and many not growing over 2 or 3 feet tall. Other clones will grow from 5 to 10 feet tall if left unburned and undisturbed, such as along a fencerow, for several decades. There is also much variation in leaf size and appearance, timing of nut production, and other characteristics. The leaves are covered below by a variable thin layer of dirty-white woolly hairs.

The other, larger forms of chinquapin occurring in Florida that do not spread widely by underground stems are best discussed using their original names. The leaves of Allegheny chinquapin and Small's *Castanea ashei* are coated below with white woolly hairs, whereas the leaves of Small's *Castanea floridana* are mostly hairless and shiny green on both sides at maturity. These three nonrunning forms of chinquapin are typically large shrubs, but a few individuals will grow into small trees, and, very rarely, an individual will become a medium-sized tree.

The largest chinquapin in Florida was a very unusual tree apparently fitting Small's description of *Castanea ashei*. Located near Interlachen in Putnam County, it was about 50 feet tall with a trunk diameter of 2 feet and a full crown 60 feet in diameter. There were six of these large chinquapins at this location at one time, but only two remained as of 1993, and they are all now long dead. There was similarly a tree fitting Small's description of *Castanea floridana* in Fort White in Columbia County that was about 30 feet tall with a 2-foot trunk diameter and a 60-foot crown spread. It was cut down in 1997 by the homeowner, who didn't like the sharp spines on the burrs. These larger forms of chinquapin are susceptible to the chestnut blight and are clearly in danger of going extinct, at least here in Florida.

The common, dwarfed, running form of chinquapin occurs as part of the ground cover in some longleaf pine forests on well-drained uplands. It occurs throughout northern Florida, including the panhandle and as far south on the peninsula as Citrus and Lake Counties (Wunderlin et al. 2020), but it is most common north and west from Alachua and Levy Counties. It is generally most common on the more fertile of the sandy upland longleaf pine soils, especially those with some clay in the subsoil. Its associates, in addition to longleaf pine and wire grass, include sand post oak, post oak, bluejack oak, southern red oak, turkey oak, mockernut hickory, dwarf waxmyrtle, shiny blueberry, rufous buckthorn, woolly pawpaw, bracken fern, summer farewell, blazing star, bluestar, lopsided Indian grass, and many other grasses and wildflowers.

Although formerly common in the extensive upland longleaf pine forests, chinquapin is not common today, and the nonrunning forms have become very rare in Florida. Almost all of the more fertile longleaf pine uplands have been cleared or converted to other kinds of landscapes. In 1958, Mr. B. O. Glenn, a farmer in Suwannee County, told me he had selected his farmland based on its abundance of chinquapin, which was a good indicator in his opinion of soil good enough to support farming. The best remaining populations of chinquapin are in south-central Suwannee and Columbia Counties on private lands, where the dwarfed, running chinquapin forms a major part of the ground cover on several thousand acres of longleaf pine–turkey oak–wire grass sandhill habitat that is still intact as of 2020. There are smaller populations of chinquapin at Suwannee River State Park, Ichetucknee Springs State Park, O'Leno State Park, Torreya State Park, and Blackwater River State Forest, to name a few places.

The dwarfed, running form of chinquapin is extremely fire and drought toler-ant and might best be described as fire dependent. The trunk and stems are easily killed by fires, but the plant benefits greatly from them, as they prevent hardwood trees and larger shrubs from growing up over and shading it out. Chinquapin re-sponds to fire by resprouting vigorously and producing an abundant nut crop. It will even produce an abundant nut crop the same year following a growing-season fire although the ripening time of the crop will be somewhat delayed. Chinquapin is not at all shade tolerant and does not tolerate flooding or poorly drained soils. Reproduction is both by extensive spreading of underground stems (in the case of the dwarf form) and by seed. The seeds do not remain viable for more than a few days following harvest and should be planted immediately if gathered for the purpose of growing seedlings or planting.

There has been some interest recently in attempting to determine how old clonal patches of various species of shrubs growing in native habitats might be (Takahashi et al. 2011). It seems to me that the clonal patches of the dwarfed, running form of chinquapin might sometimes be quite old. I have seen patches that are an acre in size, and I have grown this form of chinquapin from seed on typical habitat in Suwannee County and observed that it grows and spreads slowly.

The wildlife value of chinquapin is high. The nuts, which mostly ripen be-ginning in mid-September on bushes that have not burned recently and ripen mostly in October on recently burned bushes, are very appealing to most mam-mals and birds and are sweeter and tastier to us humans than chestnuts. (The nuts of the dwarfed, running form ripen earlier than the nuts of the larger forms.) Sharp spines on the burr protect the nut until it is ripe, at which time the burr opens, holding the nut out in plain view. Where the habitat is burned every one to three years, some clones of the dwarfed, running chinquapin will often produce nuts abundantly. Dwarfed, running chinquapin is not a good nut producer where fire is infrequent unless it grows in full sunlight free from competition by other plants. Large, healthy trees of the nonrunning forms of chinquapin produce nuts more abundantly and regularly and are less sensitive to competition from other bushes and trees although none of the forms are shade tolerant. Animals in the longleaf pine uplands that make extensive use of chinquapin nuts when they are available include fox squirrel, gray squirrel, flying squirrel, gray fox, raccoon, black bear, white-tailed deer, Florida mouse, old-field mouse, cotton mouse, blue jay, bobwhite quail, wild turkey, and red-headed woodpecker. White-tailed deer also browse the foliage. In addition, grasshoppers and other insects feed abundantly on the foliage, thus providing additional food for many wild creatures. It seems to me that grasshoppers are more abundant in low, extensive patches of chinquapin than almost anywhere else in the wilds of Florida. Birds in this habitat that make extensive use of grasshoppers for food are the American kestrel, loggerhead shrike, northern mockingbird, bobwhite quail, and wild turkey (Terres 1980). Other kinds of birds and mice, flying squirrels, foxes, skunks, lizards, gopher frogs, toads, and spiders also eat grasshoppers. Finally, the low thickets of the running chinquapin are ideal cover in upland pine habitats for bobwhite quail, wild turkeys, and rabbits. Two of the five wild turkey nests I have located in the wild were in chinquapin patches, the others being in sand blackberry patches.

Chinquapin is not used commercially or for landscaping although it would be suitable for a wildlife hedgerow in a rural setting, especially if the hedgerow could be burned periodically. The nuts are gathered by people familiar with their great taste although gathering them is not easy. The spines on the burrs are painfully sharp, and the weather is often hot in September when the nuts begin to ripen. The nuts do not usually last more than a day or two after the burr opens because of all the wild animals that eat them.

American Beech—*Fagus grandifolia*

Fagaceae

American beech is a common, large, deciduous tree of the hardwood forests of the eastern United States. It has smooth, light gray bark and simple leaves that have a regular pattern of parallel veins and toothed edges. In late fall, the leaves turn gold to tan, and some inner crown leaves on saplings stay on the tree through the winter. The buds, once fully formed, are strikingly long and pointed. This tree is often 80 to 100 feet tall with a 2-foot trunk diameter at maturity. As of March 16, 2015, the largest one reported in Florida, at Wakulla Springs State Park, was 126 feet tall with a trunk diameter of 3 feet 3 inches and an average crown spread of 78 feet (Florida Champion Trees Database 2020).

In Florida, American beech occurs primarily on damp slopes adjacent to the creeks and rivers of the panhandle. It is common at Florida Caverns State Park, Torreya State Park, and Wakulla Springs State Park. Its easternmost range in Florida includes trees on the slopes adjacent to the Suwannee River in Hamilton and Suwannee Counties including some at Big Shoals State Park. The southeasternmost stands occur in Alachua County just north of the town of Alachua along several different creek systems, some of which go underground and some of which are tributaries of the Santa Fe River. The easternmost, largest, and healthiest part of the Alachua County beech population was destroyed in the 1970s by a landowner to make pastureland for cattle. Most of the rest of this population is contained in Alachua County's Mill Creek Preserve.

Slope forests containing beech also often contain southern magnolia, sweetgum, spruce pine, pignut hickory, Shumard oak, water oak, basswood, Florida maple, and many other species. In the Florida Panhandle, on the upper edges of the zone supporting beech and magnolia forest, white oak (*Quercus alba*) is usually present, and in the lower parts where some moisture continually seeps to the surface, Christmas fern is often found. This damp, densely shaded type of forest has often been called the "climax forest" type for North Florida. Today, most ecologists don't believe there is one climax forest type for northern Florida. Instead, they might say that this is the climax forest for the particular conditions found on slopes that provide a continuous supply of moisture and have moderately fertile soil. Part of the reason for the composition of this forest is its protection from both fire and drought that the combination of continuously seeping moisture and a slope with a stream at its bottom affords. Different soils on flat land, even where protected from fire, will support a different forest community, no matter how much time it has to succeed toward a climax condition, primarily because of occasional severe drought.

Beech also grows on slope forests on less fertile soil, such as in the steephead ravines of the western panhandle. On the most infertile of such sites common associates may include swamp tupelo, laurel oak, sand live oak, southern magnolia, sourwood, wild olive, sparkleberry, mountain laurel, witch-hazel, sweetleaf, red star anise, serviceberry, and even silky camellia.

Beech is one of the most shade-tolerant trees in the eastern United States. It can start as a seedling and grow to maturity in the full shade of the dense slope forests that are its home. In these forests, beech usually develops a wide crown that extends to some extent out below the crowns of the other canopy trees, thus making the shade of these forests denser than forests lacking beech (Platt and Schwartz 1990). Beech is not demanding with regard to soil pH or fertility, is very wind-firm, and can withstand some minor flooding, but it is quite vulnerable to drought if planted outside its narrow niche of the slope forest. It is also easily damaged by fire.

In the northeastern United States and in the Appalachian Mountains, American beech trees are being killed in large numbers by beech bark disease, an association of fungi in the genus *Nectria* with an introduced European scale insect, the beech scale, *Cryptococcus fagisuga*. This disease complex has been slowly spreading southward from its introduction in Nova Scotia around 1890 (Houston 1998).

Beech reproduces almost entirely by seed, although both the root collar and the roots can produce sprouts. In Florida, the nuts that are produced in most years are usually hollow, and the few sound and fertile nuts are mostly eaten by squirrels. For this reason, beech does not reproduce very well here. Charles Salter, who had a native plant nursery at Monticello for many years, told me that in the Florida Panhandle, beech will produce a big crop of sound seed at irregular intervals averaging perhaps once every five years, apparently corresponding to summers without drought. These fertility problems are puzzling because, otherwise, the beech trees seem to be quite healthy and vigorous in their special habitats in Florida. I raised several from seed from the Alachua County population and planted three in Gainesville that are growing rapidly and in good health. Perhaps their reproduction is better suited to the ice age climate that occurred here about 15,000 years ago. If so, beech trees may have a long wait before party time resumes for their species in northern Florida.

The wildlife value of beech in eastern North America north of central Georgia and Alabama is high because of the importance of its mast crop. However, in Florida, the mast crop is only occasionally significant and, then, only for individual trees or small areas rather than for an entire habitat. The most significant consumers of the nuts, when they are available, are white-tailed deer, gray squirrel, black bear, wild turkey, wood duck, blue jay, rose-breasted grosbeak, and tufted titmouse (Martin et al. 1951). The red-bellied woodpecker also eats the nuts when available (Beal 1911). The value of the foliage for supporting insect populations is moderate, and the value of seedlings and sprouts for browsing animals is low. On the other hand, old beech trees that become hollow make excellent den trees.

Beech wood is hard, strong, and even-grained, making it valuable in the eastern United States well north of Florida. In Florida it is too scarce to have any special commercial value.

Beech is too restricted in its moisture requirements to be widely used as an ornamental in Florida. However, on slopes where there is continuous moisture underground, it can make a magnificent shade or specimen tree. In order to successfully grow healthy beech trees in North Florida, it is probably necessary to obtain planting stock grown from seed from the native northern Florida beech population and to provide the trees with irrigation during periods of drought.

Oaks—*Quercus* spp.: Introduction

Fagaceae

There are about 25 species of oak native to Florida. Most of these are large trees, but some are small trees, and two species are small shrubs that sprout from underground runners. The two small shrubs are runner oak and dwarf live oak (also often called "runner oak"), both of which are important mast producers for wildlife in pine flatwoods forests and dry prairie habitats.

The oaks of Florida were classified into two groups by two early monographers of the genus *Quercus:* the red oak group and the white oak group (Camus 1936–54; Trelease 1924). Unfortunately, oak classification is no longer so simple. Modern taxonomists have subdivided the white oak group into two subgroups: the white oak group and the live oak group, and worldwide, classification is much more complex (Walter Judd, University of Florida botanist, personal communication). However, for our ecological discussion of the oaks of Florida, the old classification system of Camus and Trelease works well. They used more than a dozen morphological characteristics, mostly of the flowers and fruit, to do their classifications. It is interesting that the hybridization records for the oaks of North America precisely follow these two classification systems. Nearly all the species within each of the two groups can hybridize with each other. Thus, live oak, which, according to these two monographers, is a white oak, hybridizes easily with other members of the white oak group such as white oak, post oak, overcup oak, and bluff oak. Conversely, it has never been hybridized with any red oak although many people have tried. Indeed, no hybrids of white oaks with red oaks, as defined by this classification, have ever been found or created.

The most easily observed feature that distinguishes to which subgenus the various oak species in our region belong is the length of time the acorn takes to mature. Among the members of the white oak group, acorns mature in roughly six months, whereas in the red oak group, acorns overwinter as small immature acorns and require roughly 18 months to mature. There is one oak which fails to conform to this pattern—the runner oak *Quercus pumila.* It is clearly a red oak, yet its acorns usually mature the first year (in roughly six months) although they sometimes overwinter as do the acorns of other red oaks (see *Quercus pumila* for more discussion). Two other characteristics even more consistent with this classification are the shape of the male flowers and whether or not the inside surface of the acorn shell is smooth (white oak group) or covered with pubescence (red oak group).

Compared to the red oaks, the white oaks are longer lived, have slightly stronger, more rot-resistant wood, are more firmly and deeply rooted, and are somewhat slower growing. From personal observations following hurricanes and other storms in Florida and North Carolina, I conclude that white oaks are so much more wind-firm than red oaks that severe wind storms will sometimes eliminate nearly all large red oaks while almost all of the large oaks of the white oak group, with the exception of swamp chestnut oak, will remain standing. The acorns of the white oak group, and particularly live oak, contain considerably less tannin and are therefore more palatable to many animals. On the other hand, red oaks often produce acorns more consistently and, with the exception of live oak (from the white oak group), more abundantly than the white oaks. As a result of both greater acorn production and higher tannin content, red oaks generally reproduce more rapidly and abundantly. The oaks that invade old field pine plantations are usually mostly of the red oak group.

Florida trees in the red oak group are laurel oak, water oak, willow oak, bluejack oak, blackjack oak, turkey oak, southern red oak, black oak, Arkansas oak, Shumard oak, myrtle oak, scrub oak (*Quercus inopina*), and runner oak. The white oak group includes live oak, sand live oak, bluff oak, Chapman oak, post oak, sand post oak, overcup oak, white oak, swamp chestnut oak, chinquapin oak, and dwarf live oak.

The oaks are widespread, diverse, adaptable, successful, and ecologically important. They are found throughout the northern hemisphere's temperate and tropical zones in a great diversity of habitats: from the nearly desert conditions of scrub in Florida and pinyon-juniper forests and chaparral in the southwestern United States to cloud forests in Central America; from river bottoms to mountain ridge-tops; and from prairie ecotones to coastal dunes.

Acorns are the single most important wildlife food item in northern Florida, being a staple in the diet of many species of mammals and birds, and it is this relationship with wildlife that explains part of the reason for the success of oak trees. Because they stash away massive numbers of acorns as food reserves, squirrels and jays, in particular, benefit the oak populations almost as much as the oaks benefit them. These animals often stash the acorns in widely scattered spots in the ground or on the ground under leaf litter in ways that enable germination. Jays stash acorns much farther from the parent tree than squirrels do and are therefore even better seed dispersers than squirrels. Blue jays have been observed flying as far as a mile to stash acorns, carrying three to five acorns per trip (Darley-Hill and Johnson 1981). Because they seem to prefer to stash acorns under pines, blue jays are largely responsible for the great numbers of oak seedlings often observed growing in pine plantations. Old field pine plantations that contained no trees of any kind before the pines were planted will often have hundreds of oak seedlings per acre by the time the pines are 20 years old.

Animals greatly benefiting from the acorn crop in late fall and early winter include white-tailed deer, black bear, raccoon, wild hog, domestic hogs, domestic cattle, beaver, fox squirrel, gray squirrel, flying squirrel, cotton mouse, Florida mouse, golden mouse, oldfield mouse, wood rat, eastern cottontail rabbit, wood duck, wild turkey, bobwhite quail, common grackle, blue jay, Florida scrub jay, brown thrasher, red-bellied woodpecker, and red-headed woodpecker. Many other animals such as opossum, gray fox, mallard duck, common crow, and others make use of acorns. Some of this information is such common knowledge for hunters, naturalists, biologists, ecologists, park managers, and foresters that its great importance for the wildlife food chain is often either overlooked or taken for granted when decisions are being made. Furthermore, acorn consumption by some wildlife species, such as the American beaver—a phenomenon which Axel Nail, former superintendent at Ichetucknee Springs State Park, observed and reported to me—is not well known to date.

Besides providing acorns, oak trees support a wide variety of insects that in turn provide food for birds, lizards, tree frogs, spiders, dragonflies, and other small animals. This insect part of the food chain is highly significant and is often completely overlooked. The importance of the insect and small animal community in the crowns of oak trees is great for many bird species. For instance, migrating warblers, vireos, kinglets, and bluegray gnatcatchers are especially abundant in the crowns of live oaks because of the insects and spiders they can find there. Squirrels feed extensively on the oaks' pollen-producing flowers in March. The bark, especially of live oak and the other white oaks, often supports an array of epiphytes and invertebrates. Finally, mature oak trees often provide cavities that many kinds of animals, such as honey bees, wasps (in winter), spiders, tree frogs, lizards, rat snakes, indigo snakes, bats, gray squirrels, flying squirrels, cotton mice, wood rats, raccoons, opossums, barred owls, screech owls, chickadees, titmice, wood ducks, and woodpeckers use for their homes.

Oaks are also exceptional for their value as shade trees. Laurel oak and live oak are the two most commonly used trees for this purpose in northern Florida, with live oak being one of the best landscape trees and laurel oak being one of the worst. The two trees which are probably best suited for use as street and parking lot trees in northern Florida are bluff oak and live oak.

Oak wood is famous for its utility for making furniture, flooring, railroad ties, and many other products requiring strength and physical durability. Oak logs also make superior firewood. Whiskey and wine barrels are made of white oak, and, from the perspective of history, the best wood for the keels, ribs, and other timbers of colonial-era sailing ships was live oak. Small- to medium-sized logs of laurel oak and water oak are ideal for the cultivation of shiitake mushrooms.

White Oak—*Quercus alba*

White oak is a well-known tree. Its 7-inch-long by 4-inch-wide deciduous leaves with their prominent, rounded lobes exemplify one of the classic leaf shapes. The name "white oak" may have to do with the tree's light tan-gray bark or perhaps the light brown heartwood that stands in contrast with the dark reddish-brown heartwood of many red oak species. Also, when white oak begins leafing out in spring, the new foliage is covered with a fine white fuzz, giving it, at first, a distinctive silvery whitish-pink hue that soon turns silvery green and finally bright green. White oak can grow to be a large and long-lived tree, reaching a maximum height of 150 feet with a trunk diameter of 8 feet and, on rare occasions, reaching an age of 600 years (Rogers 1990). More typically, however, white oaks grow to be about 100 feet tall with 3-foot trunk diameters and attain a maximum age of 300 years. Some trees on poor soil or at high elevations in the Appalachian Mountains are much smaller.

White oak is a common and important denizen of the hardwood forests of the eastern United States. It occurs in Florida primarily on the clay and sandy clay soils of the northern half of the central panhandle region. The southeasternmost population of about 100 trees occurs in San Felasco Hammock Preserve State Park in Alachua County. Another group of perhaps four trees grows at Robinson Sinks in northern Alachua County, and a few white oak trees grow along the Suwannee River in northern Suwannee County.

White oak in the Florida Panhandle often grows in the ecotone between upland pine forest and the hardwood forest that grows on the side of a slope going down to a stream. Common tree associates in Florida include pignut hickory, white ash, sweetgum, blackgum, basswood, American beech, winged elm, hophornbeam, southern magnolia, loblolly pine, shortleaf pine, southern red oak, upland laurel oak, and water oak.

White oak has a moderate growth rate that is faster than that of American beech, southern magnolia, and pignut hickory but slower than the growth rate of sweetgum, basswood, winged elm, white ash, southern red oak, laurel oak, and water oak (Rogers 1990). It is deeply rooted with a large taproot; its heartwood is rot resistant; and it heals wounds well. It is drought tolerant, resistant to wind-throw, and able to outlive most other species of hardwood trees. In October 1954, Hurricane Hazel brought 120-mile-per-hour winds to Raleigh, North Carolina. In 1964, I had a look at the forests around Raleigh, noting that most mature red oaks and many other hardwood trees had been blown down and lay in the same direction. Most of the mature white oak trees were still standing, however. White oak is moderately shade tolerant, which enables it to grow up under less shade tolerant trees such as pines, sweetgum, water oak, laurel oak, southern red oak, and white ash and to eventually gain access to the upper canopy by outliving them. White oak suffers less damage from ice storms than most other trees because of its strong branches and rot resistance.

The wildlife value of white oak is quite high. It is a good acorn producer, and its acorns contain less tannin than red oak acorns, rendering them more palatable to many wildlife species (see the list of wildlife species that eat acorns in "Oaks—*Quercus* spp.: Introduction"). White oak makes a great den tree in old age if it becomes partially hollow.

The timber value of white oak is high. It often has a tall, straight, solid trunk, and the strong, heavy wood is ideal for making barrel staves, flooring, furniture, and, historically, wooden ships. Most of the old wooden school desks and chairs were made of white oak; most wine and whiskey barrels are still made of white oak; and, launched in 1797, the famous naval frigate the USS *Constitution* ("Old Ironsides"), the oldest warship still afloat, is built from a combination of live oak and white oak.

White oak is a great landscape tree with its dense, rounded crown of attractive leaves that display pretty fall colors: red to purplish-red to violet. It is sold in most nurseries in the central and northern parts of the eastern United States but not often in northern Florida. White oak trees derived from Florida seed sources do well on fertile upland soils with some clay content as far south as Gainesville, Florida, but they are not well adapted to sandhill soils or pine flatwoods soils. It is a safe and dependable tree near houses and streets in part because of its large, deep, strong taproot. The downside of this large taproot is that it makes white oak difficult to transplant or grow in a container. By growing it from seed in extra-deep pots like those used for starting pecan trees and then planting it in the ground when the sapling is no more than 1 year old this difficulty can be overcome.

Bluff Oak—*Quercus austrina*

Bluff oak is a rare white oak that is becoming common in the nursery trade. Its shiny, hairless leaves are not nearly as deeply lobed as those of white oak and are quite variable in size and shape from tree to tree. With their few, irregular, rounded lobes, bluff oak's leaves are most similar in shape to those of sand post oak, but the lack of hairs under the leaves of bluff oak distinguish it. The acorn is a moderately sized white oak acorn with a bowl-shaped cup that comes about one quarter of the way up the side of the acorn. Bluff oak bark is similar to that of white oak: light brownish-gray with vertical cracks, plates, and strips.

Bluff oak grows at a moderate pace, eventually becoming quite tall but not nearly as massive as white oak. The average mature size is about 100 feet tall with a trunk 2 feet in diameter. The largest one currently known is 129 feet tall with a trunk diameter of 3 feet 4 inches (American Forests 2018). The maximum age attained by bluff oak is unknown but is probably at least 300 years. One remarkable thing about bluff oak is its taproot, which is massive and deep, even on seedlings that are only 2 or 3 years old. Seedlings I have transplanted with trunk diameters of half an inch had taproots 1 to 2 inches in diameter that gradually tapered like big carrots, extending down many feet well below a depth I could comfortably dig.

Bluff oak normally grows on soils that have an abundance of available calcium, such as soils overlying limestone or phosphate rock deposits. As a result, it has a spotty distribution that includes widely scattered places in eastern Mississippi, Alabama, Georgia, South Carolina, and northern Florida as far south as Alachua County. It is listed as vulnerable by the International Union for the Conservation of Nature's (IUCN) Red List of Threatened Species, because of its rarity, spotty distribution of very small, isolated populations, and the dearth of seedlings and saplings in these populations (Beckman 2017). On the northern Florida peninsula, it occurs in San Felasco Hammock, the River Rise State Preserve section of O'Leno State Park, and the hardwood forest that borders the Ichetucknee River in and below Ichetucknee Springs State Park, but it does not occur in Gulf Hammock. Perhaps the reason for its absence in Gulf Hammock and similar forests near the Gulf of Mexico is that these low-lying forests are subject to occasional severe flooding with slightly brackish water during extreme storm events, such as Hurricane Elena in September 1985. Bluff oak is neither especially flood tolerant nor salt tolerant. It shows some flood tolerance, however, as evidenced by its occasional occurrence on river banks.

Bluff oak is slightly shade tolerant but does best in full sun. It is a tough competitor, however, and like white oak can struggle along until other trees die and allow it to eventually find a place in the sunlight of the upper canopy of the forest.

Bluff oak is very wind-firm. I surveyed an area of about 40 acres of hardwood forest in San Felasco Hammock that was flattened by the Storm of the Century in 1993. The only trees still standing in this devastated area were a few saplings of various kinds that were able to bend without breaking and a few scattered mature bluff oaks and live oaks. All the large trees of other species such as water oak, laurel oak, Shumard oak, swamp chestnut oak, sweetgum, pignut hickory, winged elm, white ash, and red maple were lying on the ground in a broken, tangled mess.

Bluff oak is a valuable wildlife tree, though perhaps somewhat less valuable than some of the other oak species. This is because its acorn crop is somewhat irregular, it is an uncommon tree, and it is less likely to provide hollows for cavity-seeking animals. On the plus side, its acorns are less bitter than red oak acorns and are therefore highly sought after by many wildlife species. (See "Oaks—*Quercus* spp.: Introduction" above for specific species that eat acorns.)

Bluff oak produces valuable wood similar to white oak. However, there is now so little of it remaining that it is no longer economically important.

Bluff oak is now often planted as a street and parking lot tree in Gainesville. It is perhaps the best deciduous tree to use in northern Florida for urban landscaping of this kind or, for that matter, to plant near a house. It withstands heat, drought, repeated pruning, abuse, and poor air quality better than perhaps any other tree, and it is safe during storms. Bluff oak is very drought tolerant, heals wounds well, tolerates the excessive lime often present in urban soils, and remains a healthy and attractive tree when other species would either die or look stressed. A good example of bluff oak's adaptability to the extremities of urban situations can be found at the gymnasium parking lot at P. K. Yonge School in Gainesville. Here, at least 40 years ago, the late landscape designer Noel Lake planted a number of bluff oaks in narrow tree island strips in the middle of a large parking lot without an irrigation system. These trees are still healthy and attractive. Noel Lake is largely responsible for the introduction of both bluff oak and Shumard oak into the nursery trade.

Swamp Chestnut Oak—*Quercus michauxii* (including a note on Chinquapin Oak [*Q. muehlenbergii*])

Swamp chestnut oak is also called "white oak," "basket oak," and "cow oak." It is a large tree with large, wide, scalloped-edged, deciduous leaves up to 11 inches long and 6 inches wide that give it a distinctive appearance. It has the largest acorns of any oak in Florida and large, robust acorn cups. The acorn size, varying from 1 to 1½ inches in length, is quite variable from tree to tree. The bark is rather thin, tan colored, and scaly with a vertical texture. (There is a related, somewhat smaller, more northern species—the Chinquapin oak [*Quercus muehlenbergii*]—with narrower leaves and smaller acorns that occurs sparingly on the calcareous soils of bluffs and glades in a few spots in the central panhandle of Florida.) A good place to see swamp chestnut oak is San Felasco Hammock State Preserve Park in Alachua County.

Swamp chestnut oak is found as an occasional tree in upland hardwood forests on fertile soil extending northward and westward from Citrus, Marion, and Putnam Counties in the northern part of the Florida peninsula to reach as far north as New Jersey and southern Indiana and as far west as eastern Texas (Edwards 1990, "*Quercus michauxii*"). Its distribution is somewhat sporadic, however. It is restricted to fertile soils with at least a moderate amount of available calcium and a good supply of moisture based on the soil's good moisture-holding capacity, or subsoil seepage, or on the presence of adjacent streams or wetlands. Although moisture loving, and, in spite of its name, this tree is not found in swamps and is not especially flood tolerant, being found only on the slightly higher ridges in Gulf Hammock. It grows in mesophytic (moderately moist), calcareous hardwood forests, slope forests, high floodplains of spring-fed rivers, and on the edges of seepage communities where calcium is available. Its common canopy tree associates are pignut hickory, sweetgum, white ash, winged elm, laurel oak, water oak, Shumard oak, Carolina basswood, and southern magnolia. It often has grape vines growing up into its crown and provides good habitat for resurrection fern on its largest branches.

Swamp chestnut oak becomes a large canopy tree in old-growth forests where it is very competitive, in part because of its somewhat greater shade tolerance than the other species of oaks growing in Florida. It is often massive in old age, the largest one reported in Florida in 2013 from Ichetucknee Springs State Park being 143 feet tall with a trunk diameter slightly over 5 feet at 4½ feet above ground and having an average crown spread of 122 feet (Florida Champion Trees Database 2020). The national champion in Maryland has a trunk diameter of nearly 8 feet (American Forests 2015). Like most oaks in the white oak group, swamp chestnut oak lives a long time, exceeding 200 years of age on occasion. It is not as deeply rooted and not as wind-firm as the other Florida white oaks, being more similar in this regard to laurel oak.

The wildlife value of swamp chestnut oak is high. Its large acorns are actively sought by deer, wild turkeys, wild hogs, and squirrels, a fact not lost on northern Florida's hunters, who often locate deer stands or hog traps under large, healthy trees of this species, knowing that these animals will come every day (or night) while the acorns are falling. Once it reaches old age, swamp chestnut oak can become a valuable cavity tree.

The wood of this tree had considerable commercial value in the past for its many uses. It has been sold as white oak for the making of whiskey barrels (hence the replacement of "swamp chestnut oak" with the name "white oak" in the timber business), and it has been used for flooring, furniture, tool handles, and the making of split wooden objects such as baskets (hence the name "basket oak"). The acorn crop often contributes to the diet of domestic cattle and hogs (hence the name "cow oak").

Swamp chestnut oak can become a magnificent shade tree with a distinctive appearance and has been increasingly planted, especially in Gainesville. It is easy to grow and transplant and does quite well as long as the soil is suitable and there is enough space for the crown and root system to fully develop. It withstands the high lime content of urban and suburban situations well. However, it does not do well on deep, infertile sands or on acidic pine flatwoods soils, and it needs more room, both for its root system and its crown, than most other trees. It does not adapt as well as bluff oak to confined spaces in parking lots, along streets, or next to buildings.

Post Oak—*Quercus stellata*

Post oak is a moderately large oak of the white oak group with deciduous, lobed leaves. The principal lobes often form a distinctive cross shape. The leaves are from 4 to 6 inches long and often nearly as broad. The new twigs are clothed with tawny- to olive-colored fuzz (pubescence). At maturity, post oak has a sturdy trunk, a few large, horizontal branches, and a wide, rounded, rather open crown. Post oak is common on upland soils throughout most of the eastern United States from the eastern seaboard to the edge of the Great Plains, ranging from New Jersey, southern Ohio, and southeastern Iowa southward into Florida and eastern and central Texas. Although more common in the Florida Panhandle, it reaches as far south onto the Florida peninsula as Alachua County. (Some range maps show post oak reaching farther south in Florida, because they do not separate post oak from sand post oak.) A good place to see post oak in Florida is in Blackwater River State Forest in the western panhandle. It is slow growing and long lived, obtaining a maximum age of about 400 years. Large post oaks are often 60 to 80 feet tall with trunk diameters of about 2 feet, but they sometimes reach heights of 100 feet with trunk diameters of 3 feet. The three largest ones reported to American Forests (2000, 2004, 2008) were all around 84 feet tall with trunk diameters a bit over 6 feet.

Post oak is a fire-adapted tree occurring in association with pines in fire-adapted forests and also occurring in either pure stands or in association with blackjack oak on the ecotone between the eastern forests and the prairies of the Great Plains in a region spanning Texas, Oklahoma, and Kansas known as the "Cross Timbers" (Carey 1992, "*Quercus stellata*"). In Florida, where post oak is not particularly abundant, some of the more common associates are longleaf pine, southern red oak, blackjack oak, mockernut hickory, and wire grass. Post oak grows on soils that have a clay subsoil and is not found on the deep sands that support its close relative, sand post oak.

Post oak is wind-firm, highly drought tolerant, moderately fire tolerant, but intolerant of shade and competition. It does not abide prolonged flooding, wet soils, or excessive watering. It is also not amenable to soil disturbance in its root zone. Post oak usually grows where fire, poor soil fertility, or dry climate reduce competition from other hardwood species.

The wildlife habitat value of post oak is high. Its acorns are a valuable food source for a wide range of species including white-tailed deer, fox squirrel, wild turkey, bobwhite quail, and red-headed woodpecker. (For a more complete list, see "Oaks—*Quercus* spp.: Introduction" above.) Because large post oaks are often isolated, large hardwood trees in a pine-dominated landscape, they are particularly valuable to many wildlife species. Unfortunately, it is precisely because of this that they are often killed in southwest Georgia and the Florida Panhandle by managers of quail plantations, who, eager to maximize quail production, don't want to maintain trees that might harbor predators such as hawks or owls or support the prey base of such predators.

Post oak got its name because the highly rot-resistant heartwood was once used extensively for making fence posts. It is also used for making railroad ties and as firewood. It is not often used for lumber, except for the variety of post oak known as "delta post oak," which grows in the lower Mississippi River drainage.

Post oak makes an excellent shade tree in places where it grows naturally, and most of the trees currently serving this purpose existed before people constructed their housing developments. Post oak is available from some nurseries but is not often planted. There are many reasons for this. One is that post oak starts out with a massive taproot, making it poorly adapted for growing in a pot and difficult to transplant. Another is that it is slow growing and not readily adaptable to some soil conditions and maintenance practices. It needs a clay subsoil and does not like watering, soil compaction, or root disturbance. Where it grows naturally and is left undisturbed, it does great.

Sand Post Oak—*Quercus margaretta* (*Q. stellata* var. *margaretta*)

Sand post oak is the sandy soil cousin of post oak. It is a smaller tree, with smaller, less deeply lobed leaves than post oak, and unlike post oak, its new twigs are hairless. Like post oak's leaves, sand post oak's leaves are fuzzy underneath. Sand post oak occurs on the southeastern coastal plain from southeastern Virginia into Texas, occurring throughout northern Florida and as far south as central peninsular Florida (Godfrey 1988). It is much more common than post oak in Florida.

Sand post oak demonstrates considerable variability in terms of its leaf shape, tree size at maturity, and the degree to which it spreads by root suckers. There are clones of sand post oak that have spread widely by root suckers, and there are other sand post oaks that are isolated single stems without any root suckers. In some clones, all of the trees are small in stature, sometimes no more than 5 feet tall with stems no more than a few inches in diameter. This is especially true in the extreme western edge of the native range in central Texas but also occurs sometimes in Florida. Mostly, though, sand post oak is a small- to medium-sized tree with large, mature trees growing to heights of 50 or 60 feet with trunk diameters of 1 to 2 feet. Sand post oak is slow growing, often nearly ceasing growth once it reaches a particular size, which may be anywhere from very small to moderately large. Mature sand post oaks can remain healthy over a life span of more than 200 years (Greenberg and Simons 2000).

Sand post oak is a fire-adapted species that lives in longleaf pine savannas on sandy soil in association with longleaf pine, turkey oak, and bluejack oak on the poorer sites and with longleaf pine, southern red oak, bluejack oak, and mockernut hickory on more fertile sites. Other plants in these open woodlands include wire grass, showy pawpaw, chinquapin, and a wide assortment of sandhill wildflowers. Places where sand post oak is common are often places where chinquapin occurs. Sand post oak does not occur in the poorest, driest sandhill sites, and it is replaced by post oak on the more fertile soils that have clay near the surface. Sand post oak is much less common than turkey oak, but there are sandhills where sand post oak is the most common oak under the longleaf pines.

Sand post oak is very drought tolerant, perhaps even more so than longleaf pine and turkey oak. Similarly, it is quite fire tolerant, though perhaps not as fire tolerant as longleaf pine. Sand post oak is not tolerant of flooding and is only slightly more shade tolerant than turkey oak and bluejack oak. It cannot withstand competition from the large, fast-growing upland hardwood forest trees such as laurel oak, water oak, and black cherry, which quickly overtop and kill it in the absence of fire.

The wildlife value of sand post oak is good. It provides acorns for a wide variety of mammals and birds to eat (see "Oaks—*Quercus* spp.: Introduction" above) as well as a nesting structure for some bird species.

The wood of sand post oak is similar to post oak and can be used for the same purposes. It is most commonly used for firewood.

Sand post oak trees left over from the original forest are attractive small shade trees at some rural homesites, but this species is almost never planted as a landscape tree, although it might be a good choice on some sandy sites where a small- to medium-sized shade tree is desired.

Chapman Oak—*Quercus chapmanii*

Chapman oak is usually a shrub or very small white oak tree that is found intermixed with other shrubs and small trees on dry, sandy soils throughout most of Florida (except for the northern central panhandle counties). It also occurs in extreme southern Alabama and South Carolina and a small area in southeast Georgia (Kartesz 2015). Its 2- to 3-inch-long, tardily deciduous leaves are slightly wavy or very slightly lobed. Its acorns are about two thirds of an inch in diameter with cups that cover about half of the nut. Chapman oak occasionally reaches small tree size, mostly in xeric hammocks. An exceptionally large one located in a xeric hammock in the Ocala National Forest near Sweetwater Spring and Juniper Creek measured 45 feet tall with a 26-inch trunk diameter (American Forests 2008).

Chapman oak is mostly found in sand pine scrub, coastal scrub, or scrubby flatwoods forests. It is a part of the shrubby undergrowth in these environments, mixed with other shrubs and small trees such as sand live oak, myrtle oak, crookedwood, scrub palmetto, and saw palmetto. It also occurs in nearby xeric hammocks where it attains greater size. One of the best places to observe Chapman oak is in the scrub and xeric hammock habitats of the Ocala National Forest.

Chapman oak is exceptionally drought tolerant and is always found on very sandy soils. It is neither shade nor flood tolerant but is fire adapted by its ability to resprout vigorously from its root collar following a fire.

The wildlife value of chapman oak is high because of its acorns, which, being white oak acorns, are less tannic and more palatable than red oak acorns. It is a good mast producer, and the acorns are a valuable part of the diet of scrub jays, bears, deer, mice, squirrels, and many other species.

Chapman oak is not used commercially or as an ornamental.

Overcup Oak—*Quercus lyrata*

Overcup oak is a medium- to large-sized white oak that grows on river floodplains and the banks of streams and rivers from the Santa Fe and Suwannee River systems north and west throughout the Florida Panhandle and the rest of the southeastern United States coastal plain and Piedmont. It occurs naturally as far west as eastern Texas and as far north in the Mississippi River floodplain as the southern ends of Illinois and Indiana (Solomon 1990). Its deciduous leaves are, on average, about 6 inches long with irregular lobes that, while not as rounded on the ends as those of most white oaks, are not as sharply pointed on the lobal ends as those of most red oaks. The leaves are widest toward the tip, narrow in the middle, and tapered toward the base. The bark resembles that of post oak and sand post oak with light gray to dark gray irregular and deeply furrowed bark plates. Overcup oak's most distinctive feature is its large acorn, which on average measures a bit over 1 inch in length and width, often being wider than long and having an acorn cup that encircles most of the nut.

The size, growth habit, and growth rate of overcup oak varies according to variations in habitat and growing conditions. Its maximum size is quite large, with one in Congaree National Park in South Carolina measuring 123 feet tall with a 7-foot trunk diameter (Godfrey 1988). At maturity, overcup oak is typically 100 feet tall with a trunk diameter of 2 to 3 feet and a spreading crown composed of a few large branches that often support resurrection fern. Its growth is usually rather slow, and trees can attain a maximum age of about 400 years (Solomon 1990).

Remarkable in its ability to withstand flooding, overcup oak inhabits sites that experience flooding too prolonged for the survival of any other oak species. It is not as flood tolerant as bald-cypress, swamp tupelo, or water tupelo but withstands flooding better than laurel oak, water oak, and sweetgum. In Florida, it seems to prefer the floodplains of major rivers that have fertile soil containing an abundance of available calcium. Overcup oak is found in various places on the Santa Fe River floodplain, even though neither Godfrey (1988) nor Solomon (1990) report its presence there, and it frequently grows along the Suwannee River and the major rivers of the Florida Panhandle. Common tree associates include bald-cypress, water hickory, sweetgum, river birch, laurel oak, water oak, water-elm, water locust, red maple, and Florida elm.

The wildlife value of overcup oak is high. As a white oak, its acorns are less tannic than the acorns of red oaks, which makes them more palatable and broadly consumed by a wider range of animals. The acorns are too large for some of the avian acorn specialists, such as blue jays and common grackles, to easily ingest, but they are eaten by white-tailed deer, black bear, wild hog, raccoon, gray squirrel, and wild turkey.

Where overcup oak is abundant, the wood has been used commercially. Though it *is* a white oak, its wood is not considered as valuable as that of most other white oak species because of defects resulting from insect borer damage and because the wood often checks (splits) excessively during seasoning (Solomon 1990). Overcup oak is rarely used as an ornamental, although it would make a handsome shade tree if planted in a landscape located in the floodplain of a major river.

Live Oak—*Quercus virginiana*

Live oak is in a class by itself. Among Florida's oaks, it is by far the largest, longest lived, and most adaptable, and it has the widest-spreading crown and the strongest, heaviest, and most rot-resistant wood. It grows throughout Florida, but its natural range is restricted to a narrow band along the coast north and west of Florida until it reaches Texas, where it grows inland as well. It is called "live oak" in part to because of its nearly evergreen leaves that drop in the spring just as the new leaves begin to grow.

It is not always easy to distinguish live oak from some of the other oaks, particularly laurel oak. Live oak's leaves are shiny on top but dull (not shiny) below (the result of a normally dense covering of tiny hairs), and they have an undercurled edge. By contrast, laurel oak's leaves are nearly as shiny below as they are on top and lack the undercurled edge. Another difference between these two is the rough bark that covers live oak's trunk and all branches larger than about 3 inches in diameter. Laurel oak bark is smoother and is very smooth on the upper trunk and on branches smaller than 6 inches in diameter.

The acorns of these two oaks also differ from one another. Live oak produces oblong acorns about three quarters of an inch long that are relatively low in tannins and mature after six months of growth, whereas laurel oak produces acorns about one third of an inch long and wide that are high in tannins and take 18 months to mature. The greatest differences between these two species are not immediately visible, however. Whereas laurel oak generally lives only 30 to 80 years and is easily uprooted by high winds, live oak can live 200 to 300 years or more and is the most wind-throw resistant of Florida's trees.

In 1990, there were three trees tied for the honor of being named the largest live oak in Florida. Each had a trunk diameter of about 9 feet at 4½ feet above the ground, and their heights varied from 72 to 89 feet, with average crown spreads ranging from 138 to 154 feet. These trees were located in Alachua, Duval, and Lake Counties. In 2016, the one in Alachua County, the Cellon Oak, was 87 feet tall with a trunk diameter of 10.3 feet 4½ feet above ground and a crown spread of 160 feet (Florida Champion Trees Database 2020), making it the undisputed champion. Ring counts on a large severed branch indicate that the Cellon Oak was slightly less than 200 years old in 1997 (Ward and Ing 1997). The tallest live oak I've measured was 115 feet tall. Two live oaks that I measured in San Felasco Hammock both had trunk diameters of about 8 feet and were 100 feet tall. Average size for a mature live oak is probably from 60 to 80 feet in height with a trunk diameter of 2 to 5 feet. When growing right next to the coast or in other stressful locations, live oak will be somewhat smaller and often may not exceed 50 feet in height. Also, live oak is very diverse genetically. Some individuals will not grow tall, whereas others will grow quite tall.

Live oak grows naturally in several kinds of situations. Most of the natural forests where live oak is abundant are hammock (hardwood) forests that are either next to the coast or on low ground next to a river, lake, or prairie and subject to occasional flooding. The most common associates near the coast and along the St. Johns River are cabbage palm and southern red-cedar. In other low hammocks, laurel oak, water oak, sweetgum, persimmon, Florida elm, sugarberry, American holly, slash pine, and loblolly pine are common associates. This seems contradictory to many people, who think of live oak as an upland tree, and, indeed, live oak also thrives on well-drained soil in upland hardwood forests in association with laurel oak, sweetgum, pignut hickory, red bay, sugarberry, and many other species, although it is less abundant in these situations. Some of this confusion comes from the fact that many people do not separate live oak from sand live oak, which is often abundant on very dry, deep sands in scrub and xeric hammock habitats. Both live oak and sand live oak are common invaders of disturbed habitats, growing along old fencerows, in pioneer forests on old fields, and invading upland longleaf pine forests that no longer burn at frequent intervals.

Live oak grows in a wide diversity of habitat conditions. It grows well on both sandy and clay soils and on both acidic soils and limestone outcrops. Live oak, southern red-cedar, slash pine, and cabbage palm withstand the salt spray and the salty soil of coastal forests better than all other trees with the exception of the mangroves. Live oak can withstand periodic severe flooding on otherwise well-drained sites better than other upland trees, hence its ability to grow beside rivers, lakes, and prairies. This is a complex situation, for although it does not grow on continually wet soil as readily as swamp laurel oak and water oak, it can withstand periodic severe flooding better than they can. It can also withstand drought better than many other kinds of trees although probably not as well as sand live oak, myrtle oak, bluff oak, post oak, and sand post oak. Live oak does not thrive in shady situations and will gradually die in dense shade at any stage of its life, from seedling to mature tree. For this reason it is often found on forest edges, in places subject to occasional severe floods, or near the coast where hurricanes and saltwater flooding periodically kill some of its competition.

Many people are surprised that live oak often grows fast. It does not grow tall as quickly as some other trees, but overall growth in size is often rapid for young live oaks. My father planted a live oak in our yard in 1953, at which time it was about

15 feet tall with 3 inches in basal trunk diameter. Fifty years later, in 2003, it was 80 feet tall, with a 4-foot trunk diameter and a crown spread of over 100 feet.

One of the special qualities of mature live oaks is that they usually provide support for a whole community of other plants. The upper surface of their massive branches is often covered by a dense growth of resurrection fern. Smaller branches often support greenfly orchids, mosses, liverworts, and lichens of many kinds. Spanish moss and some of its larger relatives seem to prefer live oak as well. Live oaks also usually support several large grape vines and other vines such as poison ivy, Virginia creeper, and trumpet creeper. This complex community provides additional habitat for insects, spiders, lizards, and such insectivorous birds as warblers, wrens, vireos, titmice, and woodpeckers.

Live oak is probably the single most important tree in Florida for wildlife because of its combined exceptionally high values as a mast producer, den tree, and provider of small insects for birds, plus its abundance both in the wild and in urban and suburban situations. Live oak is often a heavy acorn producer, and its acorns are the most palatable among Florida's oaks. Live oak acorns ripen early, usually dropping to the ground beginning in late September or early October, thus expanding the time when this most important of all wildlife food crops is available. Also, live oak is more variable genetically than most species, and, as this variability includes the timing of acorn ripening, the crop from this one species is available from mid-September to mid-December.

A list of animals that depend on acorns for a significant part of their diet is given in "Oaks—*Quercus* spp.: Introduction." In Florida, live oak probably provides more food for wildlife than any other species. There are some species that strongly prefer live oak acorns because of their low tannin content. Such species include raccoon, gray fox, coyote, common grackle, common crow, and red-headed woodpecker. Even so, most of the crop is eaten by deer, wild hogs, wild turkeys, black bear, blue jays, gray squirrels, flying squirrels, and fox squirrels. Martin et al. (1951) estimate that acorns make up 25 to 50 percent of the diet of these species. In fall and early winter, the bellies of these animals are often full of acorn mush.

Live oak has outstanding value as a nest and den tree. Most ancient live oaks are partially hollow, and their hollow trunks and branches provide homes for many different wildlife species. Radio tracking studies of indigo snakes in Gulf Hammock discovered that, in this coastal hardwood forest habitat, these animals almost always use the basal part of large, old live oaks for their winter dens (Moler 1985). While squirrel hunting, I have often seen gray squirrels travel long distances to get to the safety of a live oak den tree. Some pairs of broad-headed skinks appear to spend their entire lives on a single large live oak cavity tree. I have also seen raccoons, opossums, rat snakes, tree frogs, bluebirds, tufted titmice, downy woodpeckers, red-headed woodpeckers, barred owls, barn owls, screech owls, and honey bees using live oak cavities. I'm sure the complete list of animals that use these den trees would fill an entire page. The resurrection fern–covered surfaces of large branches of live oaks are also used for nesting. Of the four great horned owl nests I have seen in Florida, three were built on the large fern-covered branches of live oak trees.

The insect and spider populations on large live oak trees are probably the most overlooked aspect of their wildlife value. There are several kinds of gall-forming wasps that inhabit live oak in large numbers, and there are many kinds of insects on their leaves, twigs, bark, and epiphytic plants. I had a medium-sized live oak in my front yard, and it almost always had several birds in it. Warblers, vireos, wrens, titmice, chickadees, or woodpeckers were almost always there searching its trunk, branches, and foliage for insects.

Live oak is the most frequently planted and the most desirable large shade tree for use in Florida. That is not to say it is the only tree to use. There is much to be said for diversity. But because of its adaptability, reliability, durability, strength, beauty, fast growth, and high wildlife value, it is second to none in situations where live oak is not already an abundant tree in the immediate area and where a large tree is desired. It is easy to grow and transplant, survives well in parking lots and other tough situations, heals wounds easily, survives disturbances to its root zone, survives the frequent pruning done by maintenance people, stands up to strong winds better than any other kind of tree, and makes a magnificent tree in almost any situation where it has enough room. This last point, however, is an important one. To thrive, a live oak must have plenty of open space. It does not do well crowded in among other trees. One common mistake in landscaping is to plant or leave other tall-growing trees near an established, open-grown live oak. A large live oak will be damaged by young trees near it that grow tall, gradually killing the branches of the live oak by shading them.

The value of live oak for humans, other than for landscaping, is mostly historical. The live oak acorn crop was a very important staple food for the Native Americans of northern Florida prior to European colonization of the New World. Later, live oak was the most valuable timber tree in Florida during the colonial era when the large sailing ships were made of wood. Because live oak provided the heaviest, hardest, strongest, and most durable large timbers of any tree in North America or Europe, and because the heartwood is very rot and weather resistant, the large branches of live oak were individually selected, based on their shape and soundness, to make the keels, ribs, and other structural timbers of these ships. To secure a future supply of live oak timber for ship building, the United States Navy, in 1828, purchased an area of live oak forest near Pensacola called the "Naval Live Oaks Reservation," which is now part of the Gulf Islands National Seashore. The oldest commissioned warship afloat in the world, the USS *Constitution*, was constructed mostly of live oak timber, which made its hull exceedingly strong. In battle, this hull repelled cannon fire so effectively, the ship was nicknamed "Old Ironsides" (Encyclopedia Britannica 2020).

Today, the wood of live oak is still used to some extent to make industrial-strength pallets. It is also used as a source of firewood, though compared to other oaks it is very difficult to split, hard to ignite, and burns very slowly. One good way to use it for firewood is to mix it with other, faster-burning woods. The wood of live oak, and especially the heartwood, can be fashioned into mallets and tool handles although there is currently no market for this, and the wood tends to warp and split as it dries. The heartwood can also be used where great rot resistance is desired. A study done on the Austin Carey Memorial Forest found that 4 × 4s of live oak heartwood placed as posts in the ground outlasted all other kinds of wooden timbers, including various kinds of chemically treated timbers (Jake Huffman, retired wood technology professor at the University of Florida, personal communication).

Sand Live Oak—*Quercus geminata*

Sand live oak is a close relative of live oak that occurs near the coast from southeastern North Carolina to southeastern Louisiana and throughout all of Florida except for the southern tip (Kartesz 2015). It is smaller and slower growing than live oak and is quite variable in its size and leaf characteristics. Its leaves are more densely scruffy pubescent on the undersides and are often strongly curled (canoe shaped), although there are some sand live oaks in Suwannee County that have flat leaves with a strongly curled-under edge. Also, the acorns of sand live oak are usually produced two per stalk instead of one per stalk as on live oak. Finally, sand live oak blooms and leafs out nearly a month later than live oak.

The size of sand live oak is quite variable, with some of those that occupy scrub habitats or oceanside dunes rarely getting more than 10 or 20 feet tall with 6-inch trunk diameters. However, on better soil and growing apart from the intense fires of the scrub habitats or the salt spray of the ocean, some individual sand live oak trees can grow much larger, often up to 50 feet tall with 2-foot trunk diameters. There are two exceptionally large sand live oak trees in Alachua County, one in the center of Gainesville and one southeast of Gainesville in Flamingo Hammock, both measuring 71 feet tall with trunk diameters of nearly 6 feet.

Sand live oak is very drought tolerant and moderately salt tolerant, and on deep sandy soils it withstands wind storms well. On shallow, wet soils it is not as wind-firm. Sand live oak is moderately tolerant of short durations of occasional flooding but is not tolerant of shade, although it *will* grow under a closed canopy of sand, slash, or longleaf pine.

Sand live oak is common in several situations. It is the only live oak located on the deep sands of the sand pine scrub and other Florida scrub habitats where it is often abundant. Similarly, it is the most common live oak on sand dunes along the coast and on the deep sands of the longleaf pine–turkey oak–wiregrass forests of the interior. It occurs as a common tree in pine flatwoods forest areas, sometimes in rather wet situations, and is the dominant tree in many xeric hammocks. Common associates in scrub habitats include sand pine, myrtle oak, Chapman oak, crookedwood, scrub bay, scrub palmetto, and saw palmetto. Common associates in longleaf pine forests include longleaf pine and turkey oak. Its most common associate on coastal dunes and in the pine flatwoods is saw palmetto.

One difference between live oak and sand live oak is that sand live oak spreads aggressively by root suckers, often producing large patches that are all derived from one tree. In the longleaf pine sandhills of the Ocala National Forest, these sand live oak clones create what are sometimes called "oak domes." These oak domes are not as noticeable now as they were in 1980, before the Forest Service reintroduced growing-season burning to these forests. In fact, many of these clones have now been reduced to patches of oak sprouts and shrubs. The domes were probably an unnatural aspect of these forests that grew up as the result of their protection from the natural fire cycle beginning about 1930 and ending in 1985.

The wildlife value of sand live oak is high. It is a good acorn producer, and its acorns are highly palatable. In scrub habitats, it provides an important food source for scrub jays along with the same wide assortment of other animals that favor live oak's acorns. Similarly, ancient sand live oak trees often provide long-lasting cavities for cavity-nesting wildlife species.

Sand live oak was never as valuable as live oak for ship building or other commercial manufactures. Not only is it smaller than live oak, but it does not produce the same massive, long, spreading and arching branches that were so valuable in ship building. It is also not as valuable for ornamental use but is sometimes planted, mostly by accident, by people who do not notice the difference between sand live oak and live oak. It is often present in residential areas as a tree that was left there during building construction. It survives well in these situations and makes an interesting and valuable landscape tree, especially where space is limited. It has a less dense and much smaller canopy than live oak and is typically much smaller and more crooked.

Dwarf Live Oak—*Quercus minima*

This is one of the two kinds of oak commonly called "runner oak." It is a low shrub spreading by underground runners to form extensive patches of stems usually less than 3 feet tall. It is usually shorter growing and less dense than *Quercus pumila*, which is also called "runner oak." The leaves of dwarf live oak are evergreen, dully lustrous on top, dully pubescent to glabrous below, and variable in shape (see the species account for *Q. pumila* for a description of features that distinguish the two runner oaks from each other). Its leaves most closely resemble those of live oak or sand live oak but are thicker than live oak's and broader, flatter, and more often lobed than those of sand live oak. Dwarf live oak occurs naturally in suitable habitat throughout Florida and near the Atlantic coast in Georgia and South Carolina (Kartesz 2015).

The habitat of this little oak is the same as that of the other runner oak. It grows primarily in the drier parts of pine flatwoods forests and, to a lesser extent, in the moister parts of upland and sandhill pine forests in association with longleaf and slash pine, saw palmetto, gallberry, dwarf waxmyrtle, shiny blueberry, dwarf huckleberry, huckleberry, hairy laurel, bracken fern, wire grass, and an assortment of wildflowers. It is somewhat more widely distributed in Florida than *Q. pumila*, getting farther south and closer to the coast and competing somewhat more successfully with the dense thickets of saw palmetto and gallberry that strongly dominate so much of these forests, but it does not extend as far into the edges of sandhill forests. It is abundant in the Apalachicola and Osceola National Forests, in the Blackwater River, Goethe, Jennings, and Withlacoochee State Forests, at Camp Blanding, and at Morningside Nature Center in Gainesville.

This plant is highly fire dependent, doing best where growing-season fires occur every two to three years on average. The abundance of both types of runner oaks is largely dependent on the fire cycle being frequent enough to prevent saw palmetto and gallberry from completely closing the openings that occur in the shrub thicket of the drier parts of the pine flatwoods forests.

The wildlife value of this plant is high. The two runner oaks produce most of the acorn crop in extensive areas of pine flatwoods forest, supplying an essential part of the diet of the deer, bear, raccoon, fox squirrel, turkey, bobwhite quail, blue jay, red-headed woodpecker, and small rodents that live there.

Runner Oak—*Quercus pumila*

This is one of the two species of oak called "runner oak." It spreads by underground runners and puts up stems that are usually from 1 to 2 feet tall, forming large patches. Some individual plants will grow as tall as 3 to 6 feet in spots protected from fire. The narrow, unlobed, tardily deciduous leaves are usually shiny on top with white pubescence below. This little oak shares an unusual trait with willow oak. The leaves of both of these oaks are rolled in from the sides when they first come out of the bud, appearing like small needles. An even more unusual trait of this runner oak is that, although it is in the red oak group, its acorns usually (but not always) mature the first year, six to seven months following flowering. All other red oak acorns usually take 18 months to grow to maturity. The acorns are attached one by one on the stems rather than in groups of two as is often the case with the other kind of runner oak, dwarf live oak (*Q.*

minima). The other obvious difference between the two runner oaks is that *Q. pumila* has narrow, unlobed leaves whereas *Q. minima* has broader leaves, some of which usually have small, stiff, sharp-pointed lobes. Some other less obvious but distinguishing traits are that the winter buds of *Q. pumila* are sharp pointed and conical and the inner surface of the acorn shell is pubescent, whereas the buds of *Q. minima* are rounded and blunt at the tip and the inner surface of the acorn shell is glabrous. In habitats that burn frequently, several other oak species sometimes spread by underground runners to form low thickets similar to those created by the two runner oaks. These are myrtle oak, sand live oak, Chapman oak, and sand post oak.

Runner oak is abundant in much of the higher, drier parts of pine flatwoods forests that still have some native ground cover. It also occurs to a lesser extent in the lower and moister parts of sandhill and other upland pine forests where native ground cover remains. Common trees in these habitats are longleaf pine and slash pine. Common shrubs and ground-cover associates include saw palmetto, dwarf waxmyrtle, gallberry, several species of pawpaw, dwarf live oak, bracken fern, wire grass, several kinds of bluestem grasses, shiny blueberry, several kinds of huckleberry, hairy laurel, and a wide assortment of wildflowers.

Runner oak occurs in suitable habitat throughout Florida (Wunderlin et al. 2020) and as far north on the coastal plain as southeastern North Carolina (Godfrey 1988). It is quite abundant in the Apalachicola and Osceola National Forests. It also occurs at Camp Blanding, Blackwater River State Forest, Jennings State Forest, the Austin Cary Forest, and, in Gainesville, at Morningside Nature Center.

Runner oak is very fire dependent. It does best with frequent fires that occur mostly in the late spring and early summer. The ideal fire frequency for runner oak is about once every other year. It is interesting that this is the only red oak that consistently grows acorns in a single season. (Myrtle oak will sometimes do the same.) This unique feature is almost certainly an adaptation to the frequent fires that shape runner oak's natural habitat. If it took two growing seasons to mature its acorn crop the way other red oaks do, it would lose at least one year's acorn crop each time its habitat burned. This would put it at a great reproductive disadvantage.

The wildlife value of runner oak is high. The two runner oak species, *Q. pumila* and dwarf live oak, produce the main acorn crop of the pine flatwoods forests, thus supplying a very important part of the diet of the deer, bear, fox squirrel, wild hog, turkey, bobwhite quail, and small rodents that live there. Dense thickets of runner oak can serve as escape and nesting cover for quail and rabbits.

Although not used in landscaping currently, *Q. pumila* could be used as a low border hedge, given its attractive, shiny leaves with their white undersides and its ability to form a dense, low thicket.

Myrtle Oak—*Quercus myrtifolia*

Myrtle oak is a small red oak with thick, roundish, evergreen leaves that occurs mostly as a shrub or small tree in scrub and scrubby flatwoods habitats. The thick, leathery leaves are shiny green on top, slightly or moderately curved or undercurled on the edges, and about 1 to 2 inches long. The half-inch-long roundish acorns, as wide as they are long, are produced in abundance and can take one or two seasons to mature. (This variable maturation timing is unusual, having been observed in only one other oak species: *Quercus pumila.*) Myrtle oak occurs throughout Florida on suitable soils, except for some of the counties adjacent to the Georgia and southeastern Alabama border, and ranges into extreme southern Alabama, southeastern Georgia, and coastal South Carolina (Kartesz 2015). In xeric hammocks that form when scrub is left unburned for a long time, myrtle oak sometimes grows to be a small tree. The largest one ever reported measured 41 feet tall with a trunk diameter of about 22 inches (American Forests 2018).

Myrtle oak is largely confined to the deep, sandy soils of coastal dunes and interior scrubs. It grows on former coastal dunes such as the sand pine scrub of the Ocala National Forest and the scrub communities on the Lake Wales Ridge. It forms dense thickets intermixed with other shrubs and small trees such as sand live oak, Chapman oak, crookedwood, Florida rosemary, scrub palmetto, and saw palmetto. It is also present in xeric hammocks near these dunes and interior scrubs.

Myrtle oak is a short-lived tree with a slow to moderate growth rate. It reproduces both by root sprouts and by seed, and some of the clones produced by years of spreading by root suckers may be fairly old. Myrtle oak is very drought tolerant and fairly salt tolerant. It prefers to grow in full sunlight but can withstand some shade, growing well in the shade of sand pine forests. Myrtle oak withstands fire by sprouting back vigorously from the root collar and its root system.

The wildlife value of myrtle oak is high because of its abundant production of acorns, which form an important part of the diet of scrub jays, black bears, white-tailed deer, beach mice, gray squirrels, and many other species. It is also one of the trees frequently used by scrub jays for nesting.

Myrtle oak is sometimes planted as an ornamental, especially on sandy soil near the coast. Its compact, round crown of dark green foliage is attractive, and its ability to thrive on deep sand near the coast makes it a good choice in such situations.

Upland Laurel Oak—*Quercus hemisphaerica* and Swamp Laurel Oak—
Quercus laurifolia (*Q. obtusa*)

This is either one variable species of oak or, more likely, two closely related species. If considered as two species, the one growing commonly on uplands is called
"laurel oak," "upland laurel oak," or "Darlington oak" and has been given the Latin
name *Quercus hemisphaerica*. The laurel oak commonly growing in wetlands is
called "laurel oak," "swamp laurel oak," or "diamond-leaf oak" and has been given
the Latin name *Quercus laurifolia* or, historically, *Quercus obtusa*.

These two forms may sometimes be distinguished by the timing of leaf and
flower formation, which is a bit earlier in swamp laurel oak. Similarly, they may
often be distinguished by both the leaf shape (broader with a less acute leaf tip in
swamp laurel oak) and the presence of hairs in the leaf vein axils on the underside
of the leaf of swamp laurel oak (Godfrey 1988).

Although these two forms are fairly distinct in their most extreme expressions
of natural variation, they are nonetheless very similar and difficult to distinguish
when growing together in mesic hardwood forests in such places as Gulf Hammock, San Felasco Hammock, or along the Silver River in Silver Springs State
Park. Both forms, but especially upland laurel oak, have expanded their populations greatly by invading former longleaf pine ecosystems and by being planted
or coming up on their own in human-created landscapes (Carey 1992, "*Quercus
hemisphaerica*"). In common usage, both forms are called "laurel oak." Both range
along the Atlantic and Gulf coastal plains from Virginia into eastern Texas including all of Florida except for the Florida Keys.

Laurel oak is also sometimes called "water oak" by people in the timber industry who do not care to distinguish laurel oak from willow oak and water oak. The
group of closely related oaks growing in the southeastern United States that have
similar characteristics such as small, mostly unlobed leaves, fast growth, short life
span, and easily rotting, defect-ridden wood includes willow oak, water oak, both
forms of laurel oak, and bluejack oak.

Swamp laurel oak is better adapted than upland laurel oak for growing in wetlands and can be found in its most distinctive and identifiable form in such places
as the continuously wet organic soils along Juniper Creek, Mormon Branch, and
Alexander Springs Run in the Ocala National Forest and on alluvial soils in the
floodplain of the Apalachicola River. In these places, the leaf form is distinctly
diamond shaped for most of the leaves, and the leaves of seedlings and sprouts
lack the tendency to produce the irregular dentation along the leaf margin so
commonly seen in the leaves of upland laurel oak's seedlings and sprouts. Mature
swamp laurel oak has a tendency to have a somewhat buttressed base, especially
on very wet soils.

Upland laurel oak is better adapted to well-drained sandy soils and has widely and extensively invaded most of the untended uplands of Florida, forming forests of laurel oak on former longleaf pine uplands, on former agricultural lands, in rural subdivisions, and along road rights-of-way. On sandhill sites, upland laurel oak has hybridized with bluejack oak to such an extent that, on some sites, young hybrids between these two oaks are often as common as are young bluejack oaks (personal observation).

The rest of this discussion will consider both forms together as laurel oak. Laurel oak is a large tree, commonly reaching 100 feet in height with a trunk diameter of 2 to 3 feet. Maximum size is represented by one in South Carolina that in 2006 was 130 feet tall with a 7-foot trunk diameter (American Forests 2008). The small, narrow leaves are nearly evergreen, typically dropping just before the new leaves begin to appear. There is considerable variation from tree to tree and from year to year, however, with some trees dropping leaves somewhat earlier. The three eighths–inch-long acorns are quite small and nearly round with shallow cups and contain enough tannin to make them bitter. The root system tends to be very shallow although on deep sands it can be deep.

Laurel oak is a fast-growing, short-lived tree. It can attain a trunk diameter of 4 feet in as little as 35 years and seldom lives more than 70 years. The wood is not rot resistant, and in old age it often has rotten and/or hollow places inside the main trunk and main branches. Given the high frequency of rot at the base of the trunk and the often shallow root system, laurel oak is one of the most likely of our large trees to be blown down during wind storms. Large branches are prone to breaking as well because of the high incidence of internal rot.

Laurel oak is a very adaptable tree, growing well in many kinds of wetlands and uplands and on a wide variety of soils. It is especially common in bottomland forests, floodplain forests, and all sorts of upland hardwood forests. It is often more common on the upland and lowland edges of upland hardwood forests than in the mesic middles of these forests. This is the result of a combination of factors including the invasion of hardwood forest on the upland edge into what was formerly upland pine forest and the transition on the lower edge into wetland forest. Laurel oak is very well adapted to both situations.

Laurel oak is only slightly shade tolerant. Though more adaptable to shade than live oak, persimmon, or loblolly pine, it is less so than most of its hardwood associates such as sweetgum, pignut hickory, winged elm, basswood, and swamp chestnut oak. It is also less tolerant of drought than most of its associates and has a shorter life span. The result is that in the center of old, long-established hardwood forests, laurel oak is much less common than in hardwood forests that are the result of recent invasion by hardwood trees into other habitats. Laurel oak's most common upland associates are sweetgum, pignut hickory, water oak, live oak, sand live oak, loblolly pine, and black cherry; its most common lowland associates are sweetgum and red maple.

If all the seedlings, saplings, and young trees were counted, laurel oak might be the most abundant tree in Florida. It is a champ at invading new territory. In upland pine forests, it often invades in great abundance well away from any parent tree. A half mile from the nearest adult laurel oak, there are often more than 100 young laurel oaks per acre in 20-year-old pine plantations that were established on agricultural fields initially containing no trees when the pines were first planted (personal observation). One might wonder how this is possible as acorns do not blow around in the wind and squirrels do not travel long distances to bury them. But there is a good explanation. Blue jays stash acorns in the fall to secure a food supply for the rest of the year, and they will fly up to a mile or more to do so (Darley-Hill and Johnson 1981). They also fail to retrieve all of the acorns they stash, thus providing a very effective means of seed dispersal.

The wildlife value of laurel oak is high. It produces an abundance of acorns almost every year that provides food for many species of mammals and birds (see "Oaks—*Quercus* spp.: Introduction" above for details). It is also one of the best producers of small and large cavities that are used for denning and nesting by a wide assortment of wild creatures including honey bees, wasps, bats, raccoons, opossums, barred owls, barn owls, chickadees, titmice, gray squirrels, flying squirrels, rat snakes, skinks, and more. As with any kind of tree, the wildlife value depends to a large extent on the overall situation and the condition of the tree. A solid, healthy laurel oak in a forest or landscape dominated by laurel oaks may have very little wildlife value, whereas an old, hollow laurel oak in a forest or landscape not so dominated is apt to have a very high wildlife value.

The commercial value of laurel oak is low to moderate. It is abundant, and the hard, heavy, easily split wood makes great firewood. It is therefore one of the most commonly used trees for fuel wood and firewood. Laurel oak logs are also particularly well suited for use in the growing of shiitake mushrooms. The sawn lumber is suitable for constructing furniture, flooring, and other items, although there is not currently a good market for it, and laurel oak is considered less desirable for the making of these products than most other oaks. Unfortunately, laurel oak is the most common invader of pine plantations and, as such, creates a negative value by increasing the cost of plantation management. More significant, because of its aggressive tendencies and the hazard large, partially rotten or hollow laurel oaks pose to buildings and power lines, it also greatly increases the cost of managing the urban forests of cities, towns, and subdivisions.

The landscape value of laurel oak in human-created landscapes is a mix of good and bad. Laurel oak is one of the most common large shade trees in cities, towns, subdivisions, and rural housing developments. Sometimes it is planted intentionally but, more often, it is either situated in its locale before building construction or comes up on its own afterward. In any case, it makes a beautiful and desirable shade tree when young, but, as it gets large and old, it often becomes a liability.

If it is within reach of a building, parking area, power line, or human-use area, it can be a hazard. This is a consequence of mature laurel oaks' large size combined with their higher-than-average chance of being storm-toppled given both their shallow and often-weak root systems and the high probability that their interior trunk bases or major branches are hollow or rotten.

Several factors cause laurel oak interiors to rot and become hollow. The first is to some extent driven by people who, when they prune large, live branches from a mature laurel oak invariably end up introducing rot into the trunk of the tree. Once this process begins, there is no stopping it: the rot will progress throughout the interior of the trunk, eventually causing a rotten or hollow tree increasingly at risk of breaking apart or falling over. Rot at the base of the tree also sometimes begins with a human-caused impact injury to the lower trunk or by cutting or damaging the roots. Trees left on a building lot are often damaged this way. Other factors that contribute to laurel oak rot and hollowing include the fact that its wood is not rot resistant and that laurel oak is a short-lived tree that does not heal wounds well once it reaches maturity.

Water Oak—*Quercus nigra*

Water oak is a medium- to large-sized red oak with tardily deciduous leaves quite variable in shape. The leaves on mature unshaded trees are mostly unlobed and oblong or spatula shaped, being much broader toward the end of the leaf. However, on shaded seedlings and sprouts, the leaves are often highly variable, with many leaves being lobed and larger than normal. The bark, gray and smooth on small trunks and branches, is ridged on larger trunks but never becomes as rough as live oak bark. The bark of water oak, including both the outer and inner bark, is about twice as thick as the bark of laurel oak. The acorns are roundish and about half an inch in diameter. Water oak occurs in the wild from Delaware southward into the central Florida peninsula and across the Carolinas, Georgia, Alabama, Mississippi, Louisiana, and Arkansas into southeastern Oklahoma and eastern Texas (Vozzo 1990).

Water oak is a common tree of wetlands, floodplains, flatwoods streams, unburned or infrequently burned pine flatwoods forests, and hardwood forests. It occurs most often and grows best on clay or sandy clay soils but is adaptable. Water oak is short lived and excels at invading new territory such as old fields and unburned pine forests. As a consequence, the age distribution of the water oak population in Florida is strongly skewed toward seedlings, saplings, and young adult trees. It is similar to laurel oak in many ways, but has larger acorns and thicker bark. It is about the same size as laurel oak, typically getting 100 feet tall with a 2- to 3-foot trunk diameter at maturity, although on soils not well suited for it, such as pine flatwoods soils or deep, sandy soils, it is usually smaller. The largest one reported so far, located in Mississippi, has about an

8-foot trunk diameter and is 122 feet tall (American Forests 2015). The largest one I have seen in Florida was 100 feet tall with a trunk diameter of about 4 feet. Like laurel oak, water oak rarely gets older than 70 years, and the trunk is often partially rotten and/or hollow in old age. Water oak is particularly susceptible to being colonized by mistletoe.

The wildlife value of water oak is high. Its acorns are a preferred food of gray squirrels, blue jays, and many other animals, and old trees with hollows supply cavities used by a wide variety of wildlife species for dens and nesting (see "Oaks—*Quercus* spp.: Introduction"). I have observed gray squirrels ignoring laurel oak acorns while eagerly harvesting water oak acorns.

Farther northwest in Alabama and Mississippi, where water oak is a more abundant, larger, and better-formed tree, it is an important timber resource, supplying low-value wood for fuel, flooring, furniture, and railroad ties. In Florida it is mostly used as firewood and fuel wood and is considered a nuisance when invading commercial pine forests. Small- to medium-sized logs are suitable for culturing shiitake mushrooms.

The landscape value of water oak is low. It is rarely planted but is sometimes left in place when homes or businesses are constructed or comes up on its own after construction. It is rarely an attractive tree in these situations although it can have considerable wildlife value. Water oak is reported to be highly susceptible to air pollution (Vozzo 1990). As water oak gets large and old, it often becomes a liability given the probability of its coming down or losing large branches in wind storms.

Bluejack Oak—*Quercus incana*

Bluejack oak is a medium-sized deciduous tree with a leaf shape similar to that of laurel oak. However, the leaves are covered underneath with minute hairs so that they are gray and not shiny. The young leaves in the spring are pinkish in color, giving the tree an unusual and beautiful cast. When the leaves mature, they are gray-green, again giving the tree a distinctive, slightly frosty appearance. The bark is thick and rough. This is often a crooked tree with an open, rounded crown, and it is rather similar in size to its most common associate, turkey oak. Bluejack oak occurs from southeastern Virginia southwestward through the Atlantic and Gulf coastal plains of the Carolinas, Georgia, Alabama, Mississippi, and Louisiana into eastern Texas and southern Arkansas; it ranges throughout the Florida Panhandle and as far south on the peninsula as Sarasota and Highlands Counties (Kartesz 2015). The largest bluejack oaks reported in Florida have trunk diameters from 2 to 3 feet and are between 50 and 70 feet tall (Ward and Ing 1997). The normal mature size of a bluejack oak is between 40 and 50 feet tall with a trunk diameter of about 1 foot.

Bluejack oak usually grows in association with longleaf pine, turkey oak, sand post oak, southern red oak, or mockernut hickory in the sandhills or clayhills. It will withstand somewhat poorer drainage than these other hardwoods, growing farther into habitats intermediate between sandhills and pine flatwoods. It cannot survive on the most infertile of the deep sand ridges, where turkey oak is often the only hardwood beneath the longleaf pines.

Bluejack oak is moderately fire tolerant, but less so than the associates mentioned above. It is also less shade tolerant than turkey oak and its other hardwood associates. It resprouts well after being killed to the ground by fire, and it sometimes forms small clones from root suckers. It does not spread rapidly this way, however, and most bluejack oaks are of seedling origin.

Bluejack oak is unusual in several ways. In the spring, it often forms smooth, round green galls the size of golf balls on its new growth. This gall, specific to bluejack oak, is caused by the gall wasp *Amphibolips quercuscinerea*. Another oddity is that on lands that are being invaded by laurel oak, and especially in old pine plantations that have been heavily invaded by oaks, hybrids between bluejack and upland laurel oak are sometimes more abundant than bluejack oak alone. It appears that the bluejack oak population is decreasing rapidly throughout Florida as a combined result of habitat destruction, altered fire cycles, and massive interbreeding with upland laurel oak.

The altered fire-cycle problem has several consequences. Lack of fire in bluejack oak's sandhill habitat allows laurel oak and other hardwoods to invade and eventually overtop and eliminate bluejack oak and turkey oak. The common fire schedule on managed land is one burn every two to four years or longer. This allows time for sufficient fuel to accumulate to produce fairly hot fires, which bluejack cannot withstand. The natural fire cycle was probably more on the order of one fire every year or two in most of this oak's former habitat. In addition, most of these historically more frequent fires would have been lightning-sparked fires occurring during high humidity conditions associated with thunderstorms. The fires to which bluejack oak is adapted were low-intensity fires controlled by limited fuel accumulation and high humidity. This phenomenon is also true, although to a slightly lesser extent, for all the hardwood associates of longleaf pine and even, to some extent, for longleaf pine as well.

The wildlife value of bluejack oak is quite high. It is a good acorn producer and often alternates with turkey oak in its bumper crop years so that a combination of these two oaks in the longleaf pine–wire grass uplands is highly beneficial to fox squirrels, flying squirrels, red-headed woodpeckers, white-tailed deer, wild turkey, bobwhite quail, and many other species. And, like turkey oak, bluejack oak is often partially hollow when mature, thus providing cavities for the many wildlife species that need them for shelter and/or nesting.

The main human use of this oak is for firewood. It is occasionally used as an ornamental, often by mistake. It makes a better parking lot tree than laurel or

Shumard oak because it is more drought tolerant, better adapted to that kind of hot, sunny environment, and also because it is a smaller tree that needs less root space to do well. It should be planted more often than it is.

Blackjack Oak—*Quercus marilandica*

This is an uncommon to locally common tree on the clayhills of the Florida Panhandle. It becomes more abundant farther north and west, where it extends as far north as southeastern Pennsylvania and southeastern Iowa and as far west as west-central Texas (Godfrey 1988). Blackjack oak has large, broad, stiff, slightly lobed leaves that are dark green and shiny on top and dull yellowish- to olive green below. By contrast, turkey oak has deeply lobed leaves that are shiny and light to medium green on both sides. The bark of blackjack oak is dark gray to nearly black and is thick and blocky. The dark, often dense foliage and black bark give this tree an overall dark appearance. It is a small- to medium-sized tree similar in size and shape to turkey oak.

One of the difficulties in discussing blackjack oak is that turkey oak, which is the most common oak of the sandhills of Florida, has sometimes been called "blackjack oak," especially in the past.

Blackjack oak occurs in sunny situations on well-drained, infertile clayhills. Sometimes there is a sandy surface layer to these soils. The trees associated with it are longleaf pine, shortleaf pine, southern red oak, post oak, bluejack oak, turkey oak, blackgum, sassafras, and mockernut hickory. Common shrubs include chinquapin, waxmyrtle, shiny blueberry, deerberry, highbush blueberry, and sparkleberry. The ground cover is a mix of fire-adapted grasses, wildflowers, and other herbs. Blackjack oak grows intermixed with post oak in some places on the ecotone between the western edge of the eastern forests and the eastern edge of the grasslands of the Great Plains in an area known as the "Cross Timbers" that spans Kansas, Oklahoma, and Texas (Carey 1992, "*Quercus stellata*").

Blackjack oak is similar ecologically to turkey oak. It is adapted to frequent, mild fires, which it needs for its long-term survival. It requires full sunlight and disappears from habitats where fire no longer keeps larger- and faster-growing oaks and other hardwoods from invading. Blackjack oak is very drought resistant but is not flood tolerant.

The wildlife value of an individual blackjack oak is probably similar to that of a turkey oak. Because blackjack is not nearly as common or widespread in Florida as turkey oak, its overall importance to wildlife in Florida is not nearly as great. As with turkey oak, the acorn crop and the cavities in old mature trees are important for a wide range of wildlife species (see the species account "Turkey Oak—*Quercus laevis*" that follows).

The main use of blackjack oak wood by humans is for fuel wood, for which it is well suited.

Blackjack oak is rarely used for landscaping. However, its distinctive, dramatic appearance and hardiness on poor, dry clay soil makes it a good candidate for use as a specimen tree in sunny places on dry upland clay soils. It should be planted more often than it is.

Turkey Oak—*Quercus laevis*

Although turkey oak is the correct name for this tree, blackjack oak is the original name and the name many longtime Floridians use. The reason it is called "turkey oak" is that blackjack oak is also the original name for a different species of oak that occurs on infertile, well-drained clay soil throughout most of the southeastern United States (though only rarely in Florida except in a few places in the panhandle). Consequently, only one of these trees ended up with its original name. Turkey oak occurs naturally on the Atlantic and Gulf coastal plain from North Carolina into extreme eastern Louisiana, reaching as far south on the Florida peninsula as Lake Okeechobee (Harlow 1990).

Turkey oak is a small- to medium-sized deciduous tree with large, light to medium green, shiny, hairless, deeply cut, lobed leaves that have pointed tips. It has thick, rough, gray bark and crooked branches with relatively few, stout twigs. Turkey oak stems consist of about 30 years of sapwood, inside of which lies the dark reddish-brown to gray to nearly black heartwood. The heartwood is slightly more rot resistant than that of most of the other red oaks. On very poor sand or when very crowded, it is sometimes only a few inches in trunk diameter and 10 to 20 feet tall. More typically, it gets 30 to 50 feet tall at maturity and 6 to 18 inches in diameter. The largest turkey oak I have seen was growing on Pat's Island in the Ocala National Forest. In 1964, its trunk circumference at 4½ feet above ground was 8 feet (just over 2½ feet in diameter), its height was 63 feet, and its crown spread was 50 feet (Pomeroy and Dixon 1966).

Turkey oak is restricted to well-drained, sandy soil in sunny situations. It is associated with longleaf pine and wire grass on deep, infertile, excessively well-drained sands. On deep sands of somewhat greater fertility, sand post oak and bluejack oak are common associates along with longleaf pine, wire grass, broomsedge, bracken fern, and dozens of species of wildflowers. Some shrubs occurring in this association include woolly pawpaw, chinquapin, gopher-apple, winged sumac, dwarf waxmyrtle, shiny blueberry, and deerberry. On shallow sands on clayhills, turkey oak is sometimes an associate of southern red oak, mockernut hickory, and occasionally even blackjack oak, along with the other previously mentioned plants. Blackjack oak is the ecological equivalent of turkey oak on clay soils and is mostly found farther north and west along with post oak, blackgum, and shortleaf pine, in addition to the previously listed trees. Turkey oak also occurs to some extent in sand pine scrub habitat where it occurs as an occasional small tree in association with sand pine, sand live oak, myrtle oak, Chapman oak, crookedwood, scrub palmetto, saw palmetto, and Florida rosemary.

In longleaf pine–wire grass habitat, turkey oak is well adapted to the frequent, low-intensity ground fires that shaped this community of plants and animals for many thousands of years before modern man began to control and extinguish wildfires. Indeed, turkey oak needs these very frequent, mild fires in order to survive. Turkey oak is not tolerant of shade, and when fires no longer occur in its habitat, the invasion of laurel oak, water oak, live oak, and black cherry will completely eradicate turkey oak as these larger and faster-growing oaks overtop it.

One unique adaptation of turkey oak to its dry, sunny habitat is that its large, leathery leaves are oriented vertically. The leaves therefore face the sunlight edgewise when the sun is directly overhead, which is when the air temperature is near its peak and the relative humidity is lowest. This allows turkey oak to avoid overheating and excessive water loss.

Turkey oak is able to grow on low pH soils that are very low in fertility and water-holding capacity, and it is very drought tolerant on deep, sandy soil. However, on shallower sands over a hardpan, on which it sometimes occurs, it is much less drought tolerant and is often replaced in such situations by bluejack oak and/or sand post oak.

On the deep, infertile sands of the sandhills, turkey oak is a tough competitor. It is more drought, fire, and shade tolerant than bluejack oak, its most common associate in the oak midstory, but slightly less tolerant of these stressors than sand post oak.

Turkey oak appears to be short lived. Evidence for this conclusion is its small size and the fact that one can almost always see dead and dying turkey oaks scattered about in the wild. However, part of this impression stems from its slow growth rate and part from the observable impacts of its severe environment. Turkey oaks aged by the counting of stump rings of trees harvested for firewood and by the coring of living trees indicate an average age for fully mature trees of about 80 years with a maximum age of at least 170 years (personal observation). Thus, turkey oak has a longer average life span than laurel oak, water oak, and myrtle oak.

There are several myths about turkey oak that need discussion. One of these is that turkey oak sprouts from the roots to form large clones similar to those of sand live oak and myrtle oak. This is clearly not true. Turkey oak reproduces almost exclusively by seed, producing acorns in the fall that get distributed by blue jays, squirrels, and woodpeckers. It will resprout vigorously from its root collar when young but rarely, if ever, spreads by sprouting from roots or underground stems.

A second myth is that turkey oak is fast growing (Harlow 1990). Turkey oak is observably much slower growing than longleaf pine, black cherry, laurel oak, water oak, live oak, and sand live oak when all of these trees are growing together on the same site. It is faster growing, on average, than sand post oak. One likely cause of this misconception is that it can resprout vigorously following the killing of the stem by fire or by cutting.

A third myth is that turkey oaks did not occur as trees but only as sprouts in the natural longleaf pine forests of the prelogging and pre-fire-suppression days, having only grown into trees following logging of the virgin pine forest and the beginning of widespread and active fire suppression. This is also not true although the relative abundance, historically, of turkey oak as fire-maintained sprouts versus turkey oak trees is open to debate. Harper (1915) observed turkey oak, bluejack oak, sand post oak, southern red oak, and mockernut hickory as trees associated with longleaf pine and wire grass in the north central Florida peninsula. He estimated the frequency of fire at that time to be once every two years on average for any one location, and he noted that the oaks survived the fires equally as well as did the longleaf pines. Nash (1895) observed that turkey oak and bluejack oak occurred as trees interspersed among the larger longleaf pines in the virgin sandhill forests near Eustis, Florida. William Bartram ([1791] 1942) described a sandhill located between Gainesville and the Suwannee River as follows: "we ascended a sandy ridge, thinly planted by nature with stately pines and oaks." Bernard Romans ([1775] 1961) made this observation of sand-hills in Florida: "some high pine hills are so covered with two or three varieties of quercus or oaks as to make an underwood to the lofty pines." John Bartram ([1766] 1928), Vignoles (1823), and other naturalists made similar observations, usually specifically mentioning the dominance of turkey oak (which they called "blackjack oak") on poor, deep sands (see Myers [1990] for more discussion). Finally, an increment core of a turkey oak in a large area of pure sandhill habitat in Suwannee County showed the trunk to be 160 years old in 1994, thus proving that its single trunk had been there since before the Civil War. (This tree died eleven years later at the age of 171 years.) Sand post oaks in similar situations are sometimes over 200 years old (Greenberg and Simons 2000). Obviously, turkey oak and the other hardwoods associated with longleaf pine on well-drained up-land soils have commonly occurred as trees (in addition to occurring as sprouts) since long before the logging of the virgin forest or the advent of effective wild-fire control in the pinelands of Florida. Even cattle grazing, which began in the 1500s, was not widespread in the dry pinelands until the 1800s and therefore can hardly be thought of as a cause for the oak underwood observed in the sandhills by Romans and the Bartrams in the 1700s. And, in any event, turkey oak repro-duces by seed and only produces acorns in significant amounts once it attains tree size.

One difficulty in envisioning how this pine and oak forest functioned is envi-sioning how a turkey oak can begin as a seedling and then grow to tree size in an area that regularly and frequently burns during the growing season. An example of how this can occur is provided by an area of healthy sandhill habitat with dense wire grass in Suwannee County that has been burned every two years without fail for over 50 years with hot head fires in March and April. These fires are presumably consistently hotter than natural fires would have been because they are conducted during dry weather at the driest time of year, whereas lightning fires would invariably start during weather moist enough to produce lightning storms. On this tract, which in some places has been reduced to a nearly treeless prairie, small turkey oaks, bluejack oaks, and sand post oaks continually resprout, some becoming more and more vigorous until, on rare occasions, a very few finally escape one fire that is milder than average by sending up a sprout that is exceptionally thick. A few of these then survive succeeding fires and grow to become trees. Indeed, on this several-thousand-acre sandhill site, the turkey oaks grow into trees more often than do the longleaf pines, which occur here as trees at about the same frequency as the oaks. A milder fire regime would allow both longleaf pine and the sandhill-endemic oaks to reproduce and grow to tree size in much greater abundance.

The relative fire tolerance of longleaf pine and turkey oak is more complex than the preceding discussion would indicate. Fires of low intensity rarely kill healthy mature oaks or pines, kill only a few sapling pines, but kill back most oak sprouts less than an inch in diameter. Mature longleaf pines can clearly survive moderately hot fires better than mature turkey oaks can. Repeated moderately hot fires will eventually eliminate turkey oak as a tree from an area. However, occasional very hot fires actually favor turkey oak because nearly all the young oaks and all the oaks in the sprout stage will survive by resprouting, whereas most of the pines in all stages of development, including the grass stage, will be killed. If an area of longleaf pine sandhill is subjected to a succession of occasional hot fires without frequent low-intensity fires in between, it will change into scrub habitat with upland laurel oak and sand live oak often being the first and most vigorous invaders.

The wildlife benefits of turkey oak are great. In its natural habitat of scattered pines and oaks in a savanna of grasses and wildflowers, it provides a major source of both food and shelter for fox squirrels, flying squirrels, cotton mice, and red-headed woodpeckers. The southern fence lizard is also largely dependent on turkey oak as a structural part of its habitat, and the southeastern American kestrel utilizes this lizard for food during its nesting season (Bohall 1984). Turkey oak provides most of the small natural cavities in this habitat type that are needed by such species as bluebirds, tufted titmice, chickadees, great-crested flycatchers, screech owls, and flying squirrels, to name a few. Finally, it is the main acorn producer in this community, providing a major fall and early winter food source for deer, turkey, bobwhite quail, fox squirrels, flying squirrels, oldfield mice, Florida mice, woodpeckers, blue jays, and many others.

Various wildlife experts rate turkey oak acorns as a major food source for black bear, white-tailed deer, northern bobwhite, and wild turkey in Florida (Harlow 1990). One factor that makes the turkey oak acorn crop especially valuable is that it is often available later in the winter than other acorn crops. This is because of a later-than-average maturation combined with the acorns' remaining good to eat for longer than most other acorns. Compared to the acorns of other oak species, this "lasting" ability is due in part to turkey oak's acorns being less degraded by weevils, in part to their not rotting as easily, and in part to delayed germination.

The main human use of turkey oak has been for firewood. It has also been used for making charcoal briquettes and for supplying the wood for barbecue restaurants. It is well suited for use as firewood, being much easier to split than live oak, and its crooked stems and branches make for better fires than straight logs that tend to pack together too tightly.

Turkey oak has been neglected as a landscape tree, mainly because of its abundance in the turkey oak barrens where it is not very attractive. However, when given room to grow, it makes an attractive specimen tree if planted or maintained in full sunlight on well-drained sandy soil and left unfertilized and unwatered after it becomes established. (Watering and fertilizing tend to raise the pH of the soil, which can kill turkey oak.) The large, shiny leaves of a healthy turkey oak are attractive all summer and are especially attractive when they turn red in December.

Southern Red Oak—*Quercus falcata*

Southern red oak is a common and well-known large deciduous tree that grows throughout most of the eastern United States from Kentucky and New Jersey south into the northern Florida peninsula and Florida Panhandle and from there westward into eastern Texas (Belanger 1990). Its sun leaves have sharply pointed lobes, with the terminal lobe often being long and narrow, which gives the leaves in the upper crown the unique and ornate shape of multibladed scimitars. Leaf shape is quite variable, however, with shade leaves and juvenile leaves being much broader and less prominently lobed. The leaves are smooth and shiny on top but usually have a yellowish pubescence underneath that gives a distinctive and contrastive glow to the foliage, especially on windy days. The thick, moderately roughened bark is light to medium gray in color.

Southern red oak is moderately fast growing and can become quite large. The national champion in 1971 was 128 feet tall with a nearly 9-foot trunk diameter and a crown spread of 149 feet (Godfrey 1988). Here in Florida, however, it is more commonly 100 feet tall with a trunk diameter of 2 to 3 feet at maturity. It can live somewhat longer than most other Florida red oaks, with some of the remnant old southern red oaks being well over 100 years old. In the open, it often has a distinctive overall oval shape, with a tall, straight trunk clear of branches for most of its length and a massive, broadly rounded crown.

Southern red oak is an upland tree that grows on well-drained soils that have some clay either throughout or in the subsoil underneath a sandy topsoil. It is drought resistant, moderately fire tolerant, but not shade tolerant. It most commonly grew in association with longleaf pine, mockernut hickory, and post oak on the most fertile soils of the upland longleaf pine forest areas, especially on clay soils. These large, fire-tolerant hardwood trees were most abundant near the edges of the longleaf pine forests but also occurred as scattered trees well out into the vast fire-adapted and frequently burned longleaf pine forests (Harper 1915). These fertile upland pine forests no longer exist except for a few special places in southwestern Georgia where they were protected in quail hunting preserves (and where, in many cases, the oaks and mockernut hickories have by and large been intentionally removed mechanically or with herbicides). One of the best managed, largest, most complete, and most natural remaining examples of this forest type is in the privately owned Joseph W. Jones Ecological Research Center at Ichauway Plantation in west central Georgia, which is only open to the public one day a year. The largest and best example in Florida is in the Blackwater River State Forest in the Florida Panhandle. The Conecuh National Forest in southern Alabama just north of Blackwater River State Park also protects examples of this forest type.

Elsewhere, these fertile lands, which make up much of the Piedmont and coastal plain of the southeastern United States, were cleared long ago to make way for agriculture. There are small examples of southern red oak and longleaf pine growing together scattered about here and there on rural lands and in subdivisions that serve as reminders of the forests of the past. The soil in one such patch of private land in northwestern Alachua County that was never cleared and farmed still has, as of 2020, six inches of rich, black topsoil on top of red clay. On the same property, on land that had been farmed for a long time, the black topsoil is all gone as it is throughout most of the lands that once supported this kind of upland longleaf pine–southern red oak forest.

The wildlife value of southern red oak is high. It produces acorns possessing a high food value for a wide array of mammals and birds including white-tailed deer, fox squirrel, gray squirrel, flying squirrel, oldfield mouse, Florida mouse, wild turkey, bobwhite quail, wood duck, blue jay, common grackle, red-headed woodpecker, red-bellied woodpecker, and northern flicker. As a large, old, remnant hardwood tree, southern red oak is often valuable as a nest and den tree.

Southern red oak is also a valuable timber tree. The heartwood is beautiful when used to make flooring or furniture. It is rarely planted as a landscape tree, but old remnant trees are still fairly common in subdivisions and other urbanized and rural lands in western Marion and Alachua Counties, in the Florida

Panhandle, and in southwestern Georgia. It is a tree that warrants more use than is current in the landscape business. It is fast growing, adaptable, produces fewer acorns than most other oaks, and it grows into a handsome shade tree with truly unique leaf shapes and striking seasonal color.

Shumard Oak—*Quercus shumardii*

Shumard oak is a fast-growing, large tree with large, deeply lobed, shiny green leaves that have bristles on the ends of the lobes rather like its close relative *Quercus rubra*, the red oak. The deciduous leaves, on average 6 inches long and nearly as broad, often turn a brilliant scarlet in December. The normally tall, straight trunk is often buttressed at the base and has thick bark with dark furrows and light gray ridges. The acorns are large for a red oak, being about three quarters of an inch long (1¼ inches long with the cap attached) and nearly three quarters of an inch wide. When fresh, the acorns appear to be striped because of an alternating coating of minute hairs on the surface.

Shumard oak is an uncommon tree found in scattered locations in hardwood forests on fertile, calcareous soil from the north central Florida peninsula and the Florida Panhandle north and west into North Carolina, Indiana, Missouri, eastern Oklahoma, and eastern Texas, with isolated populations as far north as Pennsylvania and Michigan (Edwards 1990, "*Quercus shumardii*"). In Florida, it occurs in the central panhandle area and grows as far south as Volusia, Marion, and Citrus Counties on the Florida peninsula (Wunderlin et al. 2020). It is now most commonly seen in Florida as an ornamental tree. It is a large tree, often more than 100 feet tall with a trunk diameter of 3 feet or more at maturity in the wild. The current national champion is 117 feet tall with an average crown spread of 100 feet and a trunk diameter of slightly more than 8 feet (American Forests 2018).

Shumard oak is adapted to moist, fertile soils with a fairly high pH (Edwards 1990, "*Quercus shumardii*"). In Florida, these conditions occur in the low hammocks along the upper Gulf Coast in places such as Gulf Hammock, on outcrops of limestone such as at Buzzard's Roost on the west side of Gainesville, and along major rivers such as the Silver River, the Ichetucknee River, and the Apalachicola River. It also occurs in San Felasco Hammock in Alachua County and Twelve Mile Swamp in St. Johns County. In Gulf Hammock, where most of the forest was clear-cut and converted to loblolly pine plantations, it occurred as a somewhat stunted medium-sized tree on the slightly higher limestone ridges in association with swamp chestnut oak and Florida maple. Other tree associates in this location included cedar elm, Florida elm, cabbage palm, loblolly pine, southern red-cedar, sugarberry, and live oak. In San Felasco Hammock, its most common associates include swamp chestnut oak, laurel oak, sweetgum, pignut hickory, spruce pine, winged elm, white ash, Carolina basswood, and southern magnolia. Along the lower Ichetucknee River, bluff oak is one of its common associates. (Initiating their popularity as ornamental trees in Florida, many young Shumard oak and bluff oak trees from the lower Ichetucknee River area were removed by tree spade and planted on the University of Florida main campus by its head landscape designer Noel Lake in the 1980s. Here, they thrived and produced large acorn crops in a location visited by many admirers.)

Shumard oak is not particularly shade tolerant. It almost always grows as an emergent tree with its crown above the height of the rest of the forest canopy. However, on sites where it is well adapted, such as at Buzzards Roost, it is highly competitive and able to grow in moderate shade. It seems to be fairly drought tolerant and is able to withstand some flooding as evidenced by its occurrence in Gulf Hammock and on the higher floodplains of the Mississippi River. Shumard oak cannot survive in the poor, acidic soils of the pine flatwoods, nor is it tolerant of the deep sandy soils of the longleaf pine sandhills.

The wildlife value of Shumard oak is high. It produces a good crop of acorns periodically, which are eaten by the same wide range of species that eat other acorns. Being large, the acorns are particularly sought after by deer, hogs, squirrels, and wild turkeys. Large, old Shumard oaks can develop into valuable den trees.

Where there is enough of it available to make logging economical, Shumard oak is a valuable timber tree. It produces red oak saw timber of the highest quality for use as flooring, furniture, interior trim, and cabinetry (Edwards 1990, "*Quercus shumardii*").

The main value of Shumard oak in Florida is for planting as a shade tree. It is well suited to this purpose, being able to withstand the high pH of the soils near buildings, sidewalks, and roads. In situations where it has room to grow, it makes a handsome, large shade tree that often has brilliant fall color. However, it does not do as well as a parking lot or street tree as bluff oak or live oak. There are two main drawbacks to using Shumard oak in such situations. One is that Shumard oak's root system needs a lot of room in order to do well. The other problem is that Shumard oak is shorter lived and not as hardy in tough situations as live oak and bluff oak, and it does not heal wounds as well. One interesting note: in central Texas, there is a variety of Shumard oak, *Quercus shumardii* var. *texana*, which is a bit smaller, better adapted to dry uplands, and probably more drought tolerant than the local Shumard oak. Grown and sold in Florida when acorns of the local population were not available, *Quercus shumardii* var. *texana* has proven to grow well here but, unfortunately, does not produce the same beautiful fall color as the local variety.

Oaks Barely Reaching into Florida: Arkansas Oak—*Quercus arkansana*; Chinquapin Oak—*Q. muehlenbergii*; Dwarf Chinquapin Oak—*Q. prinoides*; Cherrybark Oak—*Q. pagoda*; Black Oak—*Q. velutina*; Willow Oak—*Q. phellos*

Fagaceae

In addition to the oak species already discussed, there are several species that extend into Florida in limited numbers from populations that are more extensive farther north. The wildlife value of these species is similar to that of the other oak species and is well covered in the introduction to oaks that prefaces this section. None of these species are abundant enough in Florida to have significant commercial value, and they are rarely used in the landscape or nursery trades in part because they are better adapted to more northerly growing conditions. Each of these species has unique characteristics and habitat requirements.

Arkansas Oak—*Quercus arkansana*

Arkansas oak is a small- to medium-sized deciduous oak with broad leaves similar to but thinner than those of blackjack oak. Arkansas oak has dark gray to black bark with vertical ridges on older trunks. It occurs on well-drained uplands in fire-adapted habitats or formerly fire-adapted habitats on sandy soils in the western panhandle of Florida and from there to the north and west into Arkansas and Texas. It is listed as threatened in Florida (Wunderlin et al. 2020). Arkansas oak is drought tolerant and somewhat fire adapted but not shade tolerant. It is generally uncommon, with scattered, isolated populations. Some of the largest population clusters occur in southeastern Arkansas and in the Florida Panhandle. There are several populations of Arkansas oak on Eglin Airforce Base.

Chinquapin Oak—*Quercus muehlenbergii*

Chinquapin oak, a medium- to large-sized tree related to swamp chestnut oak, has deciduous leaves similar to but narrower than those of swamp chestnut oak. The acorns are smaller than those of swamp chestnut oak but are similar in that they contain less tannin and are, therefore, more palatable than most acorns. Chinquapin oak is uncommon to rare in Florida, occurring on calcareous soils on bluffs, ravine slopes, and glades in the central panhandle. It is more common farther north and west, growing as far west as central Texas and Oklahoma (with scattered, isolated populations in southeastern New Mexico, Trans-Pecos Texas, and northeastern Mexico) and as far north as Michigan and the southern tip of Canada next to the state of New York (Sander 1990). Chinquapin oak grows in association with other upland hardwood trees. It is drought tolerant and somewhat shade tolerant as a seedling and potentially long lived.

Dwarf Chinquapin Oak—*Quercus prinoides*

Dwarf chinquapin oak is a close relative of chinquapin oak but is much smaller, spreads clonally, and has wider leaves with fewer lobes (5 to 8 per side as opposed to 10 to 15 per side for chinquapin oak). It occurs in the central part of the eastern United States, ranging south to central Alabama with an isolated population in Santa Rosa County in the western panhandle of Florida. It grows on well-drained, acidic, sandy soils, is fire adapted, and is not shade tolerant.

Cherrybark Oak—*Quercus pagoda*

Cherrybark oak is a large, deciduous, red oak related to southern red oak that grows best on well-drained parts of major river floodplains. It occurs on the upper portions of the Apalachicola River floodplain in the central Florida Panhandle and from there ranges northeast into Virginia, west into eastern Texas, and up the Mississippi River floodplain into southern Illinois (Godfrey 1988). It is a timber tree of major importance farther north and west but is not common enough in Florida to be of commercial importance.

Black Oak—*Quercus velutina*

Black oak is a large, deciduous, red oak somewhat similar to northern red oak. It is common throughout most of the eastern United States but uncommon in the central part of the Florida Panhandle at the southern end of its extensive geographic range, where it occurs as an occasional tree in upland hardwood forests on clay soils near the Alabama and Georgia borders.

Willow Oak—*Quercus phellos*

Willow oak is a large, deciduous red oak similar to swamp laurel oak in its appearance and habitat preferences, with the exception that it prefers clay-based soils. It is a common tree in the Atlantic coastal plain and Piedmont from New Jersey southward into South Carolina and on the Gulf coastal plain from the western Florida Panhandle north and west into southern Kentucky, Arkansas, and eastern Texas. There is an isolated population in Duval County. Willow oak grows rapidly, is intolerant of shade, and is easily damaged by fire. It is rarely planted in Florida, but it is commonly used as a shade tree farther north.

Miccosukee Gooseberry—*Ribes echinellum*

Grossulariaceae

This is a small, thorny, partially evergreen shrub occurring only in Jefferson County, Florida, and McCormick County, South Carolina (Godfrey 1988).

It is listed as endangered by the states of Florida and South Carolina and as threatened by the United States government (Catling et al. 1998). Its alternate leaves have three serrated lobes, and the sharp, stiff thorns on the twigs and branches are often branched at the base to form two thorns. The shrub is bare of leaves in summer, growing new leaves in the fall that overwinter (Godfrey 1988). Odd, comet-like, yellowish-green, bisexual flowers with five conspicuous sepals and long-stemmed stamens hang down individually on long stalks from the branches in March, and round, three quarter–inch-diameter, densely spiny fruits ripen in April and May (Florida Natural Areas Inventory 2000, "Miccosukee gooseberry"). The bark is buff colored, splitting to reveal dark purplish-brown inner bark on larger stems (Godfrey 1988).

This 3- to 4-foot-tall shrub grows on calcareous soils in mesic hardwood forests on slopes and bottomlands under a forest canopy of American beech, pignut hickory, Florida maple, sugarberry, white ash, sweetgum, Shumard oak, basswood, and elms (Catling et al. 1998). Miccosukee gooseberry takes advantage of the deciduous forest canopy by leafing out in the fall to use the winter sunlight available under a canopy that has mostly lost its leaves. Then, it sheds its leaves to become bare in summer when the shade of the forest is heaviest. It forms thickets by reproducing asexually when the ends of branches bend to the ground and take root to form new plants.

Miccosukee gooseberry flowers are pollinated by bumblebees and southeastern blueberry bees (Catling et al. 1998). The fruit is eaten to some extent by the cotton mouse, which may digest the seeds, thus perhaps *not* aiding in seed distribution (Engstrom and Radzio 2014).

Miccosukee gooseberry is not used horticulturally or in any other way by humans.

Witch-Hazel—*Hamamelis virginiana*

Hamamelidaceae

Witch-hazel is a shrub or very small tree with 2- to 6-inch-long, alternate, deciduous leaves having undulating (wavy) margins and asymmetric bases. The buds are naked and tawny to rusty colored. The small bisexual flowers are produced in late fall, winter, or very early spring and are unusual looking, with narrowly linear yellow petals (Godfrey 1988). The flowers have a pleasant aroma reminiscent of citrus. The fruit is a capsule that ripens in the fall, at which time it opens explosively, ejecting the one to four seeds for a distance of up to 30 feet (Miller and Miller 1999). The growth habit involves an often slender stem that leans in one direction and does not produce a compact or symmetrical crown. There is usually a cluster of stems for tall individuals, or there may be a widely scattered group of smaller plants with single stems that have originated by root suckers. The individual stems are not long lived, lasting perhaps 10 or 20 years, but the plants resprout continually. Two witch-hazels planted in my yard in 1960 are alive and well and increasing gradually in size 60 years later. Witch-hazel occurs from eastern Canada and southeastern Minnesota southward into Texas, the Florida Panhandle, and the central Florida peninsula (Godfrey 1988).

Witch-hazel is an uncommon plant, occurring in widely scattered spots in upland hardwood forests, wooded slopes and ravines, floodplains, stream banks, and seepage bogs. It is shade tolerant, occurring as an understory plant in the shade of hardwood forests, but it does best in partial shade. It is usually found growing on moist, fertile soils. It does not seem to be tolerant of flooding, the acidic soils of the pine flatwoods, or poor, sandy soils.

Witch-hazel is browsed to some extent by white-tailed deer, and the seeds are eaten by wild turkey, northern bobwhite, and gray squirrel (Miller and Miller 1999). Witch-hazel is pollinated at night by hooded owlet moths (Miller and Miller 1999).

Witch-hazel has been used both as an ornamental shrub and, for a long time, as the source of witch-hazel folk medicines. Native Americans used extracts from this plant, which contain tannins and other chemicals, as topical salves for treating various skin ailments, and the European settlers to North America learned from them and have continued making and using these medications up to the present day (Wikipedia 2020, "Witch-Hazel"). Witch-hazel is rarely used in Florida as an ornamental but is sold and used for landscaping more commonly much farther north.

Climbing Hydrangea (Woodvamp)—*Hydrangea barbara* (*Decumaria barbara*)

Hydrangeaceae

Usually seen climbing on tree trunks, climbing hydrangea is a vine that has numerous short branches supporting opposite, deciduous leaves with 1-inch-long petioles. The leaves are mostly smooth margined, about 2 inches wide and 4 inches long, shiny dark green on top, and lighter and duller underneath. This vine climbs upward for 10 to 30 feet or more by means of numerous roots that grow out in all directions from the vine stem and attach to the tree bark. It also sometimes grows along the ground or on logs, rocks, or tree stumps. Small white flowers are produced in clusters at the ends of the branches 1 to 2 feet out from the tree trunk. The fruit is an urn-shaped capsule. Climbing hydrangea ranges naturally from southeastern Virginia to Louisiana, reaching as far south as Polk County on the central Florida peninsula (Kartesz 2015).

Climbing hydrangea occurs in seasonally flooded hardwood and cypress forests on river banks and river floodplains, in various other kinds of cypress–hardwood swamps, in fertile bottomland hardwood forests, and sometimes in fertile, moist upland hardwood forests. It survives seasonal flooding and is quite shade tolerant, although it also grows well in partial to full sunlight. It is a moisture-loving plant that is not drought tolerant. It will grow both as a ground cover and as a vine on tree trunks but does not bloom unless growing upward on a vertical surface. It rarely grows up into the crown of a tree and does no harm to the tree it uses for support.

The main wildlife benefit of climbing hydrangea comes from the flowers, which bloom in the summer and are attractive to bees and butterflies.

Sometimes used as an ornamental plant, several varieties of climbing hydrangea have been developed and are sold by native plant nurseries. It requires moist to wet ground to do well. Climbing hydrangea can be used as a ground cover but does best when allowed to climb walls, fences, trellises, or tree trunks.

Oak-Leaf Hydrangea—*Hydrangea quercifolia*

Hydrangeaceae

Oak-leaf hydrangea is a multistemmed understory shrub with large, deciduous, serrated, lobed leaves somewhat resembling the lobed leaves of northern red oak or Shumard oak. The smaller leaves are often not lobed. Large, cone-shaped clusters of white to pinkish flowers are produced at the ends of the stems in midsummer. These remain on the plant into autumn, gradually turning darker in color. The stems have bark that peels off in thin sheets revealing new, cinnamon-colored bark beneath. This shrub often grows 4 to 8 feet tall. Its native range includes the central and western panhandle of Florida and scattered locations throughout the states of Alabama, Mississippi, and Louisiana (Kartesz 2015).

Oak-leaf hydrangea occurs along streams, in ravines, and in upland hardwood forests on rich, often calcareous, soils. It is shade tolerant but does best with partial sunlight and can grow in full sun. It is a moisture-loving plant, doing best on moist, well-drained soils with abundant organic matter content, but it is more tolerant of drought and sunny situations than most other hydrangea species. It is somewhat salt tolerant. Reproduction is both by seed and by sprouting from the roots to form clonal patches.

Oak-leaf hydrangea's wildlife value is low to moderate. White-tailed deer will browse both the foliage and the blooms, and the seeds are eaten by some songbirds (Miller and Miller 1999). Bees and other insects visit the flowers.

Oak-leaf hydrangea is a popular ornamental shrub with several named varieties. It is easy to transplant and can be successfully planted well outside its native range, being used throughout Florida north of Orlando and as far north as Ohio and as far west as California. The flower clusters are showy, the leaves and bark are distinctive, and the foliage turns patchily wine red to maroon in autumn, especially on plants getting full sun for at least part of the day. It is fairly adaptable with respect to its soil and light requirements but does best on rich soils with good organic matter content and with supplemental watering during dry periods. The use of leaf mulch is helpful to its shallow roots, meeting its preference for organic matter in the soil.

Wild Hydrangea (Smooth Hydrangea)—*Hydrangea arborescens*

Hydrangeaceae

Wild hydrangea is a deciduous shrub, commonly 3 to 6 feet tall, with opposite, green to grayish-green, roundish leaves that are often 3 inches wide and 4 inches long with serrated edges and an elongated point at the tip. It has thick twigs, and the older stems have bark that peels off in thin strips or patches revealing multicolored new bark beneath. The white flowers produced at the ends of the stems in summer occur in terminal, flat-topped clusters 2 to 6 inches wide. It occurs in nature from New York west into Illinois and south into Louisiana and the central Florida Panhandle (Godfrey 1988).

Wild hydrangea is a moisture-loving, shade-tolerant plant occurring in scattered locations in hardwood forests along streams or at the base of slopes or bluffs, often on calcareous soils. It is not flood tolerant or drought tolerant. Reproduction is both by seed and by suckering from surface roots, which results in the formation of clonal patches.

Wild hydrangea's wildlife value is moderate. White-tailed deer browse the foliage and the flower heads, wild turkeys eat the flowers and the seeds, and some songbirds eat the seeds (Miller and Miller 1999). Native bees, wasps, flies, beetles, and other pollinators visit the flowers, and the foliage is host to the caterpillars of the hydrangea sphinx moth (Hilty 2019, "Wild Hydrangea").

Wild hydrangea is used as an ornamental mostly well north of Florida and has several named varieties. Its summer blooms are showy, and its foliage turns yellow in autumn. It grows best in light to moderate shade and requires supplemental watering and fertile soil with good organic matter content to do well in cultivation.

Scentless Mock-Orange—*Philadelphus inodorus*

Hydrangeaceae

This deciduous shrub with opposite leaves produces showy, white, four-petaled flowers in late spring or early summer, each about 1½ to 2 inches wide. The leaves are 2 to 3 inches long and 1 to 2 inches wide with pointed tips, rounded bases, and margins that can be smooth or have a few small, widely spaced teeth. The fruit is a capsule containing numerous smooth, red seeds 2 to 3 mm long (Godfrey 1988).

Scentless mock-orange is an uncommon, moderately shade- and quite drought-tolerant shrub occurring naturally in fertile upland hardwood forests from Virginia and Tennessee southward to Alabama, southwest Georgia, and the central Florida Panhandle (Godfrey 1988) with scattered populations occurring throughout most of the eastern United States (Kartesz 2015). It is even more widespread because of its escapes from cultivation. It is most common in calcareous rock outcrop areas.

Scentless mock-orange can have a single trunk or be multistemmed and can spread either by seed or by underground runners that form a clump or thicket. It is rarely over 10 to 15 feet tall with slender stems. It is somewhat shade tolerant but does best in full sun. It is not flood or salt tolerant, but it does not seem to be bothered by insect pests or diseases.

The main wildlife benefit of this species is its value to insect pollinators, especially native bees.

Planted for its showy flowers, scentless mock-orange was a more popular ornamental in the past. It is sometimes found around abandoned homesteads in rural areas where, in many cases, it may have been planted more than 100 years ago. Although not aggressive in spreading or reseeding, it can persist for a long time. It is still available from plant nurseries, and there are several named varieties.

Virginia-Willow, Sweetspire—*Itea virginica*

Iteaceae

Virginia-willow is a fairly common wetland shrub with 3-inch-long by 1-inch-wide tardily deciduous, smooth, dark green leaves that have finely serrated margins. The plant grows 3 to 6 feet tall with a slender trunk and a few arching branches. Three- to 6-inch-long drooping racemes (spires) of small, white flowers are produced in late spring on the ends of its twigs, followed by capsules containing tiny black seeds. Reproduction is both by seed and by suckering from the root system.

Virginia-willow occurs throughout the southeastern United States from New Jersey and the southern tip of Illinois into eastern Texas and the southern Florida peninsula (Kartesz 2015). In Florida, it is most common in and on the margins of the cypress domes that occur scattered throughout pine flatwoods forest areas but also occurs in bayheads, cypress hardwood swamps, and along streams. It withstands prolonged flooding and high soil acidity well, is shade tolerant, and sprouts back from the base following fires, provided the fires do not also consume peat soil where the plant is growing.

Common tree associates of Virginia-willow include pond-cypress, swamp tupelo, red maple, dahoon, slash pine, loblolly bay, and sweetbay. Other associates include fetterbush, swamp doghobble, laurel greenbriar, lizard's tail, and Virginia chain fern, to name a few of the more common ones.

The wildlife value of this slender shrub is found mainly in its flowers, which are attractive to bees and butterflies. The seeds may provide food for some songbirds.

Virginia-willow is often called "sweetspire" or "Virginia sweetspire" in the nursery trade and is sold as an ornamental. In a landscaped situation, it grows more densely than in the wild and blooms more prolifically. It has pretty fall color, the leaves turning a mix of red, orange, and gold, and it spreads by suckering, which is useful if a mass or screen of vegetation is desired.

Black Walnut—*Juglans nigra*

Juglandaceae

Black walnut is an uncommon, large, deciduous tree found in the wild in Florida only in the central Florida Panhandle but occurring north of Florida throughout most of the eastern United States (Godfrey 1988). As with all members of the walnut family, it is monoecious, with its male and wind-pollinated female flowers occurring separately on the same tree. It has large, pinnately compound leaves similar in shape to the leaves of pecan. It has dark gray bark with deep, interlacing furrows and produces a large, round, hard-shelled nut encased in a tough, dark brown husk. Maximum size for a black walnut tree is indicated by one tree that measured 130 feet tall with a 7-foot trunk diameter (American Forests 2004).

In the wild, black walnut is usually found on river floodplains and adjacent slopes with fertile alluvial soil in locations that do not stay flooded for very long. It also grows on uplands with deep, fertile soil of limestone origin (Williams 1990). In these rich, diverse hardwood forests, it is one of the tallest trees, rising above the canopy of most other trees. It never occurs in pure stands in nature but usually occurs either as a single tree or comprises a small grove of trees within a forest dominated by other hardwood trees such as oak, hickory, ash, elm, maple, sweetgum, tulip tree, sycamore, cottonwood, and others. It grows fast, usually getting started when an opening in the canopy occurs that provides full sunlight. It is not shade tolerant. It is deeply rooted, moderately wind-firm, and moderately drought tolerant.

The nuts provide black walnut's main wildlife value, and they are primarily eaten by squirrels (Martin et al. 1951) because the thick, hard shell is too difficult for other animals to open (Huegel 2010). A number of insects feed on black walnut leaves, thus providing food for some birds. Two of the most common of these insects are the fall webworm and the walnut caterpillar (Williams 1990).

Black walnut is a very valuable timber tree. The dark brown heartwood is highly prized for making furniture and gun stocks, and the flavorful nuts are also valuable, being used for making baked goods and flavoring ice cream (Williams 1990).

Because of its value to the timber and food industries, black walnut has often been planted in plantations for timber or nut production. It is also planted as individual trees in openings in hardwood forests and as a landscape tree in subdivisions and parks. Black walnut's ornamental value is due to its fast growth and large size as a shade tree. However, it is not a desirable street tree because the large, heavy nuts that fall from a considerable height can make dents in cars. Also, black walnut produces a toxin called "juglone" that makes it allelopathic to some other plants such as pines, birches, apple trees, blackberries, blueberries, azaleas, potatoes, and tomatoes (Appleton et al. 2015). Conversely, grasses grow well beneath black walnut trees, making it an ideal shade tree in a pasture or large lawn. It is not well suited for planting in most landscape situations in Florida, however, because it needs better soil than is generally found here, and it is not well adapted this far south.

Pecan—*Carya illinoinensis*

Juglandaceae

Pecan is native to the Mississippi River valley and occurs from there to the southwest in scattered populations through eastern Texas into northeastern and central Mexico (Peterson 1990). It has been widely planted throughout the southeastern United States and elsewhere and is a common tree in northern Florida, especially on old homesteads, and is very common in Georgia, Oklahoma, and Texas where there are large commercial pecan groves. The deciduous, pinnately compound leaves of pecan usually have from 9 to 17 long, narrow leaflets. Pecan leafs out late and drops its leaves early and will sometimes start leafing out again in midwinter due to its being poorly adapted to the Florida climate. The bark is light tannish-gray and is broken into narrow vertical scaly ridges. Once out of the four-partitioned husk, the sweet, thin-shelled nuts are about 2 inches long by three quarters of an inch wide.

In its native range in the lower Mississippi valley, pecan is a medium-sized to very large tree, often reaching heights over 100 feet at maturity and sometimes even reaching a height of 180 feet with a trunk diameter of 7 feet (Peterson 1990). In cultivation or as an escaped tree in northern Florida and the rest of the southeastern United States, it is not as tall but can still grow an impressively large trunk. The current national champion in Virginia is 97 feet tall, has a crown spread of 106 feet, and a trunk diameter of over 7½ feet (American Forests 2018).

Pecan grows best on rich, moist, well-drained soils with an abundant supply of water. In our region, where it is usually growing on upland soils, it prefers clay soils or sandy soils that have a clay subsoil. Pecan benefits greatly from applications of dolomite, which supplies it with calcium and magnesium and raises the soil pH, and from frequent fertilization with a complete fertilizer high in potassium and phosphorus. It often does poorly if its fertilization needs are not met. Pecan is moderately tolerant of occasional flooding of short duration and is also moderately tolerant of drought. It is not shade tolerant or fire tolerant.

The wildlife value of pecan is high. The nuts are a favorite food for gray squirrels, fox squirrels, flying squirrels, mice, wood ducks, wild turkeys, American crows, fish crows, blue jays, woodpeckers, raccoons, bears, and hogs. The leaves are eaten by a wide variety of caterpillars, with fall webworms being especially conspicuous, and by insects, which then supply birds with something to eat. Old pecan trees sometimes have hollows in their trunks or branches that supply cavities for wildlife; they also often have mistletoe in their upper crowns, which supplies berries for birds to eat.

Pecan is very important economically as a nut producer, with Georgia leading the southeastern United States in pecan production. Pecan is also used to supply wood for making paneling, furniture, cabinetry, pallets, and veneer (Peterson 1990).

Pecan is often planted as a landscape tree, especially on farm homesteads and in rural subdivisions where, with plenty of space to grow, it often does well, gets large, supplies pecans to the owners, provides shade, and lives a long time. In urban situations and dense subdivisions it does not do as well. Also, pecan is not as well adapted to Florida conditions as it is to growing conditions farther north or west. The farther south one travels on the peninsula, the less well pecan does in Florida. At least one hundred different cultivars of pecan have been developed.

Water Hickory—*Carya aquatica*

Juglandaceae

Water hickory is a tall deciduous tree closely resembling pecan in its leaf and bark characteristics. The pinnately compound leaves, which are smaller and finer in texture than those of pecan, come out late in the spring and drop early in the fall. The nuts are smaller than pecans, somewhat flattened, and bitter tasting. Another name for water hickory is "bitter pecan." It occurs abundantly in the Mississippi valley and on lake and river floodplains on the Atlantic and Gulf coastal plains from Virginia across the southern United States into Texas, getting as far south on the Florida peninsula as Lake Okeechobee (Francis 1990). It is a large tree at maximum size, with one in the Congaree Swamp in South Carolina measuring 143 feet tall with a trunk diameter over 5 feet (American Forests 2008).

Water hickory, as its name implies, is usually found near water. It is most common on the floodplains of major rivers such as the Apalachicola, Suwannee, and St. Johns Rivers. Water hickory is moderately tolerant of flooding but is not tolerant of shade, almost always being taller than surrounding trees in the forest canopy. It can grow on either sandy or clay soils but prefers soils that have a calcium source either underneath or in the water of the river. Common tree associates include overcup oak, swamp laurel oak, water oak, red maple, sweetgum, sugarberry, bald-cypress, water locust, Florida elm, and water-elm.

The wildlife value of water hickory is moderate. Gray squirrels, wild hogs, and some other wild creatures occasionally eat the nuts (Francis 1990), and the leaves are eaten by insects such as the fall webworm, which provide food for birds such as the yellow-billed cuckoo.

Water hickory wood is similar to pecan, but, because of its smaller average tree size and more frequent wood defects, it is considered less valuable than pecan (Francis 1990). It is used for paneling, firewood, and fuel wood.

Water hickory is rarely planted for ornamental purposes. It can be a handsome shade tree, but in any situation where it might be useful, pecan would likely be a better choice. The exception to this might be from Gainesville south on the Florida peninsula, where water hickory from a local seed source would be better adapted than pecan to the soils and climate.

Bitternut Hickory—*Carya cordiformis*

Juglandaceae

Bitternut hickory is a common-to-abundant tree throughout much of the eastern United States, but it does not occur on the Florida peninsula, in the western panhandle, or in most of southern Alabama and Mississippi (Smith 1990, "*Carya cordiformis*"). It occurs as an uncommon tree in southwest Georgia and the central panhandle of Florida along the Apalachicola, Ochlockonee, and Chipola River drainages (Godfrey 1988). The gray-brown bark is noticeably smoother than the bark of other hickories. The naked buds, especially terminal buds on main branches, are noticeably yellow to coppery-colored because they are covered with a mix of pubescence and resinous scales. The pinnately compound leaves have seven to nine leaflets, the terminal ones larger than the others with the last leaflets (nearest the twig) often much smaller. The undersides of the leaves are duller than the tops because of their resinous scales and minute hairs. The bitter-tasting, roundish nuts are about 1 inch in diameter. Bitternut hickory is a medium-sized to large tree, reaching a maximum height of 130 feet with a trunk diameter of 4½ feet (American Forests 2018). It is the shortest-lived of the hickories, attaining a maximum age of about 200 years (Smith 1990, "*Carya cordiformis*").

In Florida and southwest Georgia, bitternut hickory is a river-bottom tree on the floodplains of major rivers, but a bit farther north it also occurs on dry upland sites along with shortleaf pine, blackjack oak, post oak, flowering dogwood, and mockernut hickory (Smith 1990, "*Carya cordiformis*"). Bitternut hickory is only moderately shade tolerant; it is also moderately flood tolerant and moderately drought tolerant.

Bitternut hickory has less wildlife value than most other hickory species given the very bitter taste of the nuts (Smith 1990, "*Carya cordiformis*"). Also, in our area, it is an uncommon tree. Even so, the nuts are likely eaten to some extent by squirrels and various other species, and the leaves are eaten to some extent by fall webworms and other caterpillars.

Bitternut hickory wood is used for the same purposes as the wood of the other pecan hickories (Smith 1990, "*Carya cordiformis*"). It makes good firewood and is well suited for making paneling and for smoking meat.

Bitternut hickory is sometimes used as a shade tree well north of Florida (Smith 1990, "*Carya cordiformis*"), although hickories are rarely planted as ornamentals. Bitternut hickory would not be a good choice as an ornamental or shade tree in our area because of its marginal adaptation to Florida's climate and its limited wildlife value. Several other hickory species are better adapted to Florida's growing conditions, are more appealing to wildlife, and produce good-tasting, edible nuts.

Pignut Hickory—*Carya glabra*

Juglandaceae

Pignut hickory is a large forest tree with deciduous, serrated, pinnately compound leaves and gray bark with rather smooth but prominent interlocking vertical ridges. The leaves are somewhat shiny and completely smooth without any pubescence. Pignut hickory and the similar-looking white ash, which often grow together in the same forest, can be distinguished by the opposite leaf and branch pattern of the ash as compared to the alternate leaf and branch pattern of the hickory. Another distinguishing feature of hickories is their nuts, which are half to two-thirds the size of a golf ball and have husks as well as very hard shells. The remains of these can be found scattered beneath a mature hickory tree even when it is not the season for nuts to be on the tree. Pignut hickory nuts have a relatively thin husk which does not split away from the nut easily at maturity, in part because the husk sections do not split all the way down to the base, a feature which distinguishes it from most other hickory nuts.

Pignut hickory grows to heights between 80 and 100 feet with trunk diameters of 2 to 3 feet. The largest one reported in Florida in 1997 was 109 feet tall with a 4-foot 8-inch trunk diameter at 4½ feet above ground (Ward and Ing 1997). The largest one nationally, in Dooly, Georgia, is 104 feet tall with a 6-foot trunk diameter (American Forests 2018). Mature pignut hickories are usually straight trunked with a compact, symmetrical crown of sturdy branches. As young trees, they have a slender yet sturdy appearance.

Pignut hickory occurs throughout most of the oak–hickory forests in the eastern United States and is a major component of the upland hardwood forests of the Florida Panhandle and the northern Florida peninsula. For instance, pignut hickory makes up about 25 percent of the forest canopy of the upland hardwood forest of San Felasco Hammock Preserve State Park. It is also a common tree at Silver River State Park, O'Leno State Park, Torreya State Park and in the hardwood forests of Withlacoochee State Forest and the Ocala National Forest. It is not as common in the Gulf Hammock and Big Bend hardwood forests because of its salt intolerance and inability to grow in very wet or occasionally flooded soils. It gets as far south on the Florida peninsula as Brevard County on the east coast and Charlotte County on the west coast (Wunderlin et al. 2020).

The most common associated trees in the Florida forests where pignut hickory grows are sweetgum, laurel oak, water oak, swamp chestnut oak, live oak, winged elm, sugarberry, loblolly pine, spruce pine, basswood, white ash, southern magnolia, hophornbeam, hornbeam, devil's walkingstick, and one-flower hawthorn. Pignut hickory is a canopy tree in these forests but is not quite as tall, on average, as loblolly pine, spruce pine, sweetgum, white ash, laurel oak, and winged elm. It is taller on average than southern magnolia, red bay, red mulberry, and live oak.

Growing at a slow pace, pignut hickory begins nut production at about 30 years of age and is able to live and produce nuts until it's 300 years old (Smalley 1990). Its nut production is more regular and dependable than that of oak trees (Smalley 1990). This tree is moderately shade tolerant, being able to grow up under and through a forest of live oak, laurel oak, water oak, black cherry, sugarberry, winged elm, white ash, and sweetgum. It withstands drought, cold, ice storms, and wind storms better than most other trees. It is quite wind-firm (Smalley 1990) given its strong and deep root system and the great strength of its wood. In an area of upland hardwood forest of about 40 acres in San Felasco Hammock that was flattened by a tornado during the "Storm of the Century" in March 1993, most of the saplings that remained standing were young pignut hickories.

Primarily because of its regular and abundant nut production, the wildlife value of pignut hickory is high, especially in its native forest environment. In spite of the extremely hard shells of the nuts, a surprisingly large number of creatures eat them. Squirrels are well known for eating these nuts, which make up an estimated 10 to 25 percent of their diet (Smalley 1990). Flying squirrels eat the nut by neatly gnawing a round hole at its end, whereas fox squirrels and gray squirrels bite the nut in half. It is not as well known that wild hogs, wild turkeys, and wood ducks also eat pignut hickory nuts. Wild hogs eat large quantities of the nuts by crushing them with their molars (personal observation), and turkeys and wood ducks grind them in their gizzards (Terres 1980). People can also eat the nuts, which are tasty. The nuts, which are produced in the fall, have a high fat content (Halls 1977), making them ideal for helping wild animals build up fat reserves for the winter.

The leaves of hickory trees also provide wildlife value by providing a food source for insects. Caterpillars are the main consumers and are often abundant. One of the largest and strangest looking caterpillars in North America is the hickory horned devil, which prefers hickory leaves (Stiling 1989). It is the larvae of the large and beautiful regal moth (royal walnut moth). Many other insects feed on the leaves as well, providing an important food source for birds such as the yellow-billed cuckoo, red-eyed vireo, parula warbler, summer tanager, and others.

In the past, hickory wood was considered quite valuable. Its great strength makes it ideal for making hammer and axe handles. However, with modern technology, other materials such as metal and plastic have largely replaced wood for tool handles, and hickory wood is now most often shipped to chip mills where it is processed for use as fuel. Hickory wood is also used for making paneling, flooring, and furniture. It makes good firewood, burning slowly with a hot flame, and is a preferred wood for smoking meat.

The landscape value of pignut hickory is high in some cases and low in others. The large, hard nuts make this tree inappropriate as a street or parking lot tree because the nuts, when they fall, can put small dents in cars. They also make a very loud noise when they hit a metal roof or wooden deck. However, when away from pavement, wooden decks, and sheet metal roofs, pignut hickory is an excellent shade tree. It is handsome, durable, wind-firm, long lived, and good for wildlife.

Pignut hickory is rarely planted as an ornamental, mainly because of its slow growth rate and the difficulty of raising it in a pot. It is also difficult to transplant: having such a deep and strong taproot, it rarely survives transplanting. The best way to plant a hickory is to plant a hickory nut or, rather, several hickory nuts. If you want the tree to come up in the spots where you plant the nuts, it is best to cover them with chicken wire to prevent squirrels from getting them. A friend of mine, who once contemplated planting some hickory nuts, gathered a boxful and put them aside for a while on his porch. A month later, when he checked the box, the nuts were gone. A year later, there were seedling hickories scattered about in his yard in places where squirrels had buried the nuts (Noel Lake, former landscape designer for the University of Florida, personal communication).

Scrub Hickory—*Carya floridana*

Juglandaceae

Scrub hickory is a small tree with relatively small nuts. It has rusty-colored scales on the underside of its pinnately compound leaves, on its buds, and on the outside of the nut husk at least until the nut is mature. It occurs in sand pine scrub on the east side of the scrub in the Ocala National Forest in Marion County and, from there, ranges southward in the interior of the Florida peninsula into Charlotte and Highlands Counties. It also occurs on the Atlantic coastal dunes from Volusia County south into Palm Beach County (Layne and Abrahamson 2006). It is often no more than a tall, gangly, multistemmed shrub on the poor, deep sands of the scrub but can attain tree size on better soils: the largest one reported from Lake County has a height of 61 feet and a trunk diameter of close to 2 feet (American Forests 2018).

Scrub hickory is closely related to pignut, sand, and black hickories. It inhabits sand pine scrub in the north central Florida peninsula and in south Florida also inhabits scrubby flatwoods and sandhill plant communities (Layne and Abrahamson 2006). It is not flood tolerant but is fairly shade tolerant, very drought tolerant, and withstands fires by vigorously resprouting from its root collar.

Where it grows on the east side of the 200,000-acre sand pine scrub plant community in the Ocala National Forest, it is an understory shrub mixed in with other shrubs such as sand live oak, myrtle oak, Chapman oak, scrub holly, scrub red bay, crookedwood, Florida rosemary, saw palmetto, and scrub palmetto under a canopy of sand pine. At the southern end of the Lake Wales Ridge, it grows in several communities, including sandhills, where it grows in association with longleaf pine, South Florida slash pine, and turkey oak in a community adapted to frequent fire. In some locations, South Florida slash pine completely replaces longleaf pine, and scrub hickory completely replaces turkey oak (Layne and Abrahamson 2006).

Scrub hickory is highly beneficial to wildlife because of its nuts, which are higher in nutritional value and calories than are acorns or saw palmetto fruits (Layne and Abrahamson 2006). The nuts are eaten by gray squirrels, southern flying squirrels, black bears, raccoons, foxes, native mice, feral hogs, scrub jays, blue jays, red-bellied woodpeckers, and red-headed woodpeckers (Layne and Abrahamson 2006). No doubt, they are also eaten by wild turkeys and fox squirrels, and they are edible by and tasty for people.

Scrub hickory may have some potential for use in landscaping given its drought tolerance and its adaptability to the southern half of the Florida peninsula.

Sand Hickory—*Carya pallida*

Juglandaceae

Sand hickory is a small- to medium-sized hickory of poor, deep, sandy soil. It occurs in isolated spots in Georgia and the central and western panhandle of Florida, but its main distribution is in Alabama, Mississippi, Tennessee, Virginia, southern New Jersey, and the Carolinas (Kartesz 2015). Sand hickory has tufts of hairs on the stalk parts (rachis and petiole) of the deciduous, pinnately compound leaves (Godfrey 1988), being similar in this feature to mockernut hickory, which also inhabits sandy uplands. However, the silvery scales that coat the undersides of sand hickory's narrower leaflets make their coloration much lighter than the undersides of either mockernut or pignut hickory leaves. Similarly, compared with mockernut or pignut hickory, sand hickory has much rougher, deeply ridged bark and golden- to coppery-colored buds. The sweet, edible nuts of sand hickory differ from those of mockernut hickory by having thin husks and thin shells and from those of pignut hickory by being larger in diameter and having husks that split all the way to the base.

Sand hickory grows on well-drained, sandy soil, either with longleaf pine and various oak species in sandhill habitat or mixed in with other hardwoods along the tops of steephead ravines and other places where there is an ecotone between upland pine forest and hardwood forest or where there is a dry ridgetop. Sand hickory is slow growing, drought tolerant, and somewhat fire tolerant but not especially shade tolerant. It is slow to leaf out in the spring and drops its leaves early in the fall.

Sand hickory has excellent wildlife value because of its sweet, edible nuts that are not as difficult to break or chew open as the nuts of most other hickories. They are eaten by many wildlife species including squirrels, mice, wild hogs, black bears, and wild turkeys. The leaves are eaten by various insects.

Sand hickory is not common enough in our area to be economically important, but the wood is similar to other hickories and therefore useful as firewood, for smoking meat, and for making tool handles.

Sand hickory is rarely planted but would be a good choice for a situation within its native range where a deciduous nut tree adapted to poor, deep, well-drained, sandy soil is desired. It is probably not well adapted to the climate of the Florida peninsula.

Mockernut Hickory—*Carya tomentosa*

Juglandaceae

In Florida, mockernut hickory is a fire-adapted tree growing in association with longleaf pine, southern red oak, post oak, wire grass, and various other of the associates of longleaf pine on the most fertile of the well-drained soils support-ing longleaf pine forests. It also grows in the ecotones between these pine forests and adjacent hardwood forests. Farther north, mockernut hickory is the most abundant hickory growing in the oak–hickory and beech–maple forests through-out most of the east-central United States. It is reportedly long lived, sometimes reaching the age of 500 years (Smith 1990, "*Carya tomentosa*"). Mockernut hick-ory is a tall, slow-growing, moderately large tree with large, pinnately compound, alternate leaves that have a covering of woolly hairs on their undersides and along the leaf stems (petioles). It has relatively thick twigs and large buds, which are especially noticeable in winter when the tree is bare of foliage. Mockernut hickory doesn't begin growing until mid-April in Gainesville, Florida. As is the case with all hickories, mockernut hickory is monoecious, producing separate male and fe-male flowers on the same tree in spring, and pollination is accomplished by wind. The nuts of mockernut hickory are often nearly round, fairly large, and come clean from the husk (unlike those of pignut hickory, which retain their husks). The nut is difficult to open, but the meat inside is tasty, having a flavor somewhere between pecan and black walnut.

Mockernut hickory is a common, widespread, and widely adapted tree in much of eastern North America but is less common and more narrowly adapted in Florida, where it grows mostly on well-drained soils with underlying clay in the Florida Panhandle and southward on the Florida peninsula to southern Levy, Marion, Lake, and Volusia Counties (Smith 1990, "*Carya tomentosa*"). It was es-pecially noticeable in the large areas of well-drained clay soils in western Alachua and Marion Counties where it grew in association with longleaf pine, southern red oak, and wire grass in this most fertile of the upland longleaf pine forests (see discussion in the species account "Longleaf Pine—*Pinus palustris*"), but most of this area has now been converted to horse ranches and subdivisions.

In the most fertile of the longleaf pine forests, and especially along the eco-tones where this forest graded into hardwood forest, mockernut hickory grew in association with southern red oak, post oak, flowering dogwood, rusty blackhaw, yellow hawthorn, fringetree, chinquapin, sassafras, New Jersey tea, poppy mal-low, and white indigo in addition to the more common associates of the upland longleaf pine forests.

One of the interesting things about mockernut hickory is its association with longleaf pine in the fire-adapted forests of Florida. Mockernut hickory does not appear to be especially fire tolerant. Its bark is not as thick and obviously fire re-sistant as the bark of longleaf pine or turkey oak. However, the living bark of this hickory is rather thick, and its stems and twigs are much thicker than those of most other trees. In this way, it is similar to longleaf pine, which has the thickest twigs and initial trunk of any pine. Thick stems are less easily killed by fire than thin ones. Even so, one might wonder how a hickory could get started in a for-est that burns every two or three years. The answer is that mockernut hickories start out growing much like any other hickory. After they have been killed back to the ground by a fire, however, they sprout back vigorously with a thicker stem. If killed back several more times, they sprout back on each occasion with an even thicker stem if they are growing in a good, sunny spot. Eventually, a sprout with a very thick stem may survive the next fire and then rapidly grow a thick enough trunk to survive succeeding fires.

Mockernut hickory is a slow-growing, drought-tolerant, shade-intolerant, wind-firm, ice storm–resistant, and moderately fire-tolerant hardwood tree. It is bothered by a number of insects, the most damaging of which is the hickory bark beetle, which kills mature trees in much the same way that pine bark beetles kill pine trees (Smith 1990, "*Carya tomentosa*"). Another pest insect of mockernut hickory is the twig girdler *Oncideres cingulate*, a type of longhorn beetle that gir-dles the large twigs, which then die and fall to the ground. This insect action often deforms seedlings and saplings when the girdled stem is the main growing tip of the tree. Other insects such as caterpillars and boring beetles attack hickories without causing much mortality.

Mockernut hickory is a particularly beneficial tree for wildlife. The nuts have high value for fox squirrels, gray squirrels, and flying squirrels. They are also eaten by wood ducks and wild turkeys (Martin et al. 1951). (It is possible to tell if a flying squirrel has eaten a hickory nut, because they chew a neat-looking round hole in the end of the nut.) The leaves provide food for insects, especially caterpillars, many of which are then eaten by birds. Finally, old hickory trees often have hollows and cavities that are used by various species for denning and/or nesting.

Hickory wood is very hard and strong and has been used to make such things as hammer handles, ladder rungs, and gymnasium equipment (Smith 1990, "*Carya tomentosa"*). One of its drawbacks is that the cured wood is particularly suscep-tible to powderpost beetles. Hickory makes great firewood and is the preferred wood for smoking hams and other meats (Smith 1990).

The landscape value of mockernut hickories is limited by their slow growth rate and, especially, their nut production: having your automobile or your head dinged by a falling hickory nut is not usually considered desirable. However, if located away from where cars park or where people congregate, hickories in gen-eral, including mockernut hickories, make very handsome shade trees. Mocker-nut hickory does not grow well in pots because of its very large taproot, nor does it transplant easily, even as a small seedling, for the same reason. However, it is possible to establish mockernut hickories in places where they are desired by planting the nuts in the fall soon after they drop from the trees. They are great trees for planting in this manner along rural fencerows.

Beautyberry (French Mulberry)—*Callicarpa americana*

Lamiaceae

This is a small, aromatic, multistemmed shrub, 3 to 6 feet tall and equally wide. It occurs naturally throughout Florida except for the Florida Keys and ranges as far north as Virginia and as far west as Arkansas, southeastern Kansas, and Texas (Kartesz 2015). It has opposite, deciduous leaves with serrated edges. Its most no-table feature is its production of shiny, bright purple berries (actually drupes) in mid-August in evenly spaced clusters along the upper and outer stems at the leaf axels. On occasion, though rarely, the berries are bright white rather than purple. The fruits stay on the bush for several months. Beautyberry is never single trunked, having several to many stems originating at the root collar. It has a deep, strong root system. The stems are green at first, turning tan with age. Clusters of small, pink, tuft-like flowers appear in late May, evenly spaced along the outer-most stems at the leaf axels.

Beautyberry, sometimes called "French mulberry," is a common shrub in hard-wood forests on dry, sandy soil. It also occurs in hardwood forests on better soil but is less common there because it does not usually get enough light beneath the denser canopy on these better soils. It does very well in upland hardwood forests that are kept open by cattle grazing. It is common in old xeric hammocks, in young, pioneer hardwood forests on lands that formerly supported upland longleaf pine forest, in the more open areas of mesic upland hardwood forests, and in many places where hardwood trees are able to live such as the edges of fields, along the sides of sinkholes, and along old fencerows. It also grows in old fields, pine planta-tions, upland pine forests that no longer burn frequently, and urban and suburban lots and gardens. Common tree associates of beautyberry include upland laurel oak, live oak, sand live oak, water oak, black cherry, pignut hickory, persimmon, Hercules'-club, and devil's walkingstick. Shrub associates include saw palmetto, hammock pawpaw, sand blackberry, sparkleberry, and one-flower hawthorn.

Beautyberry is moderately shade tolerant but flowers and fruits best in full sun. It is moderately drought tolerant but does not survive flooding. It is not seriously bothered by insects or diseases.

Beautyberry has good wildlife value because of its flowers, which are attractive to insect pollinators, and especially because of its fruit, which remains on the bush for variable lengths of time, usually extending from August until October, November, or December. The fruit is eaten by mockingbirds, brown thrashers, catbirds, robins, towhees, bobwhite quail, armadillos, and raccoons (Martin et al. 1951). According to Rex Rowan, coauthor of *A Birdwatcher's Guide to Alachua County, Florida*, birds that feed on beautyberry fruit at the edge of Payne's Prairie include catbird, gray-cheeked thrush, Swainson's thrush, and wood thrush (personal communication). Local field biologist and birder Lloyd Davis has observed rose-breasted grosbeaks feeding on the fruit in mid-October in his yard in Gainesville, Florida (personal communication). White-tailed deer browse the twigs and leaves and make heavy use of the fruit in late November (Miller and Miller 1999).

Beautyberry has been used for a long time as an ornamental. It is a rather open, ordinary-looking, deciduous bush when not in flower or fruit. However, the flowers are pretty, and the bright, showy fruit remains on the bush for up to four months. The purple color of beautyberry fruit is incomparably rich and vivid, and stems full of the shiny berries are sometimes used in flower arrangements.

Woody Mints—Genera: *Calamintha* (*Clinopodium*); *Conradina*; *Dicerandra*; and *Stachydeoma* (*Hedeoma*); plus Wild Pennyroyal— *Piloblephis rigida*

Lamiaceae

There are a number of species in the mint family that are small, woody, aromatic shrubs or subshrubs that produce small, showy flowers. Most occur in scattered, small, isolated populations in sunny habitats on deep, well-drained sands. Several species of woody mints are endemic to the Lake Wales Ridge in the center of the south-central Florida peninsula south of the area covered by this book, but there are also woody mints that occur as far north as Marion County and even southeastern Georgia, and there are several woody mints in the Florida Panhandle. Most of these plants do not get much more than 1 or 2 feet tall with very slender and often brittle stems and branches. Most, but not all, are evergreen. Some are woody perennials while others are woody annuals. Some are grown sparingly as ornamentals, being raised from seed or, in the case of the woody perennials, also rooted from cuttings. The main wildlife value of these plants is the nectar and pollen they provide to pollinators. Discussion of some of these species follows.

CALAMINTHA (CLINOPODIUM)

Ashe's Savory—*Calamintha ashei* (*Clinopodium ashei*)

Ashe's savory, sometimes called "Ashe's calamint" or "lavender basil," is a compact, low shrub of the deep sands of fossil sand dunes in sand pine scrub habitat in the Ancient Island Scrub area at the center of the Ocala National Forest in northern peninsular Florida's Marion County. From there, it extends southward in separate, isolated populations in scrub habitat on the Lake Wales Ridge, mostly in Polk and Highlands Counties. It also occurs in scrub habitat on coarse quartzite sand dunes along the Ohoopee River in southeastern Georgia in an area called the "Ohoopee Dunes." It requires very well-drained sand and full sun. It is adapted to severe fires occurring at irregular intervals of perhaps every five to ten years or more. Such fires will kill all of the mint plants but provide ideal habitat for the abundant seed bank to produce a new crop of plants. Ashe's savory needs the fires or some other mechanism—timbering, habitat management for scrub jays, or power line, gas line, or road shoulder maintenance—to keep larger woody plants from growing over and killing it by shading. Most of these larger, competing woody plants resprout from their root collars or root systems following a fire or other disturbance, so the disturbance must be repeated continually to keep them at bay.

As a mature plant, Ashe's savory is ordinarily 1 or 2 feet tall but can be 3 feet wide. It blooms in the spring from mid-March to mid-April with a grand show of pale lavender flowers that are each about a half inch across. Sometimes, it also produces a few flowers at other times of year. The fuzzy, linear, half-inch-long, evergreen, blue-green gray leaves are strongly and pleasantly aromatic. This is a rare plant that has been listed as threatened in both Florida and Georgia, and it is (or was) under review for federal listing (Wood 1994). The main pollinators are bees, and there is a very rare mason bee, the blue calamintha bee, that may be constrained to visiting only the flowers of this plant and has, so far, only been found in Highlands County, Florida (Rightmyer et al. 2011).

Scarlet Basil (Red Basil)—*Calamintha coccinea* (*Clinopodium coccinea*)

Scarlet basil is a 2- to 3-foot-tall spindly shrub with a short, often twisted trunk and a few slender, perennial, woody, horizontal branches supporting more numerous, herbaceous, upright branches with small (quarter inch), deciduous to semievergreen aromatic leaves. It produces abundant orange-red to scarlet 2-inch-long tubular flowers in both the spring and the fall, with a few flowers produced in between. The flowers are pollinated by hummingbirds and butterflies. Red basil occurs in both scrub and sandhills in isolated populations in southern Mississippi, southern Alabama, the western Florida Panhandle, the east and west coastal scrubs on the central Florida peninsula, and on the Ohoopee Dunes in southeastern Georgia. It only occurs on deep, well-drained sands, mostly in full sun.

Toothed Savory—*Calamintha dentata* (*Clinopodium dentata*)

Toothed savory is a small, low, slender-branched, aromatic, semievergreen shrub restricted to well-drained sandy soils in the central panhandle of Florida (Godfrey 1988). It produces pale pinkish-lavender flowers in late summer and/or early fall. It is a rare plant found only in a few small spots and is listed in Florida as threatened. It is a short-lived pioneer of disturbed sites and requires full sun to do well on the deep sands where it grows. It can be abundant in the few places where it occurs (Huegel 2009, "Toothed Savory").

Georgia Calamint—*Calamintha georgiana* (*Clinopodium georgiana*)

Georgia calamint is a very rare plant in the wild in Florida, occurring only in a few spots in Gadsden, Holms, Walton, and Escambia Counties in the panhandle (Wunderlin et al. 2020), although it is more common elsewhere in the southeastern coastal plain. Georgia calamint is more adaptable than most other woody mints in terms of its soil, habitat, and light requirements, and it is sometimes sold in native plant nurseries for use as an ornamental shrub. This deciduous shrub grows to be about 2 feet tall, is relatively robust, and is pleasantly aromatic. The leaf size and shape is quite variable, with some of the slightly fuzzy leaves becoming an inch long and a half inch wide. The half-inch-long flowers are very pale purple speckled with darker purple spots. The flowers bloom in great profusion during a two-week period in the fall and are pollinated mostly by bees plus a few butterflies (Huegel 2009, "Georgia Calamint").

CONRADINA

False Rosemary (Wild Rosemary)—*Conradina canescens*

False rosemary, also called "wild rosemary," "scrub rosemary," or "seaside balm," is a small, evergreen, pleasantly aromatic shrub of mostly coastal counties in southern Mississippi, southern Alabama, and the western panhandle of Florida, with one isolated population reported in Hernando County in the central Florida peninsula (Wunderlin et al. 2020). It has small, gray-green, minutely hairy, needle-like, opposite leaves, and the flower bases have a covering of white hairs. This shrub grows to be 2 to 4 feet tall. It produces pale lavender flowers almost all year with a definite peak in the late spring, and the flowers are extremely attractive to bees (Huegel 2009, "Wild Rosemary"). Most common near the coast, it is salt tolerant and requires deep, very well-drained sand and full sun to do well. Because it is less picky in its soil requirements than some of the other woody mints, it is sometimes sold in nurseries and makes a fine ornamental, provided its need for sun and well-drained sand are met.

Apalachicola Rosemary—*Conradina glabra*

Apalachicola rosemary is a very rare, evergreen, minty aromatic shrub that occurs naturally only in Liberty County, Florida, in and next to Torreya State Park; it is a federally and state-listed endangered plant (Florida Natural Areas Inventory 2000, "Apalachicola Rosemary"). It is an attractive shrub, 1 to 3 feet tall, with deep green, shiny, small, needle-like leaves with whitish undersides (Huegel 2009, "Apalachicola Rosemary"). Growing in sandhills that were originally open longleaf pine–turkey oak savanna habitat, it is now often found in the transitional areas between this habitat and steephead ravines or along roadsides and on the edges of pine plantations. Like other woody mints, it needs sun and very well-drained sand to do well although it tolerates light to partial shade. Apalachicola rosemary blooms mostly in the spring, from March through May or June, but also sporadically throughout the summer and fall, producing pale lavender and purple flowers that are very attractive to bees (Huegel 2009, "Apalachicola Rosemary"; Kubes 2009). Apalachicola rosemary's primary pollinators in their native habitat are digger bees (*Anthophora abrupta*), carpenter bees (*Xylocopa virginica*), and bee flies (*Bombylius major*) although smaller numbers of leafcutter bees, sweat bees, digger wasps, and brown skipper moths visit the flowers as well (Kubes 2009).

Etonia Rosemary—*Conradina etonia*

Etonia rosemary is a rare woody mint endemic to an area near Florahome in Putnam County, Florida, on the north-central Florida peninsula. It grows 2 to 4 feet tall and emits a minty aroma when bruised. It was discovered in 1990, named in 1991, and added to the federal endangered species list in 1993. Originally, there were 13 known populations totaling fewer than 3,000 plants in Etoniah Creek State Forest, Dunns Creek State Park, and on private land inholdings within the boundaries of the state forest (Dziergowski 2009). The populations in Dunns Creek State Park, numbering fewer than 1,000 plants, are now considered a different species, *Conradina cygniflora* (see below). The remaining population of Etonia rosemary is confined to the Etoniah Creek State Forest and nearby private lands with a population of about 1,500 plants. Etonia rosemary grows in white sand scrub in association with sand pine, Florida rosemary, garberia, scrub palmetto, Chapman oak, myrtle oak, sand live oak, and other scrub vegetation. It prefers full sun and deep, well-drained sand. Its pale lavender flowers, larger than most other members of this genus and similar in size to those of large-flowered rosemary, are attractive to bees.

Dunns Creek Dixie Rosemary—*Conradina cygniflora*

This plant, also called "False rosemary" like *Conradina canescens*, is a very rare woody mint found only in Dunns Creek State Park in Putnam County, Florida. It closely resembles Etonia rosemary and was identified as a different species in 2009, using genetic and morphological data (Edwards et al. 2009). *Conradina cygniflora* grows in white sand scrub habitat. It has fragrant, evergreen leaves, and the pale lavender flowers are attractive to bees and butterflies.

Large-Flowered Rosemary—*Conradina grandiflora*

Large-flowered rosemary is an evergreen, aromatic shrub that grows 2 to 4 feet tall in scrub vegetation along Florida's Atlantic coastal counties from Volusia County south to Miami-Dade County (Huegel 2009, "Large-Flowered Rosemary"; Wunderlin et al. 2020). It has deep green, shiny, needle-like, half-inch-long leaves and relatively large (2 cm wide), pale lavender flowers that are attractive to bees. It was listed as endangered by the state of Florida (Wood 1994), but this status has been changed to threatened. It is not listed as threatened or endangered by the federal government. Occasionally planted as an ornamental shrub, large-flowered rosemary is available from some plant nurseries.

DICERANDRA

Longspur Balm—*Dicerandra cornutissima*, Coastal Plain Balm—*D. linearifolia*, and Florida Balm—*D. densiflora*

Dicerandra is a genus in the mint family with nine species of showy, aromatic, flowering plants occurring in Florida and two occurring in southeastern Georgia. Six of these are small subshrubs and five are annuals. Four species are listed as endangered, including three on the Lake Wales Ridge in south-central Florida and one, longspur balm (*Dicerandra cornutissima*), an annual wildflower, that occurs in southwestern Marion County (Wunderlin et al. 2020). The two most widely distributed species, coastalplain balm (*D. linearifolia*), which occurs in northern Florida and southern Alabama and Georgia, and Florida balm (*D. densiflora*), which is restricted to northern Florida, are both annual wildflowers (Wunderlin et al. 2020). All *Dicerandra* species occur exclusively on the deep sands of sandhills or scrub. They are exceptionally fussy about growing conditions and, in order to thrive, must have the excellent drainage and full sun such habitats afford. They flower profusely on slender stalks that range, according to species, from 12 to 18 inches tall, and, by dense branching, the plants can become quite wide. The typical linearity of *Dicerandra* leaves varies from species to species as does their color, which ranges from lime to emerald green. The abundant, small, orchid-shaped blooms range in color from white to multiple shades of light and deep pink and lavender, many of them speckled with purple or fuchsia flecks (Huegel 2009, "Coastalplain Balm" and "Longspur Balm"). All *Dicerandra* species bloom in the fall, and fully mature plants may bear dozens of flowers at once. *Dicerandra* flowers are pollinated primarily by bees, but the blooms are also attractive to butterflies. Although plants in this genus flower prolifically, attract pollinators, and provide beautiful color, their highly specific growing requirements make them almost impossible to maintain in home gardens, so they are not typically available commercially, even from native plant nurseries (Huegel 2009, "Coastalplain Balm" and "Longspur Balm").

STACHYDEOMA (HEDEOMA)

Mock-Pennyroyal—*Stachydeoma graveolens* (*Hedeoma graveolens*)

Mock-pennyroyal is a small, slender, rare, aromatic woody mint restricted to the central panhandle of Florida, mostly in the Apalachicola National Forest (Florida Natural Areas Inventory 2000, "Mock Pennyroyal"). It is listed as endangered by the state of Florida (Wood 1994). It grows in longleaf pine–turkey oak–wire grass sandhills and in the drier parts of longleaf pine–saw palmetto–wire grass flatwoods forests (Florida Natural Areas Inventory 2000, "Mock Pennyroyal"). Its fuzzy new stems and leaves give the plant a light grayish-green color. The half-inch-long bright pink flowers bloom from April through September and are visited mostly by native bees, primarily leaf-cutter bees (*Megachile* spp.) but also by digger bees (*Anthophora abrupta*) and most of the other pollinators listed for Apalachicola rosemary (Kubes 2009).

PILOBLEPHIS

Wild Pennyroyal—*Piloblephis rigida*

Wild pennyroyal, also called "Florida pennyroyal," is the only species in its genus. It is a small, dense, pleasantly aromatic, evergreen shrub that grows to be about 2 feet tall with tiny, opposite, needle-like leaves with entire margins and a profusion of pale lavender flowers borne in clusters at the ends of branches in late winter into spring, with some flowering at other times of year. Its natural distribution is from Duval and Alachua Counties southward throughout the Florida peninsula to Dade County in South Florida with isolated populations in extreme southeastern Georgia (Wunderlin et al. 2020). It is most abundant in scrubby flatwoods but also occurs to some extent in mesic pine flatwoods, scrub, sandhills, dry prairies, and ruderal areas. It prefers well-drained acidic sandy soil and is not salt tolerant. Drought tolerant but not shade tolerant, it requires full sun to thrive and is both fire adapted and fire dependent. The flowers are especially attractive to a variety of bees and butterflies because they bloom very early in the year when few other flowers are open. Wild pennyroyal is used as an ornamental and is available in some nurseries.

Camphor Tree—*Cinnamomum camphora*

Lauraceae

Camphor tree is native to China and Southeast Asia. It was commonly planted in central and northcentral peninsular Florida during the twentieth century and has spread into many wild areas including established forests of many kinds. It has a dense crown of evergreen leaves that are bright green and shiny on top and dull light green below. The twigs are also green. The leaves have smooth edges, pointed tips, and are typically 2 to 3 inches long and three quarters to 1½ inches wide. The bark on larger trunks is thin, tan, and roughened by flat-topped vertical ridges. The leaves, twigs, inner bark, and wood are aromatic with the scent of camphor when crushed. This distinctive scent provides a sure and easy way to identify the tree. The roots of the seedlings have the scent of anise (sometimes mixed with the camphor scent).

Camphor is potentially a very large tree. From Marion County south, where it is less likely to be damaged by hard freezes, it closely resembles live oak in its shape and upper size limit. The five largest ones reported as of 1997 were between 50 and 65 feet tall with trunk diameters between 10 and 11¼ feet at breast height and crown spreads ranging from 53 to 135 feet (Ward and Ing 1997). These are huge trees, especially considering that camphor has only been planted in Florida for about one hundred years.

Camphor tree grows well in most upland situations, including urban and suburban landscapes, fencerow forests, old field forests, pine plantations, upland hardwood forests, and upland pine and pine flatwoods forests that are not frequently burned. It is quite shade tolerant, which has enabled it to invade established hardwood forests with closed canopies, and it is able to invade over long distances because of seed dispersal by the songbirds that eat its fruits in winter. Camphor tree is listed as an invasive exotic pest plant in Florida (FLEPPC 2013).

The wildlife value of individual camphor trees is moderately high, although this is countered by the fact that this species is an invasion threat to native forests.

It provides almost no insects for insectivorous birds and other animals because its foliage and bark are toxic to most insects, with the notable exception of spice-bush swallowtail butterfly caterpillars (and perhaps laurel swallowtail caterpillars). However, its fruits, which begin ripening in October, hang on all winter and provide robins, cedar waxwings, common grackles, starlings, mockingbirds, fish crows, red-bellied woodpeckers, Baltimore orioles, and probably some other birds with a valuable food source at a time when they need it. I have seen large flocks of robins, cedar waxwings, and grackles feeding avidly on camphor berries in mid-January and mid-February. I observed 100 robins feeding at one camphor tree on January 9, 2001, and 300 cedar waxwings feeding at the same tree on January 16, 2001. The robins fed every day for a period of two weeks. A similar mass feeding had occurred 8 to 10 days after a hard freeze in 1996, just as it had after hard freezes the other three or four times I observed such an event. This prompted my speculation that the freezing of the fruit, followed possibly by some fermentation, perhaps prompted the mass feeding. Whether alcohol content in the fruit attracts the birds or whether freezing or fermentation eliminates some disagreeable or toxic chemical in the fruit to render it more palatable I do not know. After all, I have seen mockingbirds feed on camphor tree fruits in the fall, long before cold temperatures and freezes set in, and Michael Meisenburg, a former president of the Alachua County Audubon Society, has seen flocks of fish crows doing the same thing (personal communication). I have also seen gray squirrels feeding on the fruits, and this has usually been after freezes, including two days after a hard freeze on January 4, 1999, and five days after a freeze on December 19, 2000. Finally, camphor trees make good, large, long-lasting den trees when their trunks or branches eventually become hollow.

The camphor tree is one of those exotic plants that has the potential of doing harm by invading native habitats and strongly dominating them to the point that many native plants and animals may decline in abundance or even completely vanish from the habitat. So far, however, I have not seen this happen with camphor tree to such an extent that native habitats are destroyed. One reason for this is that camphor trees are big, obvious trees that are fairly easy to kill with herbicides.

Camphor tree is notable for its aromatic, light-colored, even-grained wood that has light to medium weight and hardness and an attractive appearance. It is not especially strong but is rot resistant and very resistant to insect damage. It is mostly used for firewood but can also be used to make paneling, outdoor decks, outdoor furniture, and fine turned objects such as bowls.

Camphor tree has been used extensively in the past as an ornamental tree. It is an attractive evergreen shade tree that is quite wind-firm and long lived. Nonetheless, it has some serious drawbacks as a landscape addition. One is that it is not entirely cold hardy, sometimes showing damage when temperatures drop below about 20 degrees Fahrenheit. Prolonged hard freezes will some-times kill it to the ground or back to the trunk and largest branches. Another drawback is that its powerful root system is well noted for invading sewer lines and drain fields and for cracking building foundations. Perhaps the greatest drawback is its tendency to invade native forests, which makes it a threat to na-tive ecosystems and the wildlife they support in Florida south of the Suwannee and Santa Fe Rivers, where the threat of hard freezes is less serious than farther north and west. With the current warming of our climate caused by increasing levels of carbon dioxide in the atmosphere, the zone of unfettered camphor tree growth may move northward.

Spicebush—*Lindera benzoin* (including notes on Pondberry [*L. melissifolia*] and Bog Spicebush [*L. subcoriacea*])

Lauraceae

Spicebush is a large shrub (occasionally a small tree farther north) in the laurel family with dark green, spicily aromatic, deciduous leaves 2 to 5 inches long and 1 to 2 inches wide that grows in the understory of hardwood forests on moist, fertile soil. It occurs throughout most of the eastern United States, but reaches Florida only in the central panhandle on either side of the Apalachicola River, plus a few small spots along spring runs in Putnam, Marion, Orange, and Brevard Coun-ties on the Florida peninsula (Wunderlin et al. 2020). The small yellow male and female spring flowers are borne on separate, dioecious plants, and the small red fruits mature on the female plants in the fall.

Spicebush grows best on rich soil over limestone with an abundant moisture supply, usually under a canopy of oak, hickory, elm, ash, maple, sweetgum, black-gum, tulip tree, and American beech. At Florida Caverns State Park in the Florida Panhandle, spicebush additionally grows in association with red bay, sugarberry, hophornbeam, and needle palm (Surdick and Jenkins 2010). It often occurs near a stream or on a slope going down to a stream. Spicebush is very shade tolerant. It reproduces by seed and also by root suckers and sometimes forms patches of bushes that are probably clones derived from root suckers. In the small popu-lations that are (or were) located on the central Florida peninsula, it grows on organic soil on stream floodplains and under the canopy of swamp red bay, red maple, pumpkin ash, swamp laurel oak, swamp tupelo, sweetgum, Florida elm, red mulberry, and sometimes Atlantic white cedar.

The small populations on the Florida peninsula have been disappearing be-cause of various factors such as laurel wilt disease and power line maintenance using herbicides (Surdick and Jenkins 2010). Unfortunately, the laurel wilt disease may eventually decimate the other populations of spicebush. Laurel wilt disease came into North America near Savannah, Georgia, in 2002 with the introduction of the Asian ambrosia beetle *Xyleborus glabratus*, which carries the fungus *Raffaelea laurifolia*.

Two related, very rare, wetland species, pondberry (*L. melissifolia*) and bog spicebush (*L. subcoriacea*) used to occur in the western Florida Panhandle and may still occur in southern Alabama and/or southern Georgia in small, isolated populations (Surdick and Jenkins 2010).

Spicebush fruits are eaten by various birds, especially wood thrush and veery (Martin et al. 1951). The foliage is one of the preferred foods of spicebush swallowtail butterfly caterpillars.

Spicebush is sometimes used as an ornamental in the eastern United States well north of Florida.

Pondspice—*Litsea aestivalis*

Lauraceae

Pondspice is a rare wetland shrub with small, narrow, deciduous leaves and thin, crooked branches. It has multiple, crooked stems and is usually between 3 and 10 feet tall. It has small yellow flowers that bloom December to March and small red fruits that ripen in late July and persist into September. The male and female flowers occur on separate, dioecious plants. It is listed as endangered by the state of Florida (Surdick and Jenkins 2009).

Pondspice occurs in a few, perhaps 40, widely scattered populations on the edges of depression marshes and cypress domes in pine flatwoods in northeastern Florida south to Pasco County on the Florida peninsula and in Okaloosa County in the Florida Panhandle. It also occurs as a rare plant in widely scattered spots in the southeastern United States coastal plain from Virginia into Louisiana. Most of the Florida populations contain less than 100 individual plants (Surdick and Jenkins 2009), although a population of pondspice in Price's Scrub State Park in Marion County has over 100 individual plants. The associated pines may be either longleaf, slash, or pond pine, and the cypress is always pond-cypress. Dahoon or myrtle dahoon often grows nearby. Pondspice often grows in slightly deeper water than do associated shrubs such as waxmyrtle, gallberry, sandweed, fetterbush, maleberry, and swamp doghobble, and it often grows in situations where these other shrubs are uncommon or absent. Its most common herbaceous associates are maidencane and Virginia chain fern. The wetlands it frequents are usually low in both fertility and pH and retain water most of the time. The pine flatwoods ecosystem as a whole is a fire-adapted habitat, and the wetland sites within it also burn when they are dry and the surrounding pinelands are burning. The edges where pondspice grows would have been subject to frequent fire before the advent of fire suppression. Pondspice resprouts vigorously after fires have killed the aboveground part of the plant (Surdick and Jenkins 2009).

Pondspice is susceptible to laurel wilt disease, with most of the populations in northeast Florida exhibiting symptoms of it in 2009. Besides the red bay ambrosia beetle that carries the laurel wilt fungus, pondspice is also susceptible to attack by another invasive exotic ambrosia beetle, the black twig borer (Surdick and Jenkins 2009).

The wildlife value of pondspice is not well documented. Mike Campbell, forester and native plant nurseryman, observed many songbirds feeding on the fruits of a large pondspice bush in southeastern Alachua County in 1997 (personal communication). There are two records of swallowtail butterfly larvae using this species as a larval host plant, one in Louisiana for the spicebush swallowtail and one in northwestern Alachua County for the laurel (palamedes) swallowtail. There was also a bird's nest in one bush in Alachua County. Given that few of these plants are ever seen, these observations indicate that pondspice is useful for at least some wildlife species where it occurs.

The only likely human use for this plant would be as an ornamental. It is easy enough to start from seed but is very picky about where it will grow. I tried growing it in and next to ponds and wetlands in Suwannee County without success. In those locations, it did fine for a year or two but was subsequently killed either by drought or flooding. It is an attractive bush in its natural setting.

Red Bay—*Persea borbonia*

Lauraceae

Red bay occurs naturally on upland soils throughout Florida, southern Georgia and Alabama, southeastern Texas and southwestern Louisiana, and along the coast of the Carolinas (Kartesz 2015). Before the laurel wilt disease killed all of the larger trees, red bay was a medium-sized to large tree with a maximum height of 80 feet, a maximum trunk diameter of 4 feet, and a maximum crown spread of 70 feet. The trunk usually divided into a few massive branches, giving the tree an overall appearance similar to that of live oak. The bark on large trunks is reddish-brown and ridged vertically. The pleasantly aromatic, 2- to 5-inch-long by 1- to 2-inch-wide evergreen leaves have entire margins, are shiny on top, and appear hairless to the naked eye. Silkbay and swamp bay—trees closely related genetically to red bay—can be distinguished from it by their smaller size and the clearly visible small hairs that line their lower leaf surfaces and young twigs. Other similar looking trees which are not closely related genetically are sweetbay and loblolly bay, the first having smooth bark and white pubescence under its leaves and the second having slightly toothed leaf margins.

The laurel wilt disease is caused by the fungus *Raffaelea lauricola*, which is spread by the red bay ambrosia beetle, *Xyleborus glabratus*—both recently introduced exotics from Asia—that are spreading rapidly in the southeastern United States and killing red bay, swamp red bay, silkbay, sassafras, pondspice, and pond berry (Global Invasive Species Database 2020, "*Raffaelea lauricola*"). Almost all red bay trees over 1 inch in diameter are being killed (Global Invasive Species Database 2020, "*Raffaelea lauricola*"). Fortunately, red bay reproduced prolifically in the past, producing an abundance of seedlings less than 1 inch in stem diameter, so that many small red bay trees still remain in numerous locations.

Red bay occurs as a scattered tree where limestone is near the surface in calcareous mesic and xeric hammocks (calcareous upland hardwood forests). Its most common tree associates include sweetgum, pignut hickory, live oak, laurel oak, swamp chestnut oak, bluff oak, Shumard oak, winged elm, sugarberry, persimmon, white ash, Florida maple, box elder, southern magnolia, and basswood. Other calcium-loving woody plants often associated with red bay are soapberry, Carolina buckthorn, climbing buckthorn, and virgin's bower. Red bay also occurs as an invader in upland longleaf pine habitats including sandhills.

Red bay is neither tolerant of flooding nor wet soil and does not do well in highly acidic soils. It is only moderately shade tolerant, but it seems quite drought tolerant. Its growth rate is moderate, and individual trees were able to live for more than 100 years prior to the introduction of the laurel wilt disease. Red bay foliage is often full of insect galls. The stems are attacked by insects or diseases that cause them to die back, and the larger stems and trunks of old, mature trees were usually hollow. Indeed, most mature red bays often looked as if they were having a really hard time of it. However, they seemed to be able to survive in apparently poor condition for a long time, until laurel wilt came along.

The wildlife value of red bay was once high, in large part because of its usually being hollow at maturity and thus providing cavities in which many species of wildlife could nest and den. The foliage provides food for several kinds of insects including the larvae of spicebush and laurel (Palamedes) swallowtail butterflies (Pyle 1981). The fruit is eaten to some extent by white-tailed deer, black bear, bobwhite quail, wild turkey, and other birds, especially in winter (Global Invasive Species Database 2020, "*Raffaelea lauricola*").

When dried, red bay's aromatic leaves make a savory, spicy substitute for commercial bay leaves in cooking, especially for flavoring meats and soups. Red bay wood is the best in northern Florida for making fine furniture (Kay Eoff, retired University of Florida professor, maker of sail boats and custom furniture, personal communication). It is similar to mahogany in its rot resistance, stability, and easy workability, and it takes a high polish. Its interlocking grain makes it resistant to splitting and gives it a beautiful, lustrous pattern. The color of the highly valued heartwood varies from light brown to a rich reddish-brown.

Red bay was rarely planted in the past, but it sometimes made a beautiful evergreen tree when left standing in a landscaped situation.

Silk Bay (Scrub Red Bay)—*Persea humilis* (*P. borbonia* var. *humilis*)

Lauraceae

The smallest and prettiest of the red bays is the silk bay. It is often only 10 to 30 feet tall with a trunk diameter of a few inches. The largest one reported was 38 feet tall with a trunk diameter of 18 inches. The leaves, smaller than those of the other red bays, are shiny above and pubescent below with the small hairs varying in color from a nondescript buff color to a glossy golden or rusty color. The dark gray bark on old trunks is roughened by interlocking ridges.

Silk bay occurs primarily on the central Florida peninsula from Marion County southward into Highlands County, with outliers as far north as Alachua, Levy, and Putnam Counties and as far south as Collier County. It grows mainly in one very restricted plant community, Florida scrub, which is characterized by such trees as sand pine, sand live oak, myrtle oak, crookedwood, Florida rosemary, scrub palmetto, saw palmetto, deer moss, and a whole host of rare scrub-endemic plants such as several species of woody mint, scrub morning glory, scrub hickory, scrub holly, and others. The northernmost large population of silk bay occurs in the Ocala National Forest in the largest patch of Florida scrub habitat in existence.

Silk bay is remarkably drought tolerant on deep, excessively well-drained sands. It prefers full sunlight but is one of the most shade tolerant of the scrub plants. It seems to be more resistant to whatever insects and diseases trouble the other red bays and was usually vigorous and healthy in appearance before the laurel wilt disease came along. Now, most of these little trees have died of that disease, which was introduced to this country in 2002 and is now decimating the populations of seven native plants: red bay, swamp bay, silk bay, sassafras, spicebush, pondberry, and pondspice (Spence et al. 2013).

The wildlife value of silk bay is not well studied. Birds, including mockingbirds and scrub jays, eat the fruit, and its leaves are host to some gall-forming insects and to spicebush and laurel swallowtail butterfly caterpillars. When in bloom in early May in scrub habitat, the flowers attract many bees, wasps, and butterflies, including hairstreak butterflies.

The small size of silk bay prevents its wood from being used to any significant extent, except perhaps as firewood. The leaves, like those of the other red bays, are suitable as a substitute for commercial bay leaves.

This beautiful little tree had considerable potential for use in landscaping on well-drained sandy soils before the introduction of laurel wilt disease.

Swamp Red Bay—*Persea palustris*

Lauraceae

Swamp red bay, also called "swamp bay," is similar in appearance to red bay and is closely related to it. It occurs throughout Florida and the coastal plain from southeastern Virginia into eastern Texas (Kartesz 2015). The young twigs, petioles, and lower leaf surfaces have a covering of clearly visible, fine, light-colored erect hairs, and the leaves usually lack the bright, shiny appearance of red bay leaves. Swamp red bay is also a smaller tree, at least in Florida, rarely growing over 1-foot in trunk diameter or 50 feet tall.

Swamp red bay grows in seepage areas, bayheads, white cedar bogs, shrub bogs, peat bogs, on margins of swamps, and pretty much throughout the mesic and wet pine flatwoods. It is the only member of its genus that occurs in these poorly drained habitats. It is most common in bayheads, where it is associated with sweetbay, loblolly bay, swamp tupelo, fetterbush, gallberry and, to a lesser extent, with pond pine, slash pine, and red maple.

Swamp red bay is tolerant of wet soil and acid soil, but it is not really flood tolerant. It is also not particularly drought tolerant or wind-firm. It is moderately shade tolerant, however. Like red bay, swamp red bay has disease and heart-rot problems and, like red bay, it is susceptible to laurel wilt (Spence et al. 2013). It seemed rarely to live as long as 100 years even before the introduction of laurel wilt disease, which has now killed almost all of the larger trees.

The wildlife value of swamp red bay was moderate before it was decimated by laurel wilt disease. The fruit was eaten to some extent by birds, and the trunk and branches sometimes provided nesting and sheltering cavities, though not nearly as often as did those of red bay. Because many sprouts and seedlings still survive in the wild, the leaves provide food for insects, including gall wasps and the caterpillars of the laurel and spicebush swallowtail butterflies.

Swamp red bay has little commercial value and is not used for landscaping. Its leaves appear dull compared with the shiny leaves of red bay, and swamp red bay does not grow well on the well-drained, lime-rich soil of most landscaped situations.

Sassafras—*Sassafras albidum*

Lauraceae

Sassafras is a tree in the laurel family with deciduous, sometimes mitten-shaped or three-lobed leaves. It is currently a moderately common tree throughout most of the eastern United States. It is more common and larger in the Florida Panhandle than on the peninsula but ranges southward into the central Florida peninsula, mostly as a shrub on well-drained sandy soil. Its leaf size is quite variable: leaves from 4 to 6 inches long and 2 to 4 inches wide are typical in most of its range, but on sassafras trees of the Florida peninsula the leaves are usually much smaller, often only 1 or 2 inches long. They have a dull surface that shows the pattern of the veins as slight indentations. The leaves are slightly aromatic, and the freshly cut or split roots are strongly aromatic. Sassafras is dioecious, producing greenish-yellow flowers in the early spring and single-seeded, one third–inch- to half-inch-long, dark blue drupes at the ends of bright red stalks in the fall on the female trees. The gray bark on large trees is roughened with vertical ridges and fissures. Mature tree size is also quite variable with some clones on the central Florida peninsula not exceeding 10 feet in height and a couple of inches in diameter. Farther north, sassafras can be quite large, often 60 feet tall and 2 feet or more in trunk diameter. The 2018 national champion, in Kentucky, is 62 feet tall with a 7½-foot trunk diameter (American Forests 2018).

In Florida, sassafras occurs mostly as a minor component of upland longleaf pine forests. It spreads widely by root suckers and resprouts vigorously following fires. On the Florida peninsula, it grows in association with longleaf pine, turkey oak, and wire grass or with longleaf pine, southern red oak, sand post oak, mockernut hickory, and wire grass. In the Florida Panhandle and especially farther north, sassafras is well known for its ability to invade abandoned fields and spread along fencerows. It is intolerant of shade but very drought tolerant and moderately fast growing.

Sassafras is highly susceptible to the laurel wilt disease, a recently introduced disease caused by the fungus *Raffaelea lauricola* that is carried and transmitted to the trees by the invasive exotic red bay ambrosia beetle, *Xyleborus glabratus* (Randolph 2017). This disease has been in Florida since 2002 and is likely to expand into the heart of sassafras's range in the near future, having already spread as far as North Carolina and Alabama, with an isolated spot in Arkansas (Randolph 2017).

The wildlife value of sassafras is moderate in Florida but higher farther north where the tree is more common and larger. Its fruits are eaten by small mammals, bobwhite quail, wild turkey, catbird, mockingbird, and especially the great crested flycatcher, eastern kingbird, and pileated woodpecker (Martin et al. 1951). It is also browsed by white-tailed deer (North Carolina Cooperative Extension 2018).

Sassafras is a host plant for the caterpillars of the spicebush swallowtail butterfly and the laurel (Palamedes) swallowtail butterfly.

Although sassafras wood is lightweight, soft, and brittle and therefore of little commercial importance, it has been used to make fence posts and furniture (Griggs 1990). Extracts from the roots and bark were used in the past to flavor root beer and to make sassafras tea and herbal medicines. An extract from the pith was once used to thicken gumbo, but the presence of a carcinogen (safrole) in sassafras prompted the United States Food and Drug Administration to ban its use in ingestible products (North Carolina Cooperative Extension 2018). For a brief period in the early seventeenth century, sassafras was a major export item from North America to England, second in importance only to tobacco (Wikipedia 2020, "*Sassafras Albidum*: Commercial Uses").

Sassafras is frequently used as an ornamental tree in the eastern United States well north of Florida. It has attractive leaves and often produces good fall color. While its large, deep taproot can make transplantation difficult, the tree does well when grown from seed in a deep pot and then planted. Sassafras requires full sun to thrive but grows in almost any well-drained soil and is very drought resistant. Because it is susceptible to laurel wilt disease, sassafras is now a risky choice for landscape use in the southern United States.

Coralbean (Cherokee Bean)—*Erythrina herbacea*

Leguminosae; Fabaceae

Coralbean is a woody deciduous shrub in the legume family that freezes to the ground each winter in areas that experience winter freezes. At the southern end of the Florida peninsula and along the Gulf Coast in Mexico, it grows into a large shrub or small tree. Its interesting-looking compound leaves consist of three leaflets that are pointed at the tip with a bulge in the middle. In April and May, before or at the same time its foliage begins to grow, coralbean produces its inflorescence, a spike of horizontally oriented tubular, deep red or deep pink flowers. These are followed by bean pods that start out green but eventually turn brown and split open to reveal very hard, red to red-orange beans. The plant, especially the beans, contain a poison.

Coralbean grows best in well-drained, sunny situations and is very drought tolerant. It is amenable to a wide variety of soil conditions from deep sands to clay soil and from acidic soils to shell mounds and limestone outcrops. It is also fairly salt tolerant, occurring along the coast in many places. Because of the stored energy in its swollen taproot, which grows to great size as the plant ages, it can survive fires and other disturbances. Plants that are decades old have taproots 3 or 4 inches in diameter that extend down as a large mass as far as 4 feet into the ground. Normal-sized roots then spread out and down from this massive base, and, in areas subject to occasional freezing temperatures, the aboveground stems repeatedly grow upward from the top of this base.

Coralbean occurs scattered about in coastal and xeric hammocks, sandhills, and scrub on the Florida peninsula. It is rare in the Florida Panhandle and largely absent from Georgia but occurs along the coast of South Carolina and in a broad area on the Gulf coastal plains of Texas and Mexico.

The wildlife value of this plant is moderate. When in bloom, it is attractive to hummingbirds and butterflies and is also sometimes robbed of nectar by Cape May warblers and orchard orioles, which pierce the flower on its side. Coralbean also supports two native moth species: the Erythrina stem borer and the Erythrina leafroller, the larvae of which often kill back the flower spikes, growing tips, and young stems (Sourakov 2011).

Coralbean's only commercial value relates to its horticultural use. It is sometimes planted as an ornamental in sunny situations, where its bright, deep pink to scarlet flowers are both unusual and attractive in early spring. Reproduction is by seed. Typically hardy and trouble free except for occasional attacks from the Erythrina stem borer and Erythrina leafroller, coralbean requires no watering, spraying, or fertilizer once established. Establishing coralbean is as easy as collecting the ripe beans and putting them in the ground where you want new plants. One drawback of planting coralbean is the toxicity of the beans, which hang onto the flower spikes in their pods throughout the summer and fall. Because of their bright color, these poisonous beans could easily be attractive to small children. The beans are very hard, are not dangerous to handle, and would likely be toxic only if chewed and swallowed.

Sweet Acacia—*Vachellia farnesiana* (*Acacia smallii* and *A. farnesiana*)

Leguminosae; Fabaceae

This sharply spiny shrub or small tree barely reaches into northern Florida. It has delicate, twice pinnately compound leaves and produces pretty little yellow puffball flowers and clusters of small, thick, woody bean pods. It is found along the coast and on Indian mounds from Levy County southward on the Florida peninsula, being most common in South Florida, South Texas (where it is sometimes called "huishache"), Mexico, and Central America. It also occurs in a few spots along the coast in the Florida Panhandle (Godfrey 1988; Wunderlin 1998) and is established throughout most of the tropics including much of South America, Africa, Arabia, southern Europe, southern Asia, and parts of Australia (Bell et al. 2017).

Sweet acacia prefers well-drained calcareous soils and requires full sun exposure to grow well. It is moderately salt tolerant and extremely drought tolerant but not at all amenable to flooding, fire, freezing temperatures, or shade. If killed to the ground by freezing temperatures or fires, however, it sprouts back vigorously from its strong root system. Sweet acacia is sometimes damaged by infestations of thorn bugs (*Umbonia crassicornis*) (Huegel 2010), and the last sweet acacia I had in my yard died of something that caused stems to die back and eventually killed the whole plant; I was unable to identify or treat the disease (personal observation).

The flowers, often produced in late fall and winter when other flowers are scarce, are visited by bees. The thorny branches provide nesting habitat for birds; the thorn bugs that sometimes feed on the twigs may provide food for lizards and birds; the seeds are a potential food source for birds; and the nutritious bean pods are a potential food for white-tailed deer.

This plant is used by humans in diverse ways. The flowers are processed in both India and southern Europe to produce a perfume, which, in Europe, is known as "cassie." The seeds are used as food in southern Asia, Mexico, and tropical America although all parts of this plant contain tannins and various alkaloids. The foliage and bean pods provide forage for livestock. Moreover, sweet acacia exudes a gum from stem wounds that is similar and perhaps even superior to gum arabic as an important commercial component of ceramic glazes, paints, inks, dyes, and more. It is also widely used as a thickening and binding agent in the making of sweets (Duke 1983).

Sweet acacia is used sparingly as an ornamental. The delicate foliage and yellow puffball flowers are attractive, but the sharp spines and the tree's susceptibility to freezing temperatures are serious drawbacks.

Mimosa (Silk Tree)—*Albizia julibrissin*

Leguminosae; Fabaceae

Mimosa is an ornamental tree with doubly compound, deciduous, fern-like leaves averaging about 12 inches long and 6 inches wide. In summer it produces abundant, showy flower clusters about 1½ inches wide resembling little pom-poms that are white at the base then mostly pink. The flowers are followed in the fall by 6-inch-long flattened bean pods containing hard, brown seeds. Mimosa has relatively smooth gray bark on the trunk and branches and a deep taproot. It grows very fast and develops into a rather flat-topped, umbrella-shaped small tree 10 to 30 feet tall with a trunk up to about a foot in diameter. Occasionally, trees are twice that size. The national champion is 59 feet tall with a 4-foot trunk diameter and a crown spread of 103 feet (American Forests 2018).

Mimosa was introduced into the United States in 1745 from Asia and became a highly popular landscape ornamental throughout the south. It was commonly planted in Florida in the 1950s. Its popularity began to decline when mimosa wilt began killing many of the planted trees, although wilt-resistant varieties were later produced. Mimosa is now known to be a problematic invasive exotic pest, and it is not as often planted today as it was in the past. It nonetheless remains common in many places because of its having escaped cultivation.

Mimosa will grow on almost any upland soil type. It grows rapidly in full sun and does well under a canopy of pines because of the light shade they cast. It cannot survive in the shade of a closed-canopy hardwood forest, however. Mimosa is difficult to control because it sprouts back vigorously from its root collar if cut down or killed to the ground by fire. Moreover, a large store of seeds remains in the soil beneath established mimosa trees. These seeds can sprout each year for up to 50 years in Texas after the parent tree has been removed (Texas Invasive Species Institute 2014). In Florida, with its higher humidity, the seeds can still last at least 20 years in the soil with a few coming up each year (personal observation).

The wildlife value of mimosa is primarily related to its flowers, which are attractive to butterflies and hummingbirds. In addition, there are a number of invasive exotic insects that feed on mimosa, including the mimosa webworm that feeds on its leaves, a jumping plant louse or "psyllid" (*Acizzia jamatonica*) that also feeds on the leaves, the eastern palearctic seed beetle that feeds on the seeds, and a metallic wood-boring beetle (*Agrilus subrobustus*) that bores into the wood (Texas Invasive Species Institute 2014).

Mimosa has long been valued by people as an attractive ornamental tree and is still available for sale. It was used more frequently several decades ago but is still commonly planted throughout Florida and the rest of the southern half of the United States. Its use is now problematic as it is an invasive exotic plant capable of invading and displacing native ecosystems. Its seed pods get distributed by the wind and by people dumping raked-up piles of yard trash in inappropriate places. Once established, it is difficult to eliminate given the long-lived seed bank that persists for decades in the soil.

Indigo Bush (False Indigo)—*Amorpha fruticosa*

Leguminosae; Fabaceae

Indigo bush is a slender-stemmed shrub with deciduous, pinnately compound leaves that are about 1 foot long. The attractive 6-inch-long flower spikes stick out above the foliage in the late spring or early summer, bearing conical clusters of closely packed dark purple flowers with bright yellow anthers. Indigo bush is uncommon but can be locally abundant in the right situation. It is often multistemmed and gets from 3 to 12 feet tall. Indigo bush occurs throughout most of the eastern and central United States and throughout most of Florida (Kartesz 2015).

Indigo bush is usually found in association with water. It occurs in freshwater marshes, on the edges of salt marshes, and along the edges of ponds, lakes, rivers, and swamps. However, it can also grow in upland situations. It is only slightly shade tolerant, doing best in full sun.

The flowers of indigo bush are quite attractive to bees and butterflies, and it is a larval host plant for three species of butterflies and one moth: the silver-spotted skipper, the southern dogface, the gray hairstreak, and the black-spotted prominent moth (*Dasylophia anguina*). Bees that obtain nectar and pollen from indigo bush in Illinois also occur in Florida and so probably visit our Florida indigo bushes as well: halicid bees (*Lasioglossum* spp.), masked bees (*Hylaeus* spp.), andrenid bees (*Andrena* spp.), little carpenter bees (*Ceratina* spp.), and cuckoo bees (*Nomada* spp. and *Coelioxys* spp.) (Hilty 2018, "False Indigo").

Indigo bush has been planted both in this country and abroad for its beauty, for benefiting bees and other pollinators, and for waterway bank stabilization. It has become an invasive exotic problem in some parts of the world but is quite safe here in the southeastern United States where it is native. It is a great substitute for Chinese butterfly bush (*Buddleia davidii*) (Christman 2017), a commonly planted exotic and often short-lived "butterfly plant." As an ornamental, indigo bush is adaptable, doing well in most situations where it gets plenty of sun.

Eastern Redbud—*Cercis canadensis*

Leguminosae; Fabaceae

Eastern redbud is a small tree with 3-inch by 3-inch, smooth-edged, heart-shaped, deciduous leaves that are medium green on top and pale green underneath. The dark-colored twigs have a zigzag growth pattern. Showy, bright pink flowers are produced along the twigs and stems in early spring before most trees have leafed out. Bean pods produced later in the year are green at first but soon turn brown and remain on the tree into winter. Eastern redbud grows naturally in eastern North America from southern Iowa, southern Michigan, and southern Pennsylvania south to the Gulf of Mexico and into central peninsular Florida and Mexico (Dickerson 1990), but it has been planted well beyond its native range throughout most of the United States. Redbud is normally 20 to 40 feet tall at maturity with a 6-inch to 1-foot trunk diameter. The currently listed Florida champion is 84 feet tall with a nearly 1-foot trunk diameter (Florida Champion Tree Database 2020). National champion eastern redbuds often have trunk diameters of 3 or more feet, and I have seen redbuds in Florida with 2-foot trunk diameters.

In nature, eastern redbud is an occasional understory tree in and on the edges of upland hardwood forests. It is short lived and somewhat weedy but reacts well to disturbances. It does not do well on acidic soils, wet soils, or in places that flood. It is moderately shade tolerant, especially when young, but does best in full sun.

The flowers are visited by honey bees and a variety of bees native to Florida, including the blueberry digger bee (Cane and Payne 1988), bumblebees, carpenter bees, halicid bees, mason bees, and others. These pollinators benefit from eastern redbud flowers in part because they are produced very early in the year—mid-February here in northern Florida—before most other plants bloom. The bean pods are browsed occasionally by white-tailed deer, and the seeds are eaten occasionally by gray squirrels, northern cardinals, and bobwhite quail (Dickerson 1990).

Eastern redbud is an ornamental favorite in the eastern United States, including Florida, primarily because of its display of pink flowers in early spring. It is readily available from plant nurseries, and a number of horticultural varieties have been developed. The redbuds from the northern and southern parts of the species' range differ somewhat in their growing requirements, so it is best to obtain planting stock derived from the south end of the native range to ensure that the tree is well adapted to Florida's climate and soils. Redbud grows rapidly, is trouble free when young, and needs no watering or other care once established. Its short life span may be a drawback as it only survives perhaps 20 or 30 years on average.

Water Locust—*Gleditsia aquatica*

Leguminosae; Fabaceae

Water locust in an uncommon tree, primarily occurring on river floodplains from the central peninsula of Florida westward through the Florida Panhandle into the Mississippi River floodplain where it occurs from Louisiana into southern Illinois and northward from Florida into the coastal plain of South Carolina (Kartesz 2015). It is a small-sized, small-fruited, wetland version of the honey locust. It has a light, open crown of both pinnately and twice pinnately compound deciduous leaves and usually has branched thorns on the trunk. The most distinctive difference between water locust and honey locust is the size of the fruit. Honey locust produces pods up to a foot long containing many seeds whereas water locust produces roundish pods that are 1 or 2 inches long and contain from one to three seeds. The maximum size of water locust is represented by the current national champion, which is 82 feet tall with a trunk diameter of about 2½ feet and a crown spread of 48 feet (American Forests 2018).

Almost always found on the floodplains of rivers or streams, water locust is moderately flood tolerant. In Florida, the floodplains supporting water locust generally occur either over limestone or along rivers that receive significant water inflow from springs. Water locust is not at all shade or fire tolerant, but it does have strong wood and a strong root system, making it resistant to wind damage. Common associates include bald-cypress, overcup oak, water hickory, water-elm, Florida elm, red maple, and swamp laurel oak.

The wildlife value of water locust is not well documented. Presumably, its seed pods and/or seeds can be eaten by a wide range of wildlife species such as white-tailed deer, wild hogs, gray squirrels, and wild turkeys. The small, yellowish flowers are attractive to pollinators, mostly small native bees. Water locust and honey locust foliage provide the larval food for the little underwing moth (Jeff Slotten, little underwing moth researcher, personal communication).

Water locust has strong, rot-resistant wood, but it is not common enough to be important as a wood producer. It is not used as an ornamental plant.

Honey Locust—*Gleditsia triacanthos*

Leguminosae; Fabaceae

Honey locust is a medium- to large-sized tree occurring sparsely in hardwood forests along river floodplains in the Florida Panhandle from Jefferson and Taylor Counties westward (Godfrey 1988). Its large bean pods, about 2 inches across and up to 1 foot or more long, readily distinguish mature specimens of this tree from water locust, which has much smaller pods and is much more common in Florida. The very open crown of pinnately and twice pinnately compound leaves and the large, branched thorns on the trunk make honey locust and water locust quite distinct from other native Florida trees.

In Florida, honey locust occurs naturally on soils overlying limestone outcrops, often within the upper floodplains of rivers but also scattered about in upland hardwood forests, often on the edges, and along fencerows and in old fields. It is unusual in Florida, occurring as a common tree primarily in the states adjacent to the Mississippi and Ohio Rivers (Blair 1990). It grows rapidly, is deeply rooted, very drought resistant, and was one of the trees often planted in windbreaks in the eastern Great Plains (Blair 1990). It is not at all shade or fire tolerant. Honey locust is a much larger tree than water locust: the largest one reported in the United States is 121 feet tall with a trunk diameter slightly over 6 feet and an average crown spread of 115 feet (American Forests 2015).

The wildlife value of honey locust is moderately high. The sweet, nutritious pods are eaten by cattle, hogs, squirrels, deer, opossum, bobwhite quail, crows, and starlings, while the leaves and young stems are eaten by deer and various kinds of insects (Blair 1990). The flowers are attractive to insects including honey bees (Blair 1990). Two of the more common of the many insects that feed on honey locust foliage are the invasive exotic mimosa webworm and the larvae of the native little underwing moth.

The wood of honey locust is heavy and strong, and the heartwood is rot resistant. It is not abundant enough to be used for anything other than firewood here in Florida, but elsewhere is has been used to make fence posts.

In the eastern and central parts of the United States, honey locust is often used as an ornamental and is especially well suited for planting on calcareous soils in open areas. Thornless varieties are available from commercial nurseries.

Kudzu—*Pueraria montana* var. *lobata* (*P. lobata*)

Leguminosae; Fabaceae

Kudzu is a high-climbing, rapidly growing, perennial herbaceous vine with compound leaves having three silky-haired leaflets 4 to 6 inches long and 3 to 4 inches wide. It produces half-inch-long, sweetly fragrant purple flowers in hanging clusters in late summer and hairy bean pods containing two or three hard, woody beans in the fall. It commonly freezes to the ground each winter, only to resprout vigorously the next spring with green, herbaceous vines, the most vigorous of which attain somewhat woody stems by autumn. Kudzu is famous for covering entire landscapes, including fences, power poles, abandoned buildings, and tall trees with a smothering green carpet of foliage. It has been called "the vine that ate the south" and is especially common in Georgia, Alabama, and Mississippi but is also found in northern Florida. It is an aggressive invasive exotic plant intentionally introduced from Japan in 1876 and widely promoted for use as fodder for cattle and goats and for erosion control in the early twentieth century.

Kudzu is not particularly shade, fire, or flood tolerant, but it is very aggressive on poor or depleted upland soils where its ability to fix nitrogen and grow rapidly give it a competitive advantage over other plants. Its primary means of spreading is to grow along the ground and take root at its nodes, thereby establishing new plants. It can also spread by seed, however. It currently occupies several million acres of land in the southeastern United States and continues to spread at astonishing rates (Wikipedia 2020, "Kudzu: Ecological Damage").

There are several ways of combatting kudzu. If the kudzu-infested area can be or is fenced, grazing goats or cattle on the plant is an effective and economical way of controlling it. A workable kudzu management plan might involve, for example, intense rotational grazing for several years followed by spot treatment with glyphosate and establishment of alternative vegetative cover. A good cover would be pine trees, under which, after the trees are large enough, grazing could be reintroduced in an ongoing management cycle.

There is one insect and one fungal disease that have arrived in the southeastern United States recently that may reduce the aggressiveness of kudzu. The insect is the kudzu bug *Megacopta cribraria*, which was discovered in northeast Georgia in 2009 and spread to Florida and other nearby states by 2012 (Wikipedia 2020, "*Megacopta cribraria*"). It feeds preferentially on kudzu and reduces its growth rate, but it also feeds on soybeans and other plants (Zhang et al. 2012). The fungus is the Asian soybean rust fungus *Phakopsora pachyrhizi*, which attacks the leaves of soybean and kudzu (Sikora 2014). It is able to overwinter on kudzu in the southern United States (Sikora 2014), which is problematic for soybean growers but might provide additional incentive for eliminating kudzu infestations.

The wildlife implications of kudzu are mostly negative as kudzu infestations eliminate large areas of diverse habitat for native wildlife. Deer and rabbits eat it, and it provides cover for various animals, but overall kudzu has a deleterious impact on wildlife.

Kudzu has been used by humans for erosion control and as livestock fodder. In the latter case, kudzu has little nutritional value, and by the 1920s ranchers had discovered that their cattle were getting very thin on it (Alderman and Alderman 2001). In attempts to find uses for a nuisance plant so prolific, humans have turned to making novelty foodstuffs from it, including jellies, jams, pasta, candy, and even kudzu chips. The vines can also be used to produce handsome baskets and wreaths (Alderman and Alderman 2001). Even so, the ecological damage posed by kudzu far outweighs any positive uses.

Rattlebox (Red Sesbania)—*Sesbania punicea*

Leguminosae; Fabaceae

Rattlebox is a shrub or small tree in the legume family that grows up to 15 feet tall. It has pinnately compound, deciduous leaves with seven to sixteen pairs of 1-inch-long leaflets. In spring or early summer, it bears hanging clusters of showy, bright red-orange, 1- to 1½-inch-long flowers and then produces dark brown bean pods 3 to 4 inches long. Four ridges (wings) run the length of each bean pod, which contains three to nine beans. Rattlebox, an invasive exotic native to South America, has become a common plant in Florida and most of the rest of the coastal plain of the southeastern United States, especially near the coast on the edges of disturbed wetlands. It is also a recognized invasive exotic pest in South Africa and Australia.

Rattlebox is a fast-growing plant that begins flowering and seed production at two years of age and lives up to 15 years (Hoffmann and Moran 1989). It spreads by seed, which can remain dormant in the soil for several years and can be distributed by flowing water. It appears most frequently on the edges of wetlands in open areas with abundant sunlight. Unfortunately, rattlebox has been spread around the world as an ornamental.

In an effort to gain control of this very invasive species, South Africa has imported and released three weevils: the flower- and leaf-eating opionid *Thichapion lativenter*, the seed-eating curculionid *Rhyssomatus marginatus*, and the stem- and trunk-boring curculionid *Neodiplogrammus quadrivittatus*. The combined attack of these three beetles has achieved effective control of rattlebox in South Africa (Global Invasive Species Database 2020, "*Sesbania punicea.*"). These weevils have not yet been introduced into the United States.

The wildlife value of this plant is largely negative. It is very toxic, thus providing little for native wildlife to eat. Indeed, it may kill some animals that try eating a few of the beans. Moreover, by competing with native vegetation, it can seriously degrade habitats. The flowers provide some benefit for insect pollinators, and the green fruit is eaten by the green stinkbug *Chinavia hilaris*, which, regrettably, can be a pest on cotton and other crops (Tillman 2015).

Rattlebox is sometimes planted as an ornamental in Florida, and both the plants and seeds are available for sale on the Internet. It should never be planted.

American Wisteria—*Wisteria frutescens*

Leguminosae; Fabaceae

American wisteria is a small, native, uncommon, woody vine in the legume family that is a diminutive relative of the commonly planted Asian species of wisteria. It has deciduous, pinnately compound leaves with nine to fifteen ovate to lanceolate leaflets, clusters of showy, typically bluish-purple flowers in May and June (some blooms may be partly or wholly white), and clusters of narrow bean pods in the fall, each containing three to four round beans. All wisteria beans are toxic. The vines, which climb other vegetation by twining around stems, are usually 10 to 15 feet long but can be up to 30 feet long with a basal stem diameter of perhaps half an inch (Alabama Herbarium Consortium 2018).

American wisteria occurs in scattered locations throughout most of the southeastern United States including the Florida Panhandle; it extends sparsely into the north end of the Florida peninsula as far south as Alachua and Levy Counties with an outlier in Orange County (Kartesz 2015). I have seen it growing along the Santa Fe River in Alachua County. American wisteria is found mostly on the sunny edge of the vegetation growing on the margins of rivers and streams or along the edges of ponds, lakes, swamps, or in wet thickets (Godfrey 1988). It appears to be somewhat flood and shade tolerant but does not abide drought and grows best with direct sunlight for at least part of the day.

American wisteria is not common enough to be of great importance for wildlife. Located in a good spot with plenty of sun, it flowers prolifically, thereby providing food for both honey bees and native bees. It is a larval host (as are many other legume species) to the silver-spotted skipper (Alabama Herbarium Consortium 2018).

American wisteria is sometimes grown as an ornamental with several named varieties. Unfortunately, it is not used nearly as often as the highly invasive Chinese wisteria. It is a much better choice for landscape plantings than Chinese or Japanese wisteria because it is not invasive and, therefore, does not spread aggressively, and it has denser flower clusters.

Chinese Wisteria—*Wisteria sinensis* and Japanese Wisteria—*Wisteria floribunda*

Leguminosae; Fabaceae

These two Asian species of wisteria are commonly planted as ornamentals, and both have escaped cultivation to become problem invasive exotics. Of the two, Chinese wisteria is much more often found in Florida. Both species are vigorous, fast-growing, long-lived, high-climbing vines with twining trunks, deciduous, pinnately compound leaves, and long, showy, hanging panicles of pea-type flowers that may be white, blue, or purple. Japanese wisteria's flower panicles are longer than those of Chinese wisteria, and the flowers open gradually, starting at the base; by contrast, Chinese wisteria's flowers open simultaneously (Wunderlin 1998). Both species produce long, narrow, hanging bean pods containing hard beans that are toxic if eaten. Chinese and Japanese wisteria can get quite large, climbing to the tops of tall trees with the constricting vine trunks becoming several inches in diameter.

These two Asian species of wisteria can spread along the ground, taking root along the way and thus, over time, spreading to cover a large area. They can then dominate the native vegetation to such an extent that the wildlife habitat value of the area is largely destroyed. They can also spread by seed, but, as the beans are toxic, wildlife species do not spread them about as much as they would if they were eating them.

The Asian wisteria species are able to grow on almost any upland soil. They prefer full sunlight but can spread widely, even under considerable shade. Once established, the vines are quite drought tolerant.

The wildlife value of these two species is mostly negative as they destroy native habitats. However, they *do* provide some wildlife benefits. The prolific blooms support both honey bees and native bees, and the silver-spotted skipper uses the foliage as one of its host plants.

The main human use of these vines is ornamental. The dense panicles of fragrant, showy flowers can be spectacular in the late spring. Unfortunately, even when they grow in places where they were intentionally planted, these vines tend to get out of control and start causing problems, which can include doing considerable structural damage to buildings as well as killing desirable trees and other plants. Once established, these vines are very difficult to control or eliminate.

Yellow Jessamine (Carolina Jessamine)—*Gelsemium sempervirens* and Swamp Jessamine—*Gelsemium rankinii*

Loganiaceae; Gelsemiaceae

These two very similar species of medium-sized evergreen vines produce bright yellow flowers about 1 inch wide in early spring. They have lanceolate to ovate-lanceolate, opposite leaves that are 2 to 3 inches long. The fruit is a two-valved capsule containing small, flattened seeds (Godfrey 1988). The vine stem has smooth, dark gray bark and is rarely more than a half inch in diameter. The vines climb by twining around the stems of other vegetation.

Yellow jessamine is both the more common and the more widely distributed and broadly adapted of the two species. It occurs throughout the coastal plain and Piedmont areas of the southeastern United States from southeastern Virginia into eastern Texas, ranging south into the south-central Florida peninsula (Ornduff 1970). Swamp jessamine occurs primarily in wetlands in the Florida Panhandle, southern Georgia, and extreme southern Alabama, Mississippi, and southeastern Louisiana, with a few isolated locations in eastern North Carolina and South Carolina (Kartesz 2015). One of the earliest blooming woody plants in the wild locations where it grows, yellow jessamine begins blooming in January or February, producing sweetly fragrant flowers followed by winged seeds. Swamp jessamine blooms slightly later; it begins to flower just as yellow jessamine's flowering is finishing (Godfrey 1988). The blooms of swamp jessamine are not fragrant, and its seeds are not winged (Wunderlin 1998).

While swamp jessamine is restricted to wetlands such as bogs, acidic swamps, and floodplain forests, yellow jessamine occurs either in or on the edges of almost all upland forest types, especially pine flatwoods forests, pine plantations, and upland hardwood forests. It also occurs on the shrubby edges of wetlands. In addition to crawling into and on top of fences, shrubs, and small trees, yellow jessamine will crawl along the ground, taking root and spreading as it goes and forming large clonal patches. It does not climb as high as grape vines do, rarely getting higher than 20 or 30 feet with vine stems seldom more than a half inch in diameter.

All parts of these vines contain alkaloid chemicals that are highly toxic. This toxicity extends to jessamine nectar as well as to the honey produced from it by bees; butterflies and honey bees are sometimes killed when they obtain too much nectar from jessamine flowers (Ornduff 1970). The toxic nature of yellow jessamine is aptly illustrated by the deaths of 200 goats owned by a friend of mine in Columbia County, Florida. The goats ingested the plant after they were fenced into a 40-acre pasture containing a large patch of it, and they were fatally poisoned (Frank Sedmera, personal communication). Humans have been killed by ingesting medications derived from yellow jessamine (Wormley 1870).

In spite of their toxicity, the flowers are pollinated by bees, primarily large-bodied bees such as bumblebees and southeastern blueberry bees, but also by smaller bees such as metallic sweat bees (*Dialictus*) and small carpenter bees (*Ceratina*) (Ornduff 1970).

Extracts and solutions derived from *Gelsemium sempervirens* were used as medicines in the past (Wormley 1870), and herbal remedies derived from this plant are still used today and available for sale on the Internet in spite of the high toxicity of the alkaloids they contain.

Yellow jessamine is sometimes used as an ornamental, even finding applications as a ground cover. It is adaptable, easy to grow, has showy flowers as early as February, and is evergreen. Its toxicity is its main drawback. A child who has learned to suck the sweet nectar from honeysuckle flowers might be sickened or killed by sucking the nectar from yellow jessamine flowers.

Mistletoe (Oak Mistletoe)—*Phoradendron leucarpum* (*P. serotinum*)

Loranthaceae; Viscaceae

The kind of mistletoe we have in northern Florida—usually called "mistletoe" but sometimes "oak mistletoe," "American mistletoe," or "eastern mistletoe"—is an evergreen parasitic plant that grows on the branches of hardwood trees. It occurs throughout the southeastern United States including all of Florida except for the tropical southern end of the peninsula (Kartesz 2015). (There are other kinds of mistletoe out west, in the American tropics, and in Eurasia, some of which grow on conifers. The discussion here is confined to this one local species.) Our local mistletoe has thick, green, brittle branches and opposite, thick, roundish leaves 1 to 2 inches long. Male and female flowers occur on separate (dioecious) plants. Small yellowish-green flowers appear in the fall, after which the female plants produce small, round, white and somewhat translucent fruits that are ripe from December through early April. Mistletoe grows as a round-ish clump of dense green vegetation on the stem or branch of a tree. The clump may be up to 3 feet in diameter with the basal stem being up to 1 or 2 inches in diameter.

Mistletoe is an obligate parasite, meaning that it cannot survive without being attached to the living stem of another woody plant. It derives water and nutrients such as nitrogen, phosphorus, potassium, calcium, magnesium, and iron from its host. It also gets some carbohydrate energy from its host, although it supplies most of its own energy through photosynthesis within its own foliage. Because it needs to perform photosynthesis, it needs to grow in a location within the crown of the host tree that allows it to obtain at least a moderate amount of sunlight. It uses more water than would seem normal for the amount of leaf surface it has, and it transpires large amounts of water during droughts when the foliage of the host tree is restricting water use (Coder 2016).

Mistletoe grows on a wide variety of host tree species including maples, box-elder, sweetgum, pecan, other kinds of hickory trees, black walnut, elms, ashes, tupelos, black cherry, plum trees, pear trees, sugarberry, river birch, sycamore, basswood, chinaberry, dogwood, red-cedar, and the red oak group of oak trees, which includes water oak, laurel oak, Shumard oak, southern red oak, and turkey oak. It is found on rare occasions on oaks in the white oak group such as live oak, bluff oak, sand post oak, and swamp chestnut oak. Mistletoe's sticky seeds are dis-tributed by birds that eat the berries, and, once it starts growing, its tissue grows into the branch or trunk of its host, forming a xylem-to-xylem connection which enables it to draw water and nutrients from the host tree. By releasing chemical growth regulators into the host tree, the mistletoe causes the host branch to en-large to accommodate the parasite's needs. A single clump of mistletoe in a large tree is not much of a problem for the tree, but numerous clumps in a tree will weaken it. Trees with mistletoe infestations are much more likely to die during periods of drought than are noninfested trees (Coder 2016).

Birds provide the primary dispersal mechanism for mistletoe. Mistletoe fruit contains a tangle of minute strings within a pectin jelly called "viscin" (Coder 2016). This combination of strings and viscin is an effective glue for attaching the seed to the surface of a tree branch, whether it is wiped there by a bird's beak or foot, regurgitated there by the bird, or deposited there in bird droppings. Once stuck to the branch by the dried glue, the seed germinates and grows a radicle that forms a "hold-fast" structure on the branch surface from which mistletoe tissue can then penetrate into the branch to form a permanent attachment and begin extracting water and nutrients from the tree (Coder 2016).

The bird species that reportedly eat and disperse mistletoe seeds include red-bellied woodpecker, northern flicker, yellow-bellied sapsucker, pileated wood-pecker, eastern phoebe, Acadian flycatcher, American crow, eastern bluebird, American robin, thrushes, veery, northern mockingbird, gray catbird, brown thrasher, Carolina wren, Carolina chickadee, tufted titmouse, white-eyed vireo, ruby-crowned kinglet, yellow-rumped warbler, towhee, purple finch, and cedar waxwing (Coder 2016). In addition to the value of the fruit as a winter food for these bird species, mistletoe foliage is one of the larval food plants for the hickory horned-devil (royal walnut moth) and is the only larval food plant for the great purple hairstreak butterfly, both of which can eat a significant amount of mistletoe foliage (Coder 2016).

Because mistletoe damages the trees it infests, people often want to remove it to protect their trees, especially if they have great value like large pecan trees. This can best be done by removing the branch on which the mistletoe grows, making sure that 1 or 2 feet of branch below the mistletoe's attachment point is pruned away. It may be necessary to employ a tree surgeon to do this kind of work. If the mistletoe is attached to the main trunk, then breaking off the external part of the mistletoe is an option, although this action will probably have to be repeated again in the future. Out in the country where it is possible to do, I have used a rifle on a cold day in winter, when tree branches are brittle, to shoot-prune the mistletoe-infested branches from the upper crowns of large pecan trees. There are chemical sprays that can be sprayed on mistletoe foliage to address infestation problems, but they are not as effective as pruning away the mistletoe along with its host branch because mistletoe will usually sprout back at some point after being treated chemically (Coder 2016).

Mistletoe is commonly used as a holiday decoration at Christmastime. All parts of mistletoe are toxic if consumed.

Water-Willow (Swamp Loosestrife)—*Decodon verticillatus*

Lythraceae

Water-willow is a clonal, thicket-forming shrub with slender, arching stems that bend to the ground and take root, producing new plants and gradually expanding the thicket. It is not a true willow. Its opposite or whorled deciduous leaves are 3 to 5 inches long and half an inch to 1 inch wide, and it produces clusters of small pinkish-purple flowers in the leaf axils in September, followed in autumn and winter by brown capsules containing reddish seeds. The dense, nearly impenetrable thicket it often forms is usually about 6 feet tall. The bark on the stems below water is spongy, while the bark above water exfoliates in long, thin, cinnamon-colored strips (Godfrey 1988).

Water-willow occurs in wetlands here and there throughout the eastern half of the United States, ranging north into southern Canada (Godfrey 1988) and as far south as Highlands County on the Florida peninsula (Wunderlin et al. 2020). It is always found in water or places that are often flooded including marshes, pond and lake edges, stream sides, and swamps (Godfrey 1988). Not being especially shade tolerant, it is uncommon in swamps with complete tree canopies but occurs in openings and on the edges. Water-willow is not a particularly common species in Florida or in most other states, being found in widely scattered locations.

Water-willow is strongly clonal, spreading largely by the rooting of stem tips and only occasionally reproducing by seed. This is especially true in the northernmost part of its range where some clones have been in place so long that they have accumulated genetic mutations that interfere with sexual reproduction (Eckert et al. 1999). There are many species of plants that spread out to form clonal patches, such as saw palmetto, winged sumac, persimmon, maidencane, and shiny blueberry, but almost all of these reproduce effectively by seed as well. Water-willow seems to rarely reproduce by seed, and, as a result, it has a widely scattered, irregular distribution.

The wildlife value of water-willow is high. It produces seeds that are eaten by ducks, including mallard, black duck, blue-winged teal, green-winged teal, and wood duck; the flowers are cross-pollinated by honey bees and bumblebees and are very attractive to swallow-tailed butterflies (a phenomenon I have witnessed many times at a thicket of this plant in southern Suwannee County); and there are several kinds of insects that feed on the leaves and stems (Hilty 2018, "Swamp Loosestrife"). Its thickets also provide escape cover and nesting structure for various wildlife species.

Water-willow is seldom used as an ornamental, partially as a result of its wetland requirements and partially because of its propensity to spread aggressively. However, it can be suitable for wetland restoration efforts.

Crape Myrtle (Crepe Myrtle)—*Lagerstroemia indica*

Lythraceae

Crape myrtle is one of the most frequently planted ornamentals in northern Florida and the rest of the southeastern United States. Many varieties have been produced by the nursery industry, some of which are products of hybridization with Japanese crape myrtle, *Lagerstroemia fauriei* (Christman [1997] 2008). Crape myrtle has smooth-margined, roundish, deciduous leaves that vary in length from 1 to 2 inches and typically lack a petiole (leaf stem) (Godfrey 1988). The crinkly looking blooms come in many colors from white to pink to red to purple, depending on the variety, and are produced in the summer for a rather long period, ranging from 60 to 120 days (Christman [1997] 2008). The length of the flowering period is influenced by many factors such as plant variety, pruning time and severity, plant age and, especially, how much sunlight the plant receives.

Crape myrtle is capable of reproducing from seed and by root sprouts, and it occasionally spreads to places where it was not planted, but the vast majority of existing plants have been intentionally planted. It is so popular and easy to obtain and plant that it is often overused.

Crape myrtle is adaptable to many soil and moisture situations and is fairly drought tolerant. It is not shade tolerant, clearly doing best in full sun and less well in partial shade. It is not bothered by many insect pests, except sometimes by aphids. Powdery mildew can be problematic, especially where aphids have been feeding.

The main wildlife value of crape myrtle is largely reserved for the honey bees that visit its flowers. Overall, it may have a negative wildlife value in situations where it is so overused that it takes up space where more valuable wildlife-supporting plants could grow.

The main human benefit of crape myrtle is its ornamental value, which is high. It is adaptable for use as a street and parking lot tree or shrub, and it is often used as a specimen plant in landscaped yards and gardens. It can grow into an attractive small tree with interesting bark but is often pruned back to create a large shrub. If a shrub is desired, there are many varieties that remain shrub sized without the need for excessive pruning. Extreme pruning has become such a traditional approach to this plant, that beautifully shaped, healthy, vigorously blooming specimen crape myrtles are often severely damaged by being hacked back to a few large trunks and branches. Once this is done, the chance of regaining the plant's beautiful natural structure is lost forever. Most crape myrtle experts recommend pruning only the ends of twigs once they stop blooming in mid- to late summer to prevent the maturation of the seed capsules, thus stimulating a second flush of blooms. If pruning is desired to shape a plant or to keep it to a certain size, then regular pruning of only the small branches on the outer parts of the plant is recommended.

Tulip Tree (Yellow-Poplar)—*Liriodendron tulipifera*

Magnoliaceae

Tulip tree, often called "yellow-poplar," is a tall, large tree that is common in much of the eastern United States, mainly east of the Mississippi River from Vermont westward into Michigan and south into Louisiana and northern Florida (Beck

1990). In Florida, it occurs naturally in two distinct populations, one in the Florida Panhandle from Leon County westward and the other in the St. Mary's and St. Johns River drainage basins from the Georgia border south into Orange County (Wunderlin et al. 2020). It is also common elsewhere as a planted ornamental tree. Tulip tree does especially well in the Appalachian Mountain region where it attains its largest size, sometimes growing nearly 200 feet tall with maximum trunk diameters between 8 and 12 feet in old growth stands in fertile coves (Beck 1990). More frequently, it will be 100 to 150 feet tall with trunk diameters of 2 to 5 feet (Beck 1990). Maximum age seems to be about 300 years (Beck 1990). A great place to see old growth tulip trees is the Joyce Kilmer Memorial Forest within the Nantahala National Forest in western North Carolina where there are walking trails through a magnificent virgin hardwood forest. Florida tulip trees do not get as large as the ones farther north and are usually 100 feet tall with 2- to 3-foot trunk diameters at maturity. The largest tulip tree recorded in the state of Florida as of 2018 is 110 feet tall with a trunk diameter slightly over 4 feet (Florida Champion Trees Database 2020).

Tulip tree is not easily confused with any other tree. This member of the magnolia family produces 1½-inch- to 2-inch-wide distinctive flowers in late spring (as early as mid-April in Gainesville, Florida, but in late May and early June in the southern Appalachian Mountains), unusual seed cones (sometimes called "follicles," "fruit cones," or "cones") of winged seeds (technically fruits) in autumn, and deciduous leaves that have a unique, boxy shape. These leaves typically have two lobes (sometimes three) per side with no lobe at the end, so that the end of the leaf appears to be missing. Leaf size is variable. Up to 6 inches wide, the leaves are as wide, or wider, than long (Godfrey 1988). The leaves of the tulip tree population on the Florida peninsula and northward on the Atlantic coastal plain into North Carolina often have lobes that are rounded at the ends, whereas the lobes on tulip tree leaves elsewhere are usually pointed at the ends (Beck 1990).

In its native environment, tulip tree is a fast-growing tree with a very straight trunk and a clean-looking appearance. In Florida, it grows in stream-side seepage areas and on slopes adjacent to streams. It is more widely adapted farther north but usually grows where there is reliable moisture in the soil. Tulip tree is neither especially drought resistant nor shade tolerant, and, when young, it is easily damaged by ground fires, cattle grazing, and browsing by deer and rabbits. In middle age, tulip tree is susceptible to severe damage by ice storms. Mature tulip trees have thick bark, which renders them quite fire tolerant (Beck 1990), and thick enough stems and branches to withstand ice storms.

The wildlife value of tulip tree is moderate to good. The flowers produce a lot of nectar and are important for bees, including honey bees (Beck 1990), and the fruits are sometimes eaten by purple finches, cardinals, and squirrels (Martin et al. 1951). The freshly fallen flower petals are a favorite food of white-tailed deer, dominating their diets in the southern Appalachian Region in late May and early June (Johnson et al. 1995). Seedlings and saplings are browsed by white-tailed deer and cottontail rabbits (Beck 1990). Tulip tree is a host plant for the tulip tree silk-moth, the promethean moth, the root collar borer moth, and the tulip tree beauty moth (Leckie and Beadle 2018). It is also a host plant, along with sweetbay and black cherry, for the caterpillars of the eastern tiger swallowtail butterfly (Glassberg et al. 2000).

Tulip tree is an important timber tree in the Appalachian Region, producing high quality, clear, large-dimension construction lumber, as well as smaller dimension timber useful for the making of paper, matches, furniture parts, moldings, and high-end cabinetry. The wood is also excellent for carving and whittling. Tulip tree is an important shade and ornamental tree in the landscape business throughout most of the eastern United States. However, tulip tree is not as well adapted to Florida's growing conditions as it is to those farther north and is therefore not the most suitable tree choice for landscape use in our state. It does not do well on poor, dry, sandy soils but can sometimes perform reasonably well on better soils, especially if there is supplemental irrigation or a nearby permanent water source.

Southern Magnolia—*Magnolia grandiflora*

Magnoliaceae

Southern magnolia, often referred to simply as "magnolia," is a characteristic part of the flora of the southeastern coastal plain from South Carolina southward into the central Florida peninsula and Florida Panhandle and on into eastern Texas with scattered isolated populations in North Carolina (Outcalt 1990). Its large, thick, shiny, evergreen leaves, which often, but not always, have a rusty to golden fuzz underneath, give it a distinctive appearance. The large, fragrant, white flowers that bloom from mid- to late April through May and into early June are also distinctive and showy and produce a sweet aroma that is sometimes noticeable throughout the landscapes where this tree is common. The fruit cones with their red seeds ripen in late August and September.

Southern magnolia is a slow-growing tree that is often observed as a small- to medium-sized tree in landscaped situations, but it can get quite large if allowed to survive long enough. The largest individuals recorded have trunk diameters between 4 and 7 feet and heights between 80 and 100 feet. The ages of these giants are often not recorded, but two large trees whose rings I counted were each approximately 400 years old. One was a healthy tree from Tallahassee that was removed to make way for development. The other tree, located on the north rim of Payne's Prairie, was healthy until it nearly died from a massive dose of herbicide applied to the adjacent railroad bed, after which it was severely damaged by a lightning strike. Both these magnolia trees might have lived many more years if not for the impacts of human action (and the lightning strike).

Southern magnolia grows naturally in climax hardwood forests ranging from quite dry to moist (xeric to mesic) hammocks. It is usually somewhat shorter than the other canopy trees and usually has a denser, more compact crown. It seems to grow most frequently and best on slopes where there is some sub-surface seepage or near the edge of streams, ponds, swamps, or bayheads. It is well adapted to both sandy soil and clay soil. Its most common associates include sweetgum, laurel oak, pignut hickory, spruce pine, white ash, winged elm, Florida maple, hophornbeam, hornbeam, and, from northwestern Alachua County northwestward through the panhandle and beyond, American beech. The beech and magnolia forest type is considered *the* climax forest type on fertile uplands in the areas where mesic hardwood forests grow northwest of the Florida peninsula.

Southern magnolia is quite shade tolerant although it grows best in full sunlight. It is also one of the most wind-firm of Florida's trees. This is probably largely because of its extensive and deep root system that usually includes numerous sinker roots (Outcalt 1990). However, its compact crown and relatively short stature also help it survive strong winds. Southern magnolia is not particularly tolerant of flooding, fire, or drought (Outcalt 1990). It withstands salt near the coast better than most hardwoods but not as well as live oak. Several scale insects, stem borers, and bud borers infest magnolia to varying degrees, sometimes causing minor damage. Vines, such as wild grape, supplejack, pepper vine, and Virginia creeper often compete with the crown for sunlight, and southern magnolia suffers more damage from this competition than do most kinds of trees because of its slow growth and compact crown. Even so, these vines rarely kill the tree, and some of the other vine species that climb magnolias, such as poison ivy, trumpet creeper, and crossvine, circumvent causing damage by staying below the top of the crown. Magnolia readily supports epiphytes such as greenfly orchid, resurrection fern, and needle-leaf airplant without suffering any damage.

The wildlife value of southern magnolia is good. The seeds are eaten by gray squirrel, opossum, bobwhite quail, and wild turkey (Halls 1977). Red-eyed, yellow-throated, and white-eyed vireos and thrushes such as Swainson's thrush, hermit thrush, and veery are fond of the red, flesh-covered seeds, which are an important part of their diet when they pass through Florida during fall migration. Indeed, the easiest way to see these vireos and thrushes during migration in Florida is to locate a large southern magnolia with ripe seeds hanging outside of the fruit cones in the tree crown. Woodpeckers, cardinals, towhees, and other birds also eat the seeds to some extent. Old southern magnolia trees can maintain hollow trunks for a very long time, making such individuals highly valuable as den trees. The dense foliage of large southern magnolias provides daytime roosting habitat for barred owls. Southern magnolia is not a preferred browse plant of white-tailed deer, but cattle eat the foliage and stems of magnolia seedlings, often eliminating all magnolia reproduction from forests open to cattle grazing.

The wood of southern magnolia is smooth grained and moderately hard and strong. The sapwood is light tan, and the heartwood varies from brown to dark green to black. The wood is sawn into lumber for making furniture, paneling, and other items. It is also used to make plywood and packing crates while its small-dimension logs are used for making pulpwood. Because of its smooth grain and variable colors, magnolia wood is a favored choice for making turned objects such as bowls.

Among the native landscape trees of the southern Atlantic and Gulf coast states, Southern magnolia is one of the traditional favorites. In the past, when people lived in one place for a long time and life proceeded at a gentle pace, magnolia was a beloved, often "first choice" tree to plant. There was time to watch it grow. Now, faster-growing trees are often desired. Nonetheless, magnolia remains an excellent addition to many landscapes because of its beauty and its ability to bloom while still small. It grows in shady places to which most other trees would not adapt. Southern magnolia also does well in cities, being more tolerant of air pollution than most other trees, and its resistance to wind-throw and wind damage makes it relatively safe to plant near houses and other buildings. A number of horticultural varieties have been developed and are now being used in the landscape business, and the planting of rather large nursery-grown trees is overcoming the problem of southern magnolia's slow growth rate.

Sweetbay (Sweetbay Magnolia)—*Magnolia virginiana*

Magnoliaceae

Sweetbay is a wetland tree occurring primarily on the Atlantic and Gulf coastal plains from Long Island and New Jersey southward into the southern Florida peninsula then westward through the Florida Panhandle and the Gulf coastal states into eastern Texas (Priester 1990). The variety *australis* is evergreen to semievergreen in Florida, but farther north, for instance in New Jersey and Pennsylvania, the variety *virginiana* is deciduous. The leaves in Florida fall before the new leaves appear, leaving the tree bare for a couple of weeks. Sweetbay's most distinctive feature is the silvery-white coating of hairs on the undersides of the 4- to 6-inch-long and 1- to 2-inch-wide light green leaves. In the spring and early summer, when the leaves are new and shiny silver underneath, they put on a show on sunny, windy days when the wind exposes the silvery undersides. The bark is smooth and light gray. The flowers and fruits are like smaller versions of southern magnolia flowers and fruits.

The mature size of sweetbay is quite variable. In the northern part of its range, the variety *virginiana* is a shrub or small tree, but in Florida the variety *australis* often grows to be a large tree, sometimes up to 95 feet tall with 3 feet in trunk diameter (Godfrey 1988). The largest one currently reported, in Jefferson County, Florida, is 115 feet tall with a trunk diameter slightly over 3 feet (American Forests 2018). I have seen similarly sized sweetbays along Hogtown Creek in Gainesville and in the central bayhead on Hughes Island in the Ocala National Forest.

Sweetbay is a wetland tree, growing in swamps, marshes, and, most commonly, in bayheads (baygalls) and other seepage wetland forests. It thrives in the seepage areas along streams and does best in wetlands that neither flood deeply nor for extended periods and that never dry out completely. Its common tree associates include swamp tupelo, swamp red bay, loblolly bay, red maple, swamp laurel oak, water oak, sweetgum, bald-cypress, pond-cypress, slash pine, and pond pine. It also occurs in association with Atlantic white cedar. Understory associates include gallberry, large gallberry, dahoon, myrtle dahoon, fetterbush, sweet pepperbush, Virginia-willow, coast leucothoe, red chokeberry, and various ferns such as Virginia chain fern, netted chain fern, cinnamon fern, and royal fern.

Neither shade nor salt tolerant, sweetbay is not at all amenable to drought and is only mildly flood tolerant. It stands acidic but not alkaline soils. It will invade pine forests growing on pine flatwoods soils but is prevented from doing so by fire where fires still occur on a frequent schedule. Sweetbay sprouts vigorously from its base but does not spread by root suckers. It rarely, if ever, forms pure stands.

The wildlife value of sweetbay is good. It is a preferred browse plant for both deer and cattle, the flowers are attractive to bees, and the flesh-covered red seeds are eaten by squirrels, wild turkeys, bobwhite quail, and songbirds (Priester 1990). As with southern magnolia, thrushes and red-eyed vireos are especially fond of the seeds during their fall migrations. Sweetbay is the host plant for caterpillars of the sweetbay silkmoth (Leckie and Beadle 2018) and one of the host plants, along with black cherry and tulip tree, of the eastern tiger swallowtail butterfly (Glassberg et al. 2000).

The soft, aromatic, straight-grained wood of sweetbay is easily worked, finishes well, and is used to make veneer, boxes, and containers (Priester 1990).

Sweetbay is called "sweetbay magnolia" in the nursery trade and is sold to a limited extent as an ornamental. In situations where it will grow well, it makes an attractive tree. It is not adaptable to upland landscape sites, however, and usually does very poorly when planted outside its comfort zone, which is acidic seepage areas or the edges of streams and wetlands.

Deciduous Magnolias: Pyramid Magnolia—*Magnolia pyramidata*; Umbrella Magnolia—*M. tripetala*; Ash Magnolia—*M. ashei* (*M. macrophylla* var. *ashei*); Cucumber Tree—*M. acuminata*

Magnoliaceae

Deciduous magnolias occur sparsely in southern upland hardwood forests in scattered locations, with only very limited populations occurring in the central and western panhandle of Florida.

Deciduous magnolias are uniquely beautiful trees that are sometimes grown in native plant nurseries and sold as ornamentals. All magnolias produce a cone-like fruit (often called a "follicle," a "fruit cone," or a "cone") that contains red-orange, flesh-covered seeds that hang when ripe on the outside of the fruit cone by a thread.

The wildlife value of these trees is moderate to good, with the flowers attracting bees and the seeds providing food for squirrels, wild turkeys, and migrating songbirds such as thrushes and red-eyed vireos.

Pyramid Magnolia—*Magnolia pyramidata*

Pyramid magnolia is a rare tree that occurs in scattered locations in South Carolina, Georgia, the central and western panhandle of Florida, central and southern Alabama, southern Mississippi, and near the border between Louisiana and Texas (Kartesz 2015). The largest population seems to be one on private land in eastern Texas with about 1,000 trees (Creech 2017). The largest recorded pyramid magnolia in Florida, in Gadsden County, is 102 feet tall with a trunk diameter of nearly 2 feet (American Forests 2018). One with an even larger trunk diameter grows in eastern Texas (Creech 2017). Pyramid magnolia is listed as endangered by the state of Florida (Wood 1994).

Pyramid magnolia has deciduous leaves about 9 inches long and 4 inches wide that have earlobe-like lobes at the base and are widest near the tip end. It produces large, white flowers and a fruit cone that is rose red when ripe. It has smooth gray bark and tends to grow with a single upright trunk and narrow crown.

Pyramid magnolia prefers acidic, sandy, fertile soil and is always found within southern upland hardwood forests, often on slopes that are part of a stream valley. It is shade tolerant but will sometimes grow into the upper canopy of the forest, where it obtains direct sunlight.

Umbrella Magnolia—*Magnolia tripetala*

Umbrella magnolia has large, deciduous leaves, 10 to 24 inches long and 6 to 10 inches wide, clustered at the ends of its twigs. These leaves are not lobed at the bottom ends as are the leaves of pyramid magnolia and Fraser magnolia (a species found on the lower slopes of the Appalachian Mountains). The buds at the stem tips are purple; the flowers are white with rather narrow petals; and the fruit cones turn red with red-orange seeds eventually dangling from them in the fall. Umbrella magnolia occurs primarily in West Virginia, Virginia, the Carolinas, eastern Kentucky, eastern Tennessee, and central Alabama, with an isolated population in the Ozarks of western Arkansas and with small, isolated populations in southern Alabama and the Florida Panhandle (Kartesz 2015).

Umbrella magnolia is often multitrunked with very smooth, light gray bark. It grows in the understory shade of hardwood forests, usually along streams in hilly or mountainous terrain. The largest one reported is 70 feet tall with multiple trunks, each about 18 inches in diameter at breast height but that combine at the base to be about 3½ feet in diameter (American Forests 2018).

Ash Magnolia—*Magnolia ashei* (*M. macrophylla* var. *ashei*)

Ash magnolia is a rare plant closely related to bigleaf magnolia (*Magnolia macrophylla* var. *macrophylla*). It does not get nearly as large or grow as rapidly or vigorously as bigleaf magnolia and is restricted to a few isolated populations in the central and western Florida Panhandle. There, it usually grows in the understory of hardwood forests on the sides of ravines. The largest one reported is 34 feet tall with a trunk diameter of about 8 inches (Florida Champion Trees Database 2020). Ash magnolia is listed as endangered in the state of Florida (Wood 1994). Its leaves are 12 to 24 inches long and 7 to 12 inches wide, and its flowers are up to 12 inches wide. The leaf and flower buds are silvery white to silvery green in color. Blooming in mid-April, the flowers are strikingly beautiful with their pure white petals, the three innermost ones having a spot of purple near the base. The 2-inch-long fruit cones are smaller than those of bigleaf magnolia and have red seeds that are very attractive to birds. In my yard in Gainesville, Florida, the seeds, which ripened in the first half of August, lasted no more than two days once exposed on the outsides of the cones because they were quickly eaten by mockingbirds, brown thrashers, cardinals, and other songbirds.

I planted two Ash magnolias in my yard in Gainesville in medium shade under a large live oak. They grew well for 30 years and flowered and fruited abundantly. They were moderately drought tolerant and needed no special care other than water during extreme droughts. However, when the large live oak died, leaving them exposed to full sunlight, the leaves would get sunburned each summer and the trees did poorly, declining in size and health each year.

Cucumber Tree—*Magnolia acuminata*

Cucumber tree is a medium-sized to large tree with deciduous leaves 6 to 10 inches long. It produces yellowish-green, tulip-like flowers 2 to 4 inches long and fruit cones that turn red when ripe. As with other magnolias, the fruit cones produce red-orange seeds that hang out of the cones on threads. Cucumber tree has a tall, straight trunk with rough, gray-brown, furrowed bark and can grow to large size in the southern Appalachians where it may be 100 feet tall with a trunk diameter of 2 to 3 feet. It occurs naturally from western New York south through Pennsylvania, eastern Ohio, and the Appalachians to scattered populations in Alabama, Mississippi, and Louisiana, with a population in the Ozarks and a tiny population in the west-central panhandle of Florida (Smith 1990, "*Magnolia acuminata L.*").

Cucumber tree is not shade tolerant, is fast growing, and is rather short lived, seldom getting more than 150 years old (Smith 1990, "*Magnolia acuminata L.*"). It grows in hardwood forests as a scattered tree on rich, moist, well-drained soils. Cucumber tree is the largest of the deciduous magnolias, with the largest one reported being 96 feet tall with a trunk diameter of nearly 8 feet (American Forests 2018).

Cucumber tree is the only deciduous magnolia commonly harvested for timber. Similar to but somewhat stronger and heavier than the wood of tulip-tree (yellow-poplar), cucumber tree wood is used for making furniture, plywood, crates, and pallets (Smith 1990, "*Magnolia acuminata L.*"). Cucumber tree is also planted occasionally as a shade tree in much of the eastern United States but not in Florida where it is not well adapted.

Carolina Basswood—*Tilia americana* var. *caroliniana*

Malvaceae

Carolina basswood is a medium- to large-sized tree with large, asymmetrically heart-shaped, deciduous leaves with saw-toothed edges. The leaves of mature red, white, and black mulberry trees are very similar to basswood leaves, but differ in having milky sap. More often than not, basswood has more than one trunk, and there are often additional sprouts from the base. Basswood produces small, sweetly aromatic flowers in May and June followed later by fruit clusters that hang on a branched stalk extending from the underside of a bract that looks like a modified leaf. The dark-colored bark on older stems has numerous vertical ridges that are neither usually flat topped nor scaly.

This tree, which I will refer to as "basswood" from here on, has various scientific names. Taxonomists have disagreed in the past on the scientific names for the various populations of this tree, on how many species there are in North America, and on the boundaries between the different species or varieties. There are also a number of common names for basswood trees. "Linden" is the name used for basswood by many horticulturists and nurserymen in this country and for this tree's close relatives throughout most of continental Europe and Asia, but in England it is usually called "lime tree," perhaps referring to the tea that is made using the tree's flowers. Here in North America it is also sometimes called "linn tree" or "bee tree," but it is most often called "basswood," a name derived from the fibers, called "bast," that were traditionally obtained from the tree's bark.

The three most commonly described populations of basswood in eastern North America are American basswood (*Tilia americana*) in the northern half of the United States and part of southern Canada, white basswood (*Tilia heterophylla*) in the Appalachian region, and Carolina basswood (*Tilia caroliniana* or *Tilia americana* var. *caroliniana*) in the Atlantic and Gulf coastal plains region. All of these are now largely considered part of one diverse, wide-ranging species (Godfrey 1988). There is some validity in paying attention to these three slightly discrete populations in that they are adapted to different climatic, topographic, and soil conditions and are therefore unlikely to be well adapted if moved out of the region where they grow in the wild.

Basswood occurs in the southeastern United States as an occasional tree in association with other hardwoods in upland hardwood forests and slope forests on moist, fertile soil, growing as far south on the Florida peninsula as Polk County (Wunderlin et al. 2020). Its most common associates include sweetgum, pignut hickory, laurel oak, hornbeam, and southern magnolia. It is found as an occasional tree in inland hardwood forests such as San Felasco Hammock, and it is common in parts of Gulf Hammock in Levy County and Cabbage Swamp in northeastern St. Johns County, both of which are coastal hammocks. Basswood does not occur within one mile of the coast in Gulf Hammock, however, because of its inability to withstand flooding by brackish water. Nor can it withstand more than a few days of flooding by freshwater even though it prefers moist situations with the water table near the surface. It is easily damaged by fire.

Although I have seen several large specimens in Alachua and Marion Counties, the largest basswood tree reported from Florida is in Wakulla Springs State Park in Wakulla County. In 2013, it measured 120 feet tall with a 3-foot 3-inch trunk diameter at 4½ feet above ground and an average crown spread of 75 feet (Florida Champion Trees Database 2020). The current national champion American basswood from Lexington, Kentucky, is 102 feet tall with a trunk diameter exceeding 7 feet (American Forests 2018).

In native forests, basswood always grows in competition with other hardwood trees. It is moderately shade tolerant but clearly grows best in full sunlight where it can grow rapidly. It reproduces by seed and by stump sprouts. Seedlings are uncommon, probably because they are highly susceptible to drought in their first year and because of being eaten by insects and browsing mammals. Once established, however, basswood would seem to be almost immortal as a result of its extremely vigorous sprouting ability. When the main trunk dies or breaks or gets weak, the tree sprouts vigorously from the base. These sprouts grow very fast, even under a full forest canopy.

The main trunk of basswood is often rotten or hollow inside. Sometimes this may be the result of the tree's marked susceptibility to fire damage, but it is more likely caused by a combination of the wood's very low rot resistance and the sprout origin of most trunks, which begin rotting at the center of the base almost as soon as they sprout from the rotting base of the previous trunk. This rotting process, along with the weakness of the wood and the tree's susceptibility to wood-boring insects, often causes individual trunks to be rather short lived and susceptible to breaking during storms in the native forests where basswood occurs. Basswood does much better when grown from seed and, in those instances, it is able to live at least 100 years.

The wood of basswood has several unique qualities. It is lightweight but rather strong for its weight, soft, evenly grained, and easy to whittle with a knife. It resists splitting and will not produce splinters. Also, it is nontoxic and does not impart a disagreeable taste to food. For these reasons, it has long been a favorite whittling wood and is used commercially to make wooden toys, cooking utensils, window blinds, statues, musical instruments, and various novelty items. It was used for food packaging before the advent of paper, plastic, and Styrofoam. It is also used to make plywood, packing crates, and pulpwood.

Where honey bees have access to its blooms, basswood is a source for high-quality honey. Other benefits derived from basswood trees include a tea made from its dried flowers and various herbal remedies made from its roots, inner bark, leaves, and flowers. Fibers derived from the bark can be used to make traditional artisanal items, especially baskets and cordage. Prehistoric North Americans and Europeans similarly used the bark fibers to make clothing, twine, rope, and woven baskets.

The wildlife value of basswood is moderately high for a number of reasons. First, a large proportion of basswood trunks are hollow, making available cavities for dens and nests. Second, its leaves and twigs are exceptionally edible, and basswood commonly sprouts prolifically from the root collar. Deer and rabbits browse on the sprouts, and, in native forests, insects of many kinds often eat most of the leaf material in the tree crown by the end of the summer, thus producing abundant food for insect-eating birds and other creatures. (There is much less insect consumption of basswood foliage in urban and suburban areas.) Several types of caterpillars are particularly abundant on basswood trees, which in turn provide an important food source for the yellow-billed cuckoo. Other birds commonly seen feeding in basswood trees are the summer tanager, red-eyed vireo, northern parula warbler, Carolina chickadee, tufted titmouse, blue-gray gnatcatcher, and, in suburban areas, the orchard oriole. This is also one of the trees most preferred by the yellow-bellied sapsucker that drills feeding holes in the bark during the winter. The fruits, which ripen in early September, are eaten only sparingly by squirrels, mice, and birds, but the flowers, which bloom in May and June, are very attractive to honey bees, native bees, and other insects.

Basswood is rarely used for landscaping purposes in Florida, probably, in part, for the very reasons that make it such a valuable wildlife tree—hollow trunks, insect-infested leaves, and prolific sprouting ability. However, these problems can be mostly overcome by planting seedlings, which will grow rapidly on fertile soil and do not begin rotting inside or sprouting at the base for at least 20 years. Also, the insect infestation of the leaves is much less pronounced in landscaped situations. This fast-growing, moderately large, well-shaped shade tree that is beneficial to wildlife merits more use. Seedlings are easy to grow and transplant.

Chinaberry—*Melia azedarach*

Meliaceae

This medium-sized invasive exotic tree has twice pinnately compound deciduous leaves 1 to 2 feet long, leaflets with serrated edges, thick green to purplish twigs lacking terminal buds, and rough, vertically ridged brown bark on larger stems. The leaves strongly resemble those of goldenrain tree except that chinaberry leaves have a terminal leaflet. Also, the new twigs of the chinaberry are green, which is distinctive. The small, star-shaped flowers borne in open clusters are white with pinkish-purple centers. The fruits are round, yellow, fleshy drupes about a half inch in diameter. Chinaberry trees commonly achieve 1 or 2 feet in trunk diameter and are often 50 feet tall at maturity. The largest one currently registered in North America is 65 feet tall with a trunk diameter of over 4½ feet (American Forests 2018).

This tree from Asia was often used as a shade tree in the past and often escaped into the wild. It is listed as an invasive exotic pest in Florida (FLEPPC 2013) and several other southern states. It occurs from Virginia southward throughout Florida and then westward into Arkansas and Texas (Kartesz 2015) and is often common around old rural homesites, along fencerows, and in waste places in agricultural areas. The seeds are spread by birds and, perhaps, by cattle and hogs, which eat the fruit, ingesting the seeds that then pass through their digestive systems. Chinaberry, which grows rapidly when young and out in the open, is somewhat shade tolerant as a seedling but clearly prefers to grow in full sun. Though it seems to prefer upland sites that have been disturbed in the past, it is adaptable, often coming up in a variety of situations including native forests. It comes up occasionally in San Felasco Hammock in Alachua County, Florida, in both upland hardwood forest and upland pine forest. Chinaberry trees often lose branches in storms and are rather easily blown down by high winds.

Chinaberry has yellowish-white sapwood and brown heartwood with an attractive grain. Overall its wood is soft, weak, porous, lightweight, easily split, and easy to work. It is not used commercially, other than perhaps to be mixed with other hardwoods for pulpwood or fuel wood, but it is sometimes used locally for firewood or to make furniture, turned objects, and various craft objects.

The fruit, which is produced in the summer and fall and hangs on into winter, is toxic to humans and most other mammals, but it is consumed to some extent by birds. I have observed robins eating chinaberry fruit in Alachua County, Florida, in the late fall and winter. Professional botanist David Hall has observed robins and blue jays having trouble flying and walking after eating chinaberry fruits that had begun to ferment (personal communication). A man visiting Florida from Kentucky ate some of the fruits, got sick, and was taken to a hospital in Kentucky where he died (Hunt 2001). I have not seen any pollinators visiting the flowers, but they produce nectar and are probably visited by insects.

Frequently planted as an ornamental and shade tree in the nineteenth and early twentieth centuries, chinaberry is no longer commonly planted, although it is still available for sale from some nurseries and on the Internet.

Paper Mulberry—*Broussonetia papyrifera*

Moraceae

Paper mulberry is an invasive exotic tree found commonly around old and former homesites, old limerock mines, nearby sinkholes, nearby forests, and untended areas, having originally been imported into this country with the idea of raising silkworms on the leaves. The leaves of paper mulberry have serrated edges and are mostly 4 to 8 inches long and 3 to 6 inches wide with variable shapes: unlobed, mitten shaped, or with three or more lobes. As with the other mulberry trees, it has milky sap. Paper mulberry can be distinguished from the other mulberry species by the appearance of its youngest stems or twigs, which are thick, green, and very hairy. (The mulberries in the genus *Morus* do not have noticeably hairy stems.) The leaves of paper mulberry are also fuzzy with small hairs.

Paper mulberry grows very rapidly when young. It makes a short, flat-topped, deciduous tree with a thick trunk when growing in the open but can grow tall in a shaded forest. Older, open-grown trees often develop excessively thick, gnarled trunks. Measured in 1991, a large paper mulberry in Florida was 75 feet tall with a trunk diameter slightly over 4 feet (Ward and Ing 1997).

Paper mulberry is an extremely competitive exotic invader when it is not controlled, spreading both by root suckers and by seed. Originally, most of the planted trees in our state were males (paper mulberry's male and female flowers occur on separate trees), but there are now many female trees in Florida, which flower in June and produce ripe fruit in June and July. When ripe, the fruit cluster is a 1-inch-diameter green ball with numerous bright red-orange fruits attached on the outside, each containing one seed. European starlings are especially fond of the fruit (personal observation). Although these small outer fruits are eaten by northern mockingbirds, European starlings, and probably some other birds, paper mulberry in general provides very little wildlife benefit, and, when aggressively spreading, it does great damage to wildlife habitat by displacing native plant species. The berries come at a time of year when there is no great food demand, at least in rural areas where other fruit such as blackberries and blueberries are readily available. Moreover, paper mulberry stems and leaves are not browsed by mammals or eaten by insects. It is one of the most aggressive exotic invaders of native habitats in tropical countries around the world and is a potential disaster for Florida's natural areas.

Paper mulberry, listed as an invasive exotic pest plant in the state of Florida (FLEPPC 2013), should never be planted and should be eliminated whenever possible. I killed a grove of paper mulberries on my own land with one application of hexazinone herbicide applied in March at the base of each trunk. This method is safe only if no other hardwood trees of value are close by. Hack-and-squirt applications of triclopyr in water or a soaking of the lower trunk with triclopyr mixed with oil plus a surfactant can also be effective methods of herbicidal paper mulberry removal.

Osage-Orange—*Maclura pomifera*

Moraceae

Osage-orange was originally native to the Red River Valley area of Oklahoma, Arkansas, and Texas, as well as some additional parts of eastern Texas, but it has been planted widely and become naturalized elsewhere, especially in the central and southern Great Plains but also throughout the southeastern United States (Burton 1990). It is a large shrub or small- to medium-sized tree, usually 20 to 40 feet tall, but occasionally up to 70 feet tall, with a short, often leaning and crooked trunk that may be several feet in diameter near the ground (Burton 1990). The trees are often multitrunked and often spread by root suckers to form thickets. Its deciduous, simple, smooth-margined leaves may be arranged either alternately or opposite each other on its very thorny twigs. Female trees produce a 3- to 6-inch-diameter round fruit that starts out green and eventually turns yellow. Male and female flowers occur on separate (dioecious) plants. The largest Osage-orange tree currently recorded is 72 feet tall with a trunk diameter of about 8½ feet (American Forests 2018).

A rare plant on the Florida peninsula, Osage-orange is more common in the panhandle and much more common from there northward and westward. It is often found along old fencerows and property boundaries. It was used extensively in the past as a living fence and hedgerow and as a windbreak, especially in the central and southern plains states following the dust bowl disaster of the 1930s. In such applications on the southern Great Plains, it is perhaps the very best choice (along with burr oak), as it is very adaptable and durable with regard to varying soil conditions, severe drought, heat, and occasional flooding (Burton 1990). It does poorly on wet soils, highly acidic soils, and deep, sandy soils and is not particularly shade tolerant.

Osage-orange was used for making fence posts because the heartwood is very strong and the most rot resistant of the native hardwoods (Burton 1990). Because of its great strength, the branch wood was used by Native Americans to make bows (Burton 1990).

Osage-orange is valuable for wildlife as escape cover and a place for nesting. The flowers are wind pollinated. The seeds are eaten to a limited extent by fox squirrels and bobwhite quail (Martin et al. 1951). Osage-orange fruits are reportedly an important winter food staple for fox squirrels in Kansas and also an important food for fox squirrels in Illinois (Tesky 1993).

Osage-orange is not often planted now except as an oddity. Its thorny branches, large, messy fruits, and tendency to get out of hand are negatives. A thornless, fruitless male clone that has been developed for ornamental use overcomes some of these undesirable features, but, of course, being fruitless, this clone would not serve as a winter food resource for fox squirrels.

Red Mulberry—*Morus rubra*; White Mulberry—*M. alba*; and Black Mulberry—*M. nigra*

Moraceae

Red mulberry, the only native mulberry east of Texas, is a medium-sized tree having deciduous leaves up to 6 inches long and nearly as broad. The leaves of seedlings and sprouts are often deeply lobed with up to five or more lobes, whereas the leaves of older trees are usually unlobed or have an occasional mitten-shaped or three-lobed leaf. The leaves of mature red mulberry trees, with their heart shape, long pointed tips, serrated edges, and upper and lower surfaces roughened by small hairs resemble those of basswood, but mulberry has milky sap that exudes from severed leaf parts whereas basswood does not. White mulberry lacks both the long point at the tip of the leaf and the hairs on the upper leaf surface, and its leaf-top surface is smooth and glossy. Black mulberry leaves are about half the size of the other two species and are somewhat folded upward along the midrib. The upper surfaces of its leaves are dull and rough like those of red mulberry, but it lacks the long point at the leaf tip. The bark on older trunks of all three species has flat-topped, slightly flaky ridges and shallow fissures. In Florida, the heartwood of all three species is usually yellow.

Many red mulberry trees are entirely female, some are entirely male, and some produce both male and female flowers. The fruit produced by female or bisexual trees ripens from mid-March to mid-May and is usually black when fully ripe in all three species although some varieties of white mulberry have white fruit. The name "red mulberry" presumably references the burgundy red color of the heartwood of trees in more northerly locations. Red mulberry occurs naturally as an uncommon tree throughout most of the eastern United States and throughout Florida, except, perhaps, for the extreme southern tip of the peninsula (Lamson 1990). White and black mulberry trees, both Old World species, are planted as ornamentals throughout the eastern United States and often escape into the wild.

Although not normally getting much taller than 50 feet with trunk diameters of 1 foot, white mulberry and, especially, red mulberry trees will occasionally grow very large trunks. (Black mulberry is usually smaller.) There was a mulberry of mixed red and white mulberry ancestry on the University of Florida main campus until 1999 with a trunk diameter of 3 feet; a red mulberry in Cottondale, Florida, with a trunk diameter of nearly 5½ feet; and a red mulberry in northern Marion County near Reddick with a trunk diameter a bit over 5½ feet (Ward and Ing 1997). In 2018, the largest recorded red mulberry in the country, in Arkansas, measured 75 feet tall with a trunk diameter of 8 feet (American Forests 2018).

One possible reason for these significant variations in tree size is that the mulberry population in the southeastern United States consists of the original species plus a variable hybrid mix of red and white mulberry, with perhaps a bit of black mulberry ancestry as well. In natural forests well away from cities and towns, pure red mulberry still exists, but many of the trees within and near cities, towns, and other populated areas are of mixed ancestry.

The De Soto expedition into the interior of Florida and elsewhere in southeastern North America in 1538 found red mulberry to be a common and especially important fruit tree for the Native Americans (de la Vega [1605] 1996). This observation suggests that the tree was fairly abundant in Florida during that early period. Some four centuries later, Moore and Thomas (1977) reported that red mulberry appeared to be decreasing in abundance over much of its range. One possible explanation for this decline is that red mulberry was domesticated and grown as a domestic plant for some time by Native Americans. A second possibility is that hybridization with white mulberry has produced a large number of poorly adapted and genetically inferior trees. A third factor in the decline is the introduced white peach scale, an exotic insect which commonly attacks mulberry trees, often killing small trunks and branches.

Today, red mulberry is found as an uncommon tree scattered about in hammocks and other upland and lowland hardwood forests. It seems to prefer calcareous soils and often occurs on Indian mounds. However, it also grows on acidic soils with low calcium levels. Though it prefers moist sites and can withstand minor flooding, red mulberry can also grow on dry upland sites if given some help getting established. It seems to occur most frequently near streams, wetlands, or open water. It has a large number of associates, the most typical of which are probably laurel oak and sweetgum. In cities and towns, mulberry trees of mixed red and white mulberry ancestry are often common, and there is the occasional planted white or black mulberry.

Although moderately shade tolerant and often found growing as an understory tree, mulberry does rather poorly in the shade of other trees. Mulberry trees do best in the open or on an edge where they can get direct sunlight. With sufficient sunlight, the female trees will produce an abundance of fruit resembling blackberries in shape, color, and taste. The fruits ripen over an extended period from mid-March to mid-May, with some fruit observed as late as mid-June. One young tree in my yard produced a small second crop of fruit in September. Birds and mammals distribute the seeds widely by feeding on the fruit and then depositing the seeds elsewhere in their droppings.

Mulberry trees grow rapidly and are often short lived, although the large one near Reddick was at least 80 years old (James Brown, the tree's owner, personal communication). The growth rate for mulberry trees under favorable conditions may be as much as 5 feet per year for the first few years, and red mulberry can bloom and produce fruit as early as three years from seed. On older trees, large branches often die back or break off and then regrow, and mulberry does not heal old branch scars very well. The result is that old trees are often rather awkward looking, and the trunks and main branches are often hollow. Another problem is that mulberry trees often begin leafing out so early that their new growth is sometimes nipped back by late frosts.

The wildlife value of mulberry is very high, mainly because healthy female trees continuously produce abundant fruit for a period of about two months from mid-March to mid-May, a time of year when other wild fruits and seeds are scarce. Mulberries are eaten by many kinds of birds and mammals. Martin et al. (1951) note that in this part of the country, mulberry fruit comprises at least 2 percent of the diet of cardinals, catbirds, mockingbirds, brown thrashers, common crows, fish crows, orchard orioles, red-bellied woodpeckers, red-headed woodpeckers, opossums, raccoons, and gray foxes. Other mammals that feed on the fruits include gray squirrel, fox squirrel (Lamson 1990), black bear, wild hog, and white-tailed deer. Other birds that feed on the fruit, according to Terres (1980), include wild turkey, bobwhite quail, Baltimore oriole, wood thrush, rose-breasted grosbeak, eastern towhee, scarlet tanager, great-crested flycatcher, and yellow-billed cuckoo. Hill (1993) reports house finch feeding on mulberries. Before migrating north, cedar waxwings feed extensively on mulberries when they are available, often as late as May 1st in Gainesville, Florida. Red-bellied woodpeckers, rose-breasted grosbeaks, and cedar waxwings seem to be particularly fond of the berries. Mulberry fruits are an especially important part of the diet of lingering spring migrant cedar waxwings in Texas, Arkansas, and Alabama (Witmer et al. 1997). The same is true in Florida and Maryland (personal observation) and presumably in all the areas in between these locations. The fruit, stems, and leaves, including freshly fallen leaves, are good deer browse, and the leaves are also eaten by many kinds of insects that, in turn, are eaten by birds, lizards, spiders, and other small animals. Finally, the hollow trunks and branches of old mulberry trees can be valuable den and nest sites for many cavity-nesting or cavity-dwelling species.

Red mulberry and white mulberry were commonly used in the past as ornamental trees, sometimes because of their fruit production and sometimes with an eye toward raising silk worms on the foliage. Members of De Soto's expedition related that the Native Americans of Florida gathered and ate red mulberries in large quantities (de la Vega [1605] 1996). The berries of red mulberry are sweet and tasty when fully ripe and can be eaten fresh or made into pies, jam, or wine. Red mulberry, white mulberry, and paper mulberry are all found around old homesites here in Florida. (Paper mulberry is an especially aggressive and destructive invasive exotic that should be eliminated whenever possible [see species account on paper mulberry].)

None of the mulberry species are used commercially today, although red mulberry wood was used in the past for fence posts because of the heartwood's rot resistance.

The main landscape value of red mulberry (or white mulberry, black mulberry, or a mulberry tree of mixed or unknown ancestry) is its value for wildlife. In most landscape situations in our area, where oaks and hollies are already common, a mulberry tree is the single best tree to plant to benefit wildlife. In urban and suburban situations where songbirds will be the primary fruit consumers, one female tree per city block in a sunny situation may satisfy their needs. In situations where the medium-sized or larger mammals will also have access to the fruit or where people want to enjoy it, several trees might be fully utilized. Mulberry grows rapidly in ideal situations (plenty of water, sunlight, and fertilizer). Female mulberry trees should not be planted where they will grow over a clothesline, patio, or parking area given the mess that heavy fruit production, plus bird droppings, can make in the spring. Because of the susceptibility of mulberry trees to frost damage in the early spring, it is best not to plant them in depressions that serve as frost pockets.

When growing young mulberry trees, each tree should be inspected several times during the growing season for the presence of white peach scale. Large populations of this insect can build up rapidly on twigs and stems up to 3 or 4 inches in diameter. White peach scale creates the appearance of a white cottony or woolly growth on the trunk or branches. The white covering is made of strands of wax. On close inspection, the yellow-orange scale insects can be found under the protective wax covering. Large infestations can kill branches or even an entire small tree. This scale is easy to eliminate by simply running one's hands or fingers up and down the trunk and branches to crush the tiny insects, which are soft compared to most other scale insects. Another obstacle to growing young mulberry trees is that deer love to eat them. In situations where deer are common, it is often necessary to put a chicken wire cage around the young tree until it gets tall enough to be out of reach of browsing deer.

Waxmyrtle—*Morella cerifera* (*Myrica cerifera*) and Dwarf Waxmyrtle—*Morella pumila* (*Myrica cerifera* var. *pumila*)

Myricaceae

Waxmyrtle, also known as "southern bayberry" and "southern waxmyrtle," is one of the most common and widespread shrubs in Florida. It occurs in two forms. The most common form is a large bush or small tree that grows in almost any situation from wetlands to pine flatwoods to well-drained uplands. This large form occurs naturally throughout all of Florida and the rest of the southeastern coastal plain of the United States from New Jersey into eastern Texas (Kartesz 2015). The other form is a dwarf shrub described as *Morella pumila* or *Myrica cerifera* var. *pumila* (Wunderlin at al. 2020) that forms clonal patches in upland longleaf pine forests and pine flatwoods forests subject to ground fires. Both forms usually have multiple stems. The 1- to 3-inch-long leaves of waxmyrtle, smaller and narrower on the dwarf form, are thin, light green, irregularly toothed, evergreen, dotted with tiny yellow resin glands, and pleasantly aromatic. Male and female flowers occur on separate (dioecious) plants. At casual glance, before closer examination, the bark is light gray and appears smooth, but it is, in fact, slightly rough to the touch.

The dwarf form is a component of the ground cover of some well-drained, native, upland longleaf pine forests including sandhills and mesic longleaf pine flatwoods forests. Its associates in the ground cover of the sandhill community are wire grass, bracken fern, shiny blueberry, chinquapin, woolly pawpaw, partridge pea, summer-farewell, and a great variety of other sandhill wildflowers and grasses. Its associates in the flatwoods community include saw palmetto, gallberry, dwarf live oak, runner oak, shiny blueberry, and dwarf huckleberry.

It spreads by both underground runners and by seed and is usually less than 2 feet tall.

The common form of waxmyrtle will normally be from 5 to 20 feet tall at maturity with crooked, leaning trunks up to about 6 inches in diameter. The largest individual in Florida in 1997 was 36 feet tall with a trunk slightly over 1 foot in diameter (Ward and Ing 1997). It most commonly grows in situations where there is abundant moisture and abundant sunlight. The large form, abundant in some shrub swamps and wet (hydric) hammocks, is one of the most prolific invaders of open prairie and marsh habitats and pine flatwoods forests that burn infrequently. It is common along streams and around the edges of ponds and lakes, and it rapidly invades many sites that have been severely disturbed and then abandoned, such as mine spoils and pits, borrow pits, old construction sites, and ditches.

Waxmyrtle is a symbiotic nitrogen fixer that has been found to increase the amount of total nitrogen and biologically available nitrogen in the soil where it grows as well as in the other plant species that grow in association with it (Kurten et al. 2008). Nodules on the waxmyrtle's roots contain bacteria that take gaseous nitrogen from the air or water and convert it to organic nitrogen compounds usable by the plant. Waxmyrtle's nitrogen-fixing ability enables it to invade and do well in locales where topsoil has yet to form or where soil has been removed or greatly damaged.

Although waxmyrtle cannot withstand prolonged deep flooding, it is amenable to shallow, intermittent flooding and continuously wet soils. The dwarf form of waxmyrtle is drought tolerant, occurring in the excessively well-drained deep sands of sandhill longleaf pine forests. Waxmyrtle grows well in full sunlight and in partial shade but not in dense shade. Its aboveground stems are not fire tolerant, but the plants resprout vigorously from the base and root systems after being killed to the ground by fire. Repeated ground fires can eliminate the large form, but the dwarf form is well adapted to frequent fire. The aboveground stems of waxmyrtle are short lived, rarely surviving for more than 30 years, but the clonal patches of the dwarf form can probably survive for hundreds of years.

The wildlife value of waxmyrtle is moderately high as a combined result of the fruit's high use by a few species of birds in the fall and winter and of the leaves' consumption by insects before they become, in turn, food for spiders, frogs, lizards, and birds. Martin et al. (1951) record that waxmyrtle fruits (drupes) comprise 2 to 5 percent of the diet of mottled duck, eastern bluebird, meadowlark, brown thrasher, white-eyed vireo, myrtle warbler (yellow-rumped warbler), and red-bellied woodpecker. The fruit comprises between 5 and 10 percent of the gray catbird's diet and 25 to 50 percent of the diet of tree swallows while they are overwintering along Florida's coastlines. The percentage of waxmyrtle in yellow-rumped warblers' diets is clearly much higher than 2 to

5 percent when they are migrating through or spending the winter in Florida. Kilham (1989) noted that American crows eat waxmyrtle fruits in Florida. I have seen large flocks of both tree swallows and yellow-rumped warblers feeding on waxmyrtle fruits on the northern Florida peninsula on many occasions. In addition to these two, local birder and naturalist Michael Meisenberg has observed the eating of waxmyrtle fruits in northern Florida by white-eyed vireo, blue-headed vireo, cardinal, ruby-crowned kinglet, pine warbler, and Carolina chickadee (personal communication).

Wax can be obtained from the fruits of waxmyrtle for making scented candles, and the leaves can be used for flavoring food. More commonly, waxmyrtle is sometimes used for landscaping, especially near the coast where its salt tolerance is an asset. It can make a reasonably dense, tall hedge or visual barrier and is ideal for wildlife plantings around stormwater retention basins. More use could be made of the dwarf form, of which the landscape and nursery industries seem to be largely unaware. In the sandhills and pine plantations of northern Florida, I have seen some attractive clones of the dwarf form. I have also seen clones of compact, midsized shrubs (3 to 4 feet tall) that are intermediate between the large and the dwarf form and are likely hybrids between the two.

Bayberry, Southern Bayberry—*Morella caroliniensis* (*Myrica heterophylla*)

Myricaceae

Bayberry is similar in most respects to waxmyrtle. On average, it is much smaller than the large form of waxmyrtle, the current national champion being only 14 feet tall and about 2 inches in trunk diameter (American Forests 2018): hence, the alternate name, "small bayberry." Its leaves are thicker, broader, greener, and less aromatic than waxmyrtle leaves, and its resin glands are confined primarily to the lower leaf surfaces. Male and female flowers occur on separate (dioecious) plants.

Southern bayberry occurs from New Jersey south through the Atlantic and Gulf coastal plains into eastern Texas (Kartesz 2015). It is much less common in Florida than waxmyrtle and is less widely distributed, occurring mainly in the Florida Panhandle with a few isolated populations in northeastern Florida and on the Lake Wales Ridge in the southern half of the Florida peninsula (Wunderlin et al. 2020).

Bayberry occurs primarily in wet pine flatwoods, shallow wetlands, or on dunes along the coast. It is usually found growing on soils that are acidic and low in available nitrogen. Bayberry is only moderately shade tolerant, doing best in full sun.

Bayberry is similar to waxmyrtle in that it has nodules on the roots containing bacteria that take gaseous nitrogen from the air or water and convert it to organic nitrogen compounds usable by the plant. This process is called "nitrogen fixing." It enables bayberry to do well in nitrogen-poor soils and also benefits surrounding plants of other species.

Bayberry is also similar to waxmyrtle in its value to wildlife. The fruits are eaten by the same set of birds, although not in as great a quantity given that bayberry is a less abundant and less widespread species.

Bayberry has been called "candle-berry" because its fruits can be used to make candles. The fruits are harvested and then boiled to obtain the wax covering, which is then skimmed off the top and used, often in combination with beeswax, to make scented candles.

Bayberry has potential as an ornamental plant, being somewhat greener than waxmyrtle. However, it is less adaptable than waxmyrtle, making it more difficult to grow in some landscape situations.

Odorless Bayberry, Odorless Waxmyrtle—*Morella inodora* (*Myrica inodora*)

Myricaceae

Odorless bayberry is a rare shrub native primarily to the Florida Panhandle with one population on protected land in Colquitt County in southern Georgia and several populations in extreme southern Alabama and Mississippi and perhaps extreme eastern Louisiana (Kartesz 2015; Chafin [2008] 2020, "*Morella inodora*"). It has rather thick, leathery, dark green, evergreen leaves, with slightly curled-under edges that are rounded at the tip, tapered toward the base, and about 2 to 4½ inches long. Despite having scattered white or clear resin glands on both surfaces, the leaves, when crushed, lack any significant aroma. Male and female flowers are borne on separate (dioecious) plants. The small fruits that occur individually on stalks on female plants are dark brown to blue-black in color when ripe.

Odorless bayberry is usually a large shrub but can become a small tree. The largest one reported, from southern Alabama, is 27 feet tall with a trunk diameter of about 5 inches (American Forests 2018). I saw one of similar size in 1978 on Eglin Airforce Base in the western panhandle of Florida.

Odorless bayberry occurs in wet pine flatwoods, bayheads, titi bogs, and seepage spots in steephead ravines. It is often associated with pond pine, black titi, white titi, fetterbush, and swamp azalea (Chafin [2008] 2020, "*Morella inodora*").

As with other bayberry species, the roots of odorless bayberry have nodules containing nitrogen-fixing bacteria (Chafin [2008] 2020, "*Morella inodora*").

The fruits of odorless bayberry are eaten by birds which distribute the seeds

(Chafin [2008] 2020, "*Morella inodora*"). Because it is likely that the same bird species that eat waxmyrtle fruit, such as the yellow-rumped warbler, also eat the fruit of odorless bayberry, it is similarly likely that odorless bayberry has significant wildlife value where it occurs.

Historically, odorless bayberry berries were used to treat common household illnesses such as colds and flu. Today, odorless bayberry has some potential as an ornamental shrub. Like waxmyrtle, the berries can be used to make candles, and the branches and berries make handsome dried arrangements and additions to garlands and wreaths.

Blackgum—*Nyssa sylvatica (N. sylvatica* var. *sylvatica)*

Nyssaceae

This species of tupelo is not nearly as common in Florida as the very similar looking swamp tupelo. It occurs from Alachua County northward and westward into the Florida Panhandle and from there northward throughout most of the eastern United States. It is a medium- to large-sized deciduous tree with simple, smooth, thin, glossy leaves. In Florida, it rarely gets over 100 feet tall with more than 2 feet in trunk diameter, but farther north and west it can be quite large with the diameter record set in Texas by a blackgum with a trunk diameter of about 6½ feet (American Forests 2018). As a small tree, it is similar enough in appearance to persimmon to cause confusion. One trick to telling these two species apart is that blackgum has a terminal bud at the end of each twig, whereas persimmon does not. (On a persimmon twig, the withered end of the twig sticks out past the last lateral bud.) Distinguishing blackgum from swamp tupelo is more difficult. On average, the leaves of blackgum are thinner, more flexible, and more likely to have a few teeth on the margins of some of the leaves. The flowers and fruit of blackgum are often borne in clusters of three or more, whereas the flowers and fruit of swamp tupelo are more often borne in pairs.

In northern Florida and adjacent Georgia and Alabama, blackgum is often found in association with longleaf, shortleaf, and loblolly pines, post oak, and southern red oak on upland, well-drained, sandy, sandy clay, and clay soils. It is adapted to frequent, low-intensity fires. It also occurs in upland and lowland hardwood forests.

Similar to swamp tupelo, blackgum's wildlife value is very high because of its nutritious fruits, the importance of its flowers to insect pollinators, and its potential use as a nest and den tree when hollow. (See species account on swamp tupelo for detailed information on wildlife value.)

In Florida, blackgum is not common enough to have much economic value, but it does contribute to honey bee culture and help maintain native insect pollinator populations by providing a valuable nectar source. It has some untapped potential as an ornamental tree given its ability to thrive on uplands and its beautiful fall colors.

Swamp Tupelo (Blackgum)—*Nyssa biflora* (*N. sylvatica* var. *biflora*)

Nyssaceae

Of the various kinds of tupelo that grow in Florida, this is by far the most widespread and abundant, ranging as far south on the peninsula as Lake Okeechobee (Wunderlin et al. 2020) and occurring commonly in wetlands throughout the southeastern coastal plain from Delaware and Maryland into Louisiana and eastern Texas and extending as far north in the Mississippi River floodplain as western Tennessee (Godfrey 1988). Swamp tupelo is often called "blackgum." However, the upland variety or sister species of this tupelo is also often called "blackgum." In order to talk about each of these closely related species in a way that will allow us to distinguish them, "swamp tupelo" will be used as the name of this tree and "blackgum" will be used for the upland species. Distinguishing these two similar species in the field is sometimes difficult. Swamp tupelo has, on average, thicker, stiffer leaves that are less likely to have teeth on the margins, and the flowers and fruit usually occur in pairs instead of in clusters of three or more as on blackgum. In addition to these two types of blackgum, there is a variant population (*Nyssa ursina* Small) in the Florida Panhandle that occurs as a shrub, sprouting up from a subterranean base in wet pinelands subject to frequent fires. (Two additional species of tupelo with much larger fruits and leaves, water tupelo and Ogeechee tupelo, occur in wetlands in parts of northern Florida and are discussed separately.)

Swamp tupelo can occur as a small tree or clump of trees in the middle of the marshes and wet prairies that are scattered about in wet pinelands. However, for the most part, swamp tupelo grows to be a large wetland tree, 80 to 100 feet tall with trunk diameters of 1 to 3 feet. The largest one I have seen in Florida, in the swamp on the south border of Newnan's Lake, had a trunk diameter at breast height of 4 feet 9 inches on February 19, 2000. The largest one in the country, from Virginia, is 116 feet tall with a 5½-foot trunk diameter (American Forests 2018). Swamp tupelo has simple, shiny, deciduous leaves. The trunk is often swollen at the base and has rough, gray bark that usually has vertical ridges. Sometimes, however, the bark lacks clear ridges, instead showing a bumpy texture, especially near the base. The crown is made up of many fine twigs, each of which has a terminal bud (as compared to the twigs of persimmon, which do not have terminal buds).

All tupelos are dioecious, meaning that there are separate male and female trees. Both produce flowers in spring, but only the females produce fruits, which, in swamp tupelo, are oblong, about one half to 1 inch in length, and turn from green to black in the fall when ripe.

Swamp tupelo, true to its name, grows in swamps. It grows in freshwater swamps of all sorts, from pure stands of swamp tupelo to mixed stands of ash, tupelo, red maple, and bald-cypress or, in the cypress domes of the pine flatwoods, mixed stands of swamp tupelo and pond-cypress. It also grows in bayheads (baygalls) and bogs along with sweetbay, swamp bay, loblolly bay, and, sometimes, slash or pond pine. It is associated with Atlantic white cedar, swamp laurel oak, and cabbage palm at Mormon Branch in the Ocala National Forest and with Atlantic white cedar, tulip tree, swamp laurel oak, and other trees along Deep Creek south of Interlachen, Florida. In Florida swamp tupelo can be seen along almost every stream, around most lake shores, and in almost every swamp and bayhead. It is also often seen as an isolated tree or clump of trees in marshes and wet prairies.

Swamp tupelo is not shade tolerant and does not grow very fast. It is able to survive and compete because of its ability to withstand flooding and permanently wet soil better than most other trees. In freshwater, only bald- and pond-cypress trees are better at enduring prolonged flooding. Swamp tupelo appears to be able to live at least 200 years, perhaps longer. The fruit, which is often produced prolifically, is eaten in large quantities by bears, raccoons, robins, cedar waxwings, and many other species, none of which digest the seeds, thus scattering them widely as the seeds pass through their digestive systems.

Swamp tupelo's value to wildlife is very high. The fruit is one of the most important food resources for black bears in the areas of the Osceola National Forest and the Okefenokee Swamp when the female bears are putting on fat in preparation for hibernation and giving birth. The fruits mature September through December and are an important part of the diets of black bears, raccoons, wild turkeys, wood ducks, pileated woodpeckers, flickers, robins, mockingbirds, brown thrashers, starlings, and thrushes (Martin et al. 1951). I have often observed robins and cedar waxwings feeding heavily on swamp tupelo fruits during November and December in Alachua County. Rusty blackbirds also feed on the fruits.

Two additional key features add to tupelo trees' wildlife value. First, they often have cavities that are used by cavity-nesting species and, second, they produce spring flowers that are highly attractive to pollinating insects. Bee keepers make good use of this last feature. The honey produced by bees visiting mainly tupelo trees, known as "tupelo honey," is of the highest quality.

Swamp tupelo's medium-weight wood is not rot resistant, but it is strong and tough and stands up to wear when used as a truck-bed liner or a crosstie. Moreover, it is very split resistant because of its interlocked grain. It is often used for plywood core stock and for single-ply crates and baskets. It turns well on a lathe and is easy to carve, making it suitable for making wooden bowls. It is a favorite wood for making duck decoys. Swamp tupelo is not ideal for making lumber, however, because of its tendency to warp while drying.

Swamp tupelo is rarely used in landscaping and is not well adapted for planting in upland subdivisions. However, if left on a lot where it is already well established or if planted along a ditch or stream or at the edge of a pond, it often makes an attractive tree. Its bright red autumn leaves provide the earliest and brightest fall color of any tree native to northern Florida. This tree is completely bare of leaves throughout the winter, losing its leaves earlier and leafing out later than most other deciduous trees.

Water Tupelo—*Nyssa aquatica*

Nyssaceae

Water tupelo is a large wetland tree similar in appearance to swamp tupelo. However, its leaf size, although quite variable, is much larger, ranging from 3 to 12 inches in length. The leaf margins are mostly entire, although there can be a few pointed lobes on the edges. The bark is gray and rough on larger stems and trunks, and the base of the trunk is usually swollen, often much more so than that of swamp tupelo. The fruits are much larger than those of swamp tupelo, averaging about 1 inch in length.

As its name implies, water tupelo is closely associated with water. It grows in the floodplains of major rivers in places that remain flooded for extended periods but also dry out periodically. It occurs on the Atlantic and Gulf coastal plains from southeastern Virginia into eastern Texas, ranging northward into northern Alabama and Mississippi and up into the southern tip of Illinois along the Mississippi River floodplain (Kartesz 2015). In Florida, it mainly occurs in the western panhandle and along the Apalachicola River and the lower Suwannee River. A good place to observe this tree is along the boardwalk beside the spring run at Manatee Springs State Park. A stand of these trees occupies a small depression swamp in O'Leno State Park in Columbia County.

Water tupelo is a tall tree, often reaching heights of 100 feet at maturity, usually with a straight, clear trunk. The maximum size of the trunk is larger than that of the other tupelos, in part because of the swollen nature of the lower trunk; a water tupelo can have a trunk diameter as large as 12 feet (American Forests 2018). The extent of swelling of the lower trunk can vary from 5 to 10 feet above ground. Water tupelo is intolerant of shade, growing well only in full sunlight. It prefers soils with some clay content, withstands flooding well, and is not fire tolerant. Common associates include bald-cypress, red maple, and pumpkin ash.

The fruits of water tupelo are distributed to some extent by wildlife, but the primary means of seed dispersal is by water.

The wildlife value of water tupelo is moderate to high. The fruit is not consumed by as many species or as abundantly as the fruit of swamp tupelo, but it is eaten by black bears, raccoons, wood ducks, wild turkeys, bobwhite quail, and some songbirds (LSU Agricultural Center 2021). Insects feed on the foliage, bees use the flowers to obtain nectar, and deer browse seedlings and sprouts. As with the other tupelo tree species, the trunks of old trees sometimes contain hollows usable as den and nest sites.

Water tupelo is rarely planted as an ornamental. It has some value as a timber tree, producing white wood of medium density that is tough under wear, strong, and split resistant. It is easily carved, turns well on a lathe, and is often used to make the plywood cores and single-ply objects such as crates and baskets. It is also used for making furniture, carved products such as duck decoys, and turned items such as bowls. The wood has interlocked grain, making it nearly impossible to split and causing it to warp more than most other woods when sawn into boards and dried. The wood at the base of water tupelo's swollen trunk is very lightweight, making it suitable for the crafting of fishing net floats. However, the wood is not rot resistant. Sold as tupelo honey, the honey produced by honey bees that visit the flowers is of high quality.

Ogeechee Tupelo (Ogeechee-Lime)—*Nyssa ogeche*

Nyssaceae

The least common of the tupelos, this tree is often multitrunked with an enlarged base. In occurs in northeastern Florida from Clay to Suwannee County, in the central panhandle of Florida, and from there across southeastern Georgia (Kartesz 2015), being especially common along the Ogeechee, Altamaha, and upper Suwannee Rivers (Kossuth and Scheer 1990). It appears most often as a multistemmed, large shrub to small tree with a large base on very wet, frequently flooded areas along streams. Its 3- to 6-inch-long and 1- to 2-inch-wide deciduous leaves with entire margins are dark green above and pale green with velvety pubescence below (Godfrey 1988). Ogeechee tupelo's white, male and female flowers are borne separately on the same (monoecious) tree. Male flowers are borne in compact balls, whereas the female and bisexual flowers are solitary (Godfrey 1988). The lightweight green fruits, inflated with air and with a stone at the center, turn partly reddish when ripe and are about 1½ inches long and half to three-quarters of an inch wide.

The leaves of Ogeechee tupelo are most similar to those of water tupelo, but the growth form is distinct and the bark lacks the deep vertical ridges of water tupelo. Ogeechee tupelo is most often seen as an awkward group of sprouts or small trunks growing from an ancient, thickened base. This large shrub form can be from 10 to 50 feet tall, with most of the trunks above the base not more than 1 foot in diameter but with bases that can be up to 6 feet in diameter at ground level. However, in some situations in the wet middles of forested wetlands, Ogeechee tupelo can grow to be a large, single-trunked tree. There is a grove of such trees in the center of Bradwell Bay in the Apalachicola National Forest, the largest of which was 93 feet tall with a trunk diameter of about 4½ feet in 1993 (Ward and Ing 1997).

Ogeechee tupelo is very tolerant of wet growing conditions and flooding, but it is not shade tolerant. It is almost always found near flowing water and is not at all drought tolerant. Common associates include bald-cypress, swamp tupelo, water-elm, coastal plain willow, black willow, red maple, sweetgum, pop ash, and buttonbush.

The wildlife value of this tree is probably at least moderate. It produces abundant flowers in early spring that are very attractive to bees and other insect pollinators, and the abundant fruit crop in the fall may be used to some extent by various mammals and birds. The hollow trunks and bases of old trees provide potential escape cover and nesting and denning cavities.

Many acres of Ogeechee tupelo have been planted in the lower Apalachicola River basin to provide nectar for honey bees in order to facilitate the production of tupelo honey (Kossuth and Scheer 1990). The edible, acidic fruits are sometimes gathered to produce a beverage similar to limeade or to make preserves. Ogeechee tupelo is occasionally used as an ornamental because of its unique and colorful fruit (David Hall, professional botanist, personal communication).

Wild Olive, Devilwood—*Cartrema americanum* (*C. americana; Osmanthus americanus*) and Scrub Wild Olive—*Cartrema floridanum* (*C. floridana; Osmanthus floridanus, O. megacarpus*)

Oleaceae

Wild olive is a small tree or large shrub with opposite, leathery, shiny, evergreen leaves averaging 6 inches long by 2 inches wide with entire, slightly curled-under margins. It blooms with inconspicuous, creamy white, sweetly aromatic flowers in mid-March, with male and female flowers borne on separate (dioecious) plants. The female trees subsequently produce olive-like green fruits (drupes) averaging half an inch in diameter that turn dark blue at maturity in late summer or fall and persist on the plant until spring. Wild olive often has multiple trunks and smooth gray bark on small stems, but the bark can be dark brown and vertically roughened on large trunks. Wild olive occurs naturally on the coastal plain from

North Carolina into Mississippi and down into Florida as far south as Highlands County in the central peninsula (Kartesz 2015). The largest wild olive on record in 1967 in Mayo, Florida, was 37 feet tall with a 21-inch trunk diameter (American Forests 1986). A more typical size for wild olive is 20 feet tall with a 2- to 6-inch trunk diameter.

Wild olive is adaptable, growing on sandy soils in scrub, sandhills, pine flatwoods, xeric hammocks, and among the live oaks, red-cedars, cabbage palms, and saw palmettos on coastal dunes and swales. It is moderately drought and flood tolerant, somewhat shade tolerant, and highly amenable to salt spray (Gilman and Watson 1993, "*Osmanthus americanus*"). It is rather slow growing, does best with ample sunlight, and seems not to live very long, perhaps 50 to 100 years. Wild olive does not spread by root suckers. Its only reproduction is by seed.

The closely related scrub wild olive, which differs by having larger fruit, occurs on the central Florida peninsula from Marion County south into Highlands County. It is found in scrub habitats and is typically somewhat smaller in size compared to its northern relative.

The wildlife value of these plants is moderate. Gilman and Watson (1993, "*Osmanthus americanus*") note that squirrels and birds sometimes eat the fruit. The aromatic flowers attract bees and other pollinating insects. The moderately dense, evergreen foliage provides nesting habitat for songbirds.

The only human use for wild olive is ornamental. Both species can make good hedges, visual screens, or border shrubs and are available to a limited extent from a few native plant nurseries.

Fringetree (Old-Man's Beard)—*Chionanthus virginicus* (including a note on Pygmy Fringetree [*C. pygmaeus*])

Oleaceae

Fringetree rarely gets over 20 feet tall or has a trunk thicker than 4 inches in diameter. However, the largest one reported from Florida in 1987 (and former national co-champion) was 41 feet tall with a trunk slightly over 1 foot in diameter (42 inches in circumference) (Ward and Ing 1997). The deciduous leaves are simple, 5 to 10 inches long, rather shiny on top, smooth along the edges, and arranged in pairs opposite each other on the stem. The trunk is a light tannish-gray color and smooth except for the scattered, slightly darker little bumps, called "lenticels," that are quite noticeable on this species. This tree is named for its white flower clusters that hang down rather like a white beard. It is dioecious, having male and female flowers on separate plants. It normally blooms in March, at which time the leaves also begin to grow, and the fruits ripen on female trees in late July and early August. The plump fruits are about half to three-quarters of an inch long, dark purple when ripe, and have a thin layer of juicy green flesh surrounding a single, large seed (pit).

Fringetree occurs naturally from the south-central Florida peninsula northward into Maryland and westward into eastern Texas (Kartesz 2015). There is a similar species, the pygmy fringetree, *Chionanthus pygmaeus*, which has slightly larger fruits and inhabits the sand pine scrub habitats of the Lake Wales Ridge on the south-central Florida peninsula. An exotic species, the Chinese fringetree, is also planted as an ornamental to a limited extent in Florida.

Fringetree is most common along ecotones between hardwood forests and pine flatwoods forests, high pine forests, or sand pine scrub forest. It also occurs as a widely scattered tree in upland hardwood forests. It grows on both clay and sandy soils and is usually found in well-drained situations. It never occurs in swamps or lowland hardwood forests subject to prolonged flooding. However, it is fairly common in Gulf Hammock along the hammock-pine flatwoods ecotones where the ground is poorly drained and sometimes flooded. It is common in San Felasco Hammock along the high pine–upland hardwood forest ecotones on very well-drained, sandy soil. Its most common associates are live oak, laurel oak, water oak, sweetgum, pignut hickory, and waxmyrtle. Dogwood is an occasional associate in well-drained situations.

Old-man's beard has a large, deep root system and is quite drought resistant. It is not tolerant of prolonged flooding but can withstand limited flooding. It is fairly shade tolerant, to about the same degree as dogwood, but prefers sunny situations. Fringetree's growth rate is slow (about 1 foot per year), and its life span seems to be rather short, about 20 to 40 years. However, this tree is a prolific sprouter, and the root system often survives whatever kills the aboveground part of the plant, so that a new tree may sprout up to replace the dead top. Perhaps because of its strong root system, small fringetrees are easy to transplant when dormant.

Fringetrees do not spread by root suckers. However, female trees growing in the open where they get plenty of sun produce an abundance of fruit every year. There are often some seedlings around the parent tree, and many seeds are widely distributed by birds.

Fringetree's moderate wildlife value depends primarily on its fruits, which are eaten by birds, deer, and perhaps by some rodents and insects. Halls (1977) lists deer, wild turkey, and bobwhite quail as eating the fruit. Joe Durando, who, along with his wife, owns and runs an organic farm and native plant nursery in Alachua County, has observed wild mice (probably cotton mice or oldfield mice) feeding extensively on the seeds. I have witnessed mockingbirds eating the fruits whole and guarding the tree against other competitors and have seen cardinals eating the flesh from the seeds. A species of beetle larvae consumes many of the seed kernels, thus lowering the percentage of fertile seed. The stems and leaves, however, are rarely bothered by insects. The flowers are pollinated by bees (Braman et al. 2017).

From a human point of view, the main value of this little tree is ornamental. It is a time-honored favorite and an excellent tree to use in place of, or in addition to, often-overused dogwood and redbud trees and the even more widely overused crape myrtle. It can withstand a wider range of conditions than dogwood, including full sun, wet soil, and very dry soil and is also more adaptable than crape myrtle, which does poorly in shady situations.

Privet—Genera: *Forestiera* and *Ligustrum*

Oleaceae

The various plants occurring in Florida that are called "privet" include several native species of the genus *Forestiera* and several non-native species in the genus *Ligustrum*, two of which commonly escape into the wild. All of these species produce fruits called "drupes" that are eaten by birds, which help to distribute the seeds. This is a good thing where the native *Forestiera* species are concerned, as none are very common, but it is a bad thing with regard to the non-native *Ligustrum* species, two of which have become problematic invasive exotic pests. All species of *Forestiera* have separate male and female (dioecious) plants, whereas all species of *Legustrum* are bisexual. All of the privets are easy to grow and transplant. The individual species are described below.

FORESTIERA

Swamp Privet—*Forestiera acuminata*

Swamp privet, also called "eastern swamp privet," is an uncommon river floodplain and river swamp shrub that occurs in the Mississippi River valley from Louisiana north into southern Illinois and in various places on the Atlantic and Gulf coastal plains from the southern tip of South Carolina into eastern Texas (Kartesz 2015). In Florida, it is mostly found along the Apalachicola, Suwannee, and Santa Fe Rivers (Wunderlin et al. 2020). It is spotty in its distribution, often forming 10- to 20-foot-tall dense thickets in a few areas and being absent elsewhere. The maximum recorded size of an individual plant is 41 feet tall with a trunk diameter of 9 inches (American Forests 2018). The deciduous, opposite leaves are long pointed at the tip and are from 2 to 3 inches long and up to an inch wide. The often-leaning trunks have dark brown, slightly roughened bark. The plants produce yellow, sweet-smelling flowers lacking petals in very early spring before the leaves appear. As the 1-inch-long fruits mature in April or May, they turn from green to bluish-purple with a waxy coating on the skin.

Swamp privet is almost always a leaning, sprawling, irregularly shaped plant that can spread clonally as the ends of branches lean onto the ground and take root. It also spreads by seed, which may be distributed by flowing water and by animals such as channel catfish and cedar waxwings (Adams et al. 2007). It occurs in swamps, on floodplains, and on river banks, usually where there is an abundance of available calcium either in the soil or the water.

The wildlife value of swamp privet is good. The very early flowers provide a timely nectar and pollen source for bees, and the fruits are eaten by a variety of animals including channel catfish and cedar waxwings (Adams et al. 2007) in addition to wild turkeys, wood ducks, gray catbirds, gray squirrels, raccoons, and various other species.

Upland Swamp Privet—*Forestiera ligustrina*

Upland swamp privet is a small deciduous shrub with opposite leaves 1 to 2 inches long and gray stems. In early spring, it produces small, yellow, sweet-smelling flowers lacking petals followed by small blue drupes later in the season. It is similar to swamp privet but has smaller leaves that are not long pointed toward the end, and it grows on well-drained uplands. Upland swamp privet occurs from Tennessee, Georgia, and northern Florida west into eastern Texas, ranging as far south as Hillsboro County on the Florida peninsula (Kartesz 2015). It is uncommon and spotty in its distribution and seems to prefer growing on soils underlain with limestone.

The wildlife value of upland swamp privet is probably similar to that of swamp privet in that it blooms early with flowers that are attractive to bees and produces fruits that are edible by a wide range of bird and mammal species.

Godfrey's Privet—*Forestiera godfreyi*

Godfrey's privet, listed as endangered in Florida and Georgia, is a rare, deciduous shrub with slender, leaning, tan to gray trunks, and leaves 2 inches long by 1½ inches wide. It produces downward-arching spur branches 3 to 6 inches long on vigorous stems, which serve to hook the plant onto other woody vegetation, allowing it to climb into taller plants. Without an assist from other vegetation, the stems eventually arch over to the ground and take root where they touch the soil, thus producing additional plants. A majority of the plants I have observed in the wild have gotten their start this way. Like swamp privet and upland swamp privet, Godfrey's privet also blooms very early in the spring (late February through early March) with delicate, yellow, sweet-smelling flowers lacking petals, and it produces dark blue drupes half an inch long in early April. This fruit is relished by birds, especially catbirds and mockingbirds. The seeds make their way through the birds' digestive systems and are thus scattered about to produce new plants. I have observed this in my yard, where I have a population of these plants and where new plants appear in various odd places.

Godfrey's privet occurs in a few small, widely scattered populations in the central Florida Panhandle, the north-central Florida peninsula, northeastern Florida, one spot on the coast of southeastern Georgia, and at the south end of South Carolina (Kartesz 2015). The largest population I have seen was in Sugarfoot Hammock on the west side of Gainesville, Florida, where there were about 100 plants. It has now been largely destroyed by development. Other populations occur in San Felasco Hammock and on bluffs along the Apalachicola River.

Deer browse the plant to such an extent that it has a hard time growing in forests where deer are abundant. It is also threatened by invasive exotic shrubs and vines such as Chinese privet, glossy privet, skunk vine, and cat's claw vine.

The wildlife value of Godfrey's privet is good. It is a preferred deer-browse plant; the very early blooms are a good nectar and pollen source for bees; and the fruits are actively sought out by birds such as wild turkeys, northern mockingbirds, gray catbirds, and northern cardinals. I have watched catbirds and mockingbirds fighting over access to the fruit in my yard and have planted Godfrey's privet on forest property in Suwannee County where catbirds, mockingbirds, and wild turkeys eat the fruit.

Florida Privet—*Forestiera segregata*

Florida privet is a coastal, upright, semievergreen privet with smooth, narrow, glossy leaves 1 or 2 inches long. It often grows upright with a single trunk and a compact crown. It produces small, yellow flowers followed by small, blue drupes that are about a quarter of an inch long. It is usually not more than 10 feet tall with a trunk diameter of a few inches, but an unusually large one reported in 1993 was 17½ feet tall with a trunk diameter of about 8 inches (Ward and Ing 1997). Florida privet occurs on dunes and shell mounds and near salt marshes along the Florida Gulf and Atlantic coasts from the Florida Keys northward along the Gulf Coast to Dixie County and northward along the Atlantic coast through Georgia to the southern tip of South Carolina (Kartesz 2015). It grows in association with other coastal species such as red bay, red-cedar, cabbage palm, saw palmetto, and live oak.

As with the other members of the genus *Forestiera*, the flowers are attractive to bees and the fruits are attractive to birds. Because this species is upright with a compact crown of semievergreen foliage and is adapted to coastal situations, it is used to some extent as an ornamental. It is well adapted to coastal situations but does not do well inland.

LIGUSTRUM

Chinese Privet—*Ligustrum sinense*

Chinese privet is an evergreen shrub with 1- to 2-inch-long, pale green, opposite leaves and showy clusters of small, white flowers at the ends of the new growth in spring. It is commonly planted in Georgia, Alabama, and other southern states as a hedge and ornamental plant and has escaped cultivation to become one of the most damaging invasive exotic pest plants of the southeastern United States. It reproduces both by seed and by root suckers. It is especially prolific along stream bottoms throughout most of the southeast, where it forms dense thickets that are nearly impossible to control or eliminate. Fortunately, it is not as common in Florida at present, but it is becoming more abundant. It is listed as an invasive exotic pest in most southern states, including Florida, but is still available for purchase. Pollen from Chinese privet causes serious allergy problems for some people.

Deer browse the foliage and birds eat the fruit, spreading Chinese privet far and wide. The wildlife value of the foliage, flowers, and fruit is moderate, but this is more than offset by the amount of damage done by this invasive exotic plant as it dominates wild areas, eliminating native plants and native wildlife habitat.

Glossy Privet—*Ligustrum lucidum*

Glossy privet is a tall, upright, evergreen privet with dull green, 3- to 5-inch-long, somewhat glossy, opposite leaves with pointed tips. It produces clusters of white flowers and clusters of dull bluish-purple drupes. Unfortunately, it is a good fruit producer and has escaped into the wild, causing damage to native ecosystems. It grows rapidly and gets larger than the other privets in our area. The largest one reported in Florida in 1997 was 54 feet tall with a trunk diameter of about 2 feet (Ward and Ing 1997).

Since glossy privet grows as individual, scattered plants, it is much easier to control or eliminate than the thicket-forming Chinese privet. Coating the smooth, light-gray trunk with triclopyr mixed with oil plus a surfactant will kill an individual large plant. Small plants can be manually uprooted rather easily.

Although not especially attractive and very invasive, this plant has often been used as an ornamental to provide tall, dense, visual barriers, and, unfortunately, it is still available for purchase in nurseries and on the Internet. In Florida, it is listed as an invasive exotic pest plant.

Japanese Privet (Japanese Ligustrum)—*Ligustrum japonicum*

Japanese privet is a frequently used ornamental shrub and hedge plant. It has shiny, dark green, opposite, evergreen leaves about 3 inches long and 1½ inches wide with rounded tips. It grows with arching branches and multiple stems and responds well to pruning. If left unpruned for decades, it can grow 20 feet tall with a trunk 6 inches to a foot in diameter. Japanese privet sometimes produces clusters of small, white flowers but rarely produces much fruit. Japanese privet rarely escapes into the wild and has little wildlife value other than to provide nesting cover for songbirds such as mockingbirds and brown thrashers. The pollen can be bothersome for people who are allergic to it.

White Ash—*Fraxinus americana*

Oleaceae

White ash is a tall, moderately large deciduous tree with opposite, pinnately compound leaves. The trunk is usually straight, round in cross section, and free of branches for most of its length. The bark is distinctive, having ridges that crisscross to form a diamond pattern, except near the base of the tree where the bark pattern sometimes consists of rounded knobs. Ash species are notoriously difficult to distinguish from one another, but white ash has leaf scars in the shape of a horseshoe surrounding the sides and bottom of the axillary bud. Also, white ash is the only ash species in Florida that naturally occurs on well-drained uplands, although green ash is often planted on well-drained upland situations as an ornamental (with generally poor results). Male and female wind-pollinated flowers occur on separate (dioecious) trees. The seeds (fruits) of white ash, which are produced in clusters on the female trees, are long and narrow with a narrow wing and are wind distributed.

White ash occurs throughout the eastern United States and a bit of southern Canada, ranging into Florida in the central panhandle and in the northern peninsula as far south as Marion and Citrus Counties (Schlesinger 1990). It occurs in Torreya State Park, O'Leno State Park and River Rise State Preserve, San Felasco Hammock Preserve State Park, and Silver Springs State Park. It is not especially common in Florida but was abundant in many of the hardwood forests farther north before the accidental introduction of the emerald ash borer into North America. White ash, along with all the other species of ash, is currently being decimated throughout most of the eastern United States by this introduced forest pest, which was first discovered in the United States in Michigan in 2002 (Emerald Ash Borer Information Network 2020).

White ash wood is not rot resistant, and old, mature trees often have rotten or hollow places internally. This limits their longevity, which is generally less than 200 years. Maximum size in Florida is represented by a tree beside the Suwannee River that in 2012 measured 108 feet tall with a trunk diameter of 3½ feet. The largest one reported in the country is 117 feet tall with a trunk diameter of 6½ feet (American Forests 2019).

White ash typically grows in upland hardwood forests in association with sweetgum, pignut hickory, winged elm, laurel oak, water oak, swamp chestnut oak, southern magnolia, basswood, black cherry, hophornbeam, and other upland hardwood forest species. It is not particularly shade or flood tolerant and is only moderately drought tolerant. It prefers fertile soil and is more common on calcareous soils than elsewhere. As a mature tree, white ash is usually one of the forest's taller trees, with its crown in full sun.

The wildlife value of ash trees in general is moderate. The seeds (fruits) are sometimes eaten by squirrels, wild turkeys, and wood ducks (Martin et al. 1951), and the leaves are eaten to some extent by insects. (A large, all-black, leaf-footed bug, perhaps the Florida leaf-footed bug, commonly feeds on white ash in Gainesville, Florida.) Squirrels begin feeding on the seeds in the treetops in May, and some seeds begin to drop to the forest floor in late May making them available to wild turkeys and wood ducks. The last seeds drop to the ground from October to December. The upper trunks of old, mature ash trees are often hollowed out by heart rot (*Lentinus tigrinus*) (Harms 1990). The resulting hollow trunks and branches provide a valuable habitat resource for cavity-nesting wildlife species such as gray squirrel, flying squirrel, raccoon, opossum, bats, rat snakes, barred owl, wood duck, tufted titmouse, Carolina chickadee, and prothonotary warbler.

White ash is (or was) an important commercial timber tree because of its strong, flexible, and shock-resistant wood. Uses for the wood include the manufacture of paneling, furniture, baseball bats, tool handles, canoe paddles, and archery bows. White ash wood also makes excellent firewood.

White ash was commonly planted as a shade tree and has long been well suited for this purpose. Unfortunately, there is now the threat of the emerald ash borer spreading into Florida and killing ash trees. Green ash was also commonly planted, but white ash is a much better choice for well-drained upland soils. Neither of these ash trees does well on poor, sandy soils.

Pumpkin Ash—*Fraxinus profunda* (*F. tomentosa*)

Oleaceae

Pumpkin ash and green ash are nearly indistinguishable. They are tall-growing swamp hardwoods with opposite, pinnately compound leaves and gray, moderately rough-barked trunks. The leaves of pumpkin ash often have smoother margins and are generally somewhat larger and hairier than those of green ash. The narrow, winged fruits have somewhat wider wings than those of green ash as well. Pumpkin ash trees get to 100 feet tall with trunks 1 to 3 feet in diameter.

The native distribution of pumpkin ash is unusual. It occurs from the central panhandle of Florida into Marion County on the Florida peninsula, along the coastal plain in southeastern Virginia and eastern North and South Carolina, and in various places in the Mississippi River drainage basin, often in widely isolated populations (Harms 1990).

Pumpkin ash is very unusual genetically. It is either a hexaploid, meaning it has six sets of chromosomes (Wright 1957), or an octaploid, meaning it has eight sets (Whittemore et al. 2018). Most living organisms, such as green ash, are diploid, meaning they have two sets of chromosomes. Some plants, including some white ash trees, have four sets of chromosomes. If pumpkin ash is a hexaploid, it may have come into existence, perhaps more than once, when a white ash with four sets of chromosomes crossed with a green ash with two sets of chromosomes (Wright 1965). The newly fertilized cell must then have doubled its chromosome number and begun to develop into an embryo. Without doubling the chromosome number, the new cell with three sets of chromosomes would have been doomed, because a cell with three sets of chromosomes cannot divide its chromosomes evenly during cell division. If pumpkin ash is an octaploid, it still may have arisen from hybridization between green ash and white ash, or it may have arisen because of chromosome doubling and redoubling with only green ash parentage. (Whatever the explanation, pumpkin ash seems to have been formed multiple times, giving rise to the many widely isolated populations of this species.)

Pumpkin ash grows in river floodplain swamps. It is the most abundant tree along the Ocklawaha and Silver Rivers in Marion County and also grows in the floodplains of many other rivers, including the Suwannee, Santa Fe, and Apalachicola. Its most common associates include green ash, pop ash, bald-cypress, swamp tupelo, red maple, dahoon, and, on the Ocklawaha and Silver Rivers, cabbage palm.

Pumpkin ash is more tolerant of flooding than most trees, but not quite as flood tolerant as swamp tupelo or pop ash, and not nearly as flood tolerant as bald-cypress. It is restricted to floodplains, almost never growing on upland soils. Its wood is not rot resistant, and, with age, it often becomes hollow. It does not appear to be very long lived, with its maximum age probably being between 100 and 200 years.

The native ash trees of North America are now being decimated by an introduced insect: the emerald ash borer. This insect was first discovered in Michigan in 2002 and as of 2018 had spread to all states in the eastern half of the United States except Maine, Florida, and Mississippi (Emerald Ash Borer Information Network 2020). It will no doubt show up in each of these additional states within the next few years.

The wildlife value of ash trees in general is moderate. The fruits, which are produced by the female half of the population, ash trees being dioecious, are eaten to some extent by squirrels, wild turkeys, and wood ducks (Martin et al. 1951), and the leaves are eaten to some extent by insects. Squirrels begin feeding on the fruits in the treetops in May. Some of the winged fruits begin to drop by late May, with most dropping to the ground (where they are accessible to wild turkeys and wood ducks) between October and December (Harms 1990). The upper trunks of old, fully mature trees is often hollowed out by the heart-rot fungus *Lentinus tigrinus* (Harms 1990). The resulting hollow trunks provide a valuable habitat resource for cavity-nesting wildlife. Species that frequently use pumpkin ash cavities include gray squirrel, flying squirrel, raccoon, opossum, various bat species, rat snake, barred owl, wood duck, and prothonotary warbler.

The wood of pumpkin and green ash is valuable commercially for the manufacture of tool handles, dimension stock, furniture, paneling, plywood, packing crates, and pulpwood. It also makes excellent firewood.

Pumpkin ash is not well suited for use as an ornamental tree and is rarely planted. When a wild pumpkin ash is left in a landscaped situation, it often dies or does poorly because of alterations to the soil and hydrology during development. In any case, it is no longer recommended to plant ash trees, as they will probably be doomed to early deaths by the introduced emerald ash borer.

Green Ash—*Fraxinus pennsylvanica*

Oleaceae

Green ash is a tall wetland forest tree with a straight trunk and opposite, pinnately compound deciduous leaves. It is a common tree throughout most of the eastern United States plus some of southern Canada (Kennedy 1990, "*Fraxinus pennsylvanica*"), ranging south throughout the Florida Panhandle and into the Florida peninsula as far south as Polk County (Wunderlin et al. 2020). It is similar in size and shape to white ash and pumpkin ash. The bark is not as thick and strongly ridged as that of white ash, and the leaf scars are not as obviously horseshoe shaped, occurring mostly underneath the axillary bud. The leaflets are usually narrower than those of white or pumpkin ash, and the winged fruits differ in the shapes of their seeds, the wing attachments to the seeds, and the wings themselves. The male and female flowers of all ash trees are borne on separate (dioecious) trees and are wind pollinated.

Green ash is a wetland tree, rarely getting onto well-drained soil except when planted there as an ornamental. It occurs in swamps and on the banks and in the floodplains of streams and rivers. Its associates include bald-cypress, swamp tupelo, pumpkin ash, red maple, water hickory, swamp laurel oak, water oak, and overcup oak. Additional associates in the Florida Panhandle include sycamore and cottonwood. It is flood tolerant but not particularly shade tolerant.

Green ash, along with other North American ash species, is being decimated throughout most of the eastern United States north of Florida by the emerald ash borer. The population of this invasive exotic insect is expanding rapidly and will very likely enter Florida in the near future.

The wildlife value of green ash is similar to that of pumpkin ash and white ash, and its commercial value as a timber tree is similar to that of pumpkin ash.

In the past, green ash was often planted as an ornamental, especially farther north. It is not well adapted for landscape use in Florida, and is not planted here nearly as often now as it was a few decades ago. Today, the threat posed by the emerald ash borer is enough to discourage the use of this tree as an ornamental.

Pop Ash (Carolina Ash)—*Fraxinus caroliniana* (including a note on Swamp White Ash [*F. pauciflora*])

Oleaceae

Pop ash is a small wetland tree that has opposite, pinnately compound deciduous leaves and fruits on the female trees (all ash trees are dioecious). Pop ash fruits differ from those of other ash species in that the wing forms a flat circle or disc with the seed at the center instead of extending in one direction out from the seed as with other ash species. Pop ash bark becomes scaly on older trunks but never with the prominent ridges found on other ash trees. The trunks of pop ash are often crooked and leaning, and there are sometimes multiple trunks. Pop ash is variable in average size, appearance, and chromosome number, with a larger tetraploid form sometimes described as Swamp White Ash (*Fraxinus pauciflora*). Pop ash occurs mostly on the coastal plain from southeastern Virginia into eastern Texas and as far south as the Big Cypress Swamp in southern peninsular Florida (Kartesz 2015). The maximum size attainable by pop ash is indicated by a tree measured in O'Leno State Park in 1993 that was 58 feet tall with a trunk diameter of 1½ feet (Ward and Ing 1997). Most pop ash are less than half that big.

Pop ash is quite tolerant of prolonged flooding, often growing in the wettest parts of floodplains and swamps. It is not as flood tolerant as bald-cypress but seems to be more flood tolerant than water-elm, swamp tupelo, and pumpkin ash and much more flood tolerant than red maple. Common associates include buttonbush, bald-cypress, red maple, swamp tupelo, and water-elm. A mixed stand of pop ash and buttonbush grows on Sanchez Prairie in San Felasco Hammock in an area of prolonged deep flooding. It is a fairly common tree on the banks of rivers.

Pop ash is not shade tolerant. It survives either by leaning out over a water body away from larger forest trees near it or by growing in areas of deep, prolonged flooding that prevent most large trees, other than bald-cypress, from growing there too. Pop ash is not fire tolerant, which limits its ability to grow in association with pond-cypress. It is also not especially salt or drought tolerant. As with the other species of ash native to North America, many pop ash in Florida will likely be attacked and killed by the emerald ash borer if the rapidly expanding population of this invasive exotic insect reaches the state.

The wildlife value of pop ash is similar to that of pumpkin ash but perhaps somewhat lower given its smaller size, which limits the size of the cavities it provides for cavity-nesting species. Conversely, a stand of pop ash in a deepwater wetland could support a rookery of herons and egrets.

Pop ash is not normally used commercially or as a landscape plant, but it is grown and used for wetland restoration.

Primrose-Willow, Peruvian Primrose—*Ludwigia peruviana*

Onagraceae

Primrose-willow is native to Central and South America and some of the islands in the Caribbean. It has escaped from cultivation and is causing problems in various places around the world including India, Sri Lanka, Sumatra, Java, Australia, New Zealand, and here in Florida, where it is listed as a Category I invasive exotic pest plant (FLEPPC 2013). In the United States, it is most common in Florida, especially on the peninsula (Godfrey 1988). There are some isolated populations established in the states of Washington, North Carolina, Georgia, Alabama, Mississippi, and Texas (U.S. Fish and Wildlife Service 2018). One place where it is easily viewed is on Payne's Prairie in Alachua County, Florida.

Primrose-willow is an aggressive, rapidly growing plant with densely hairy stems and leaves and pretty yellow flowers. The alternate, fuzzy leaves, which taper to a point at the end and have entire margins, are 2 to 4 inches long and up to an inch wide. The bright yellow flowers, which have four round petals, are about an inch in diameter. The capsules produce thousands of sand grain–sized seeds. The plants grow in dense patches and become 5 to 10 feet tall. The plant spreads by seed and by the rooting of the branches as they spread out from the parent plant.

Primrose-willow is a wetland plant, aggressively invading marshes and the shores of lakes and ponds. It is sun-loving and not shade tolerant. It is very tolerant of shallow flooding and can form floating mats over deep water but can sometimes be killed by prolonged deep flooding or by fires during droughts. The upper parts of the plant are not frost tolerant, and the entire plant can be killed by hard freezes (U.S. Fish and Wildlife Service 2018).

The value of this plant is almost entirely negative as it can block waterways and eliminate valuable wildlife habitat. The flowers are visited by bees, wasps, and butterflies.

This plant has been transported around the world as an ornamental for its pretty yellow flowers. It should never be planted in North America or anywhere else outside of its native range, as it so easily escapes cultivation and is so invasive when it does.

Sycamore (Plane-Tree)—*Platanus occidentalis*

Platanaceae

Sycamore is a well-known, large, fast-growing, common tree throughout most of the eastern United States but is native in Florida only along the Apalachicola, Choctawhatchee, and Escambia Rivers in the panhandle (Godfrey 1988). However, it is widely planted as an ornamental well outside its native range. It is most noticeable for its bark, which, on young stems and the upper trunk and branches of large trees, is a smooth light gray with patches of various shades of tan, green, or chalky white. The bark on the lower trunk becomes roughened with brown, square-shaped scales, thus giving the trunk a distinctive two-toned look. The large, alternate, deciduous, somewhat maple leaf–shaped leaves are as wide, or wider, than long and rather stiff, with a dozen or so prominent points around the edge. The leaves turn brown in early autumn. The 1-inch-diameter balls of flowers and fruits hang on long, slender stalks from the twigs in summer and fall. Male and female flowers occur in separate clusters and balls on the same (monoecious) tree.

Sycamore can grow to be a huge tree, especially on moist, fertile soil in the middle and northern parts of the eastern and midwestern United States, and it is considered to be the largest hardwood tree native to eastern North America. The national champion from Kentucky in 1973 was over 12 feet in trunk diameter and 96 feet tall (Godfrey 1988). The current champion, from Ohio, is 124 feet tall with a trunk diameter of 11½ feet (American Forests 2018). In Florida, on moist, fertile soil in native forests, it is normally 100 feet tall with a 2- to 4-foot trunk diameter at maturity. When planted as a shade tree on well-drained upland sites south of its native range on the northern Florida peninsula, it normally reaches heights of 90 feet and achieves a 2- to 3-foot trunk diameter, with the Florida champion, from Alachua County, being 96 feet tall with a trunk diameter a bit over 6 feet (Florida Champion Trees Database 2020).

In the wild, sycamore is most common and does best on alluvial soils along streams and on bottomlands. It tolerates wet soil but is not especially flood tolerant and will die if the entire root area is inundated for more than two weeks during the growing season (Wells and Schmidtling 1990). Sycamore is fast growing and not particularly shade tolerant, although it bears shade better than cottonwood or black willow (Wells and Schmidtling 1990). Neither is sycamore particularly drought tolerant, doing best near permanent water sources. It withstands high winds fairly well because of its moderately strong wood and strong root system.

The wildlife value of sycamore is low, but the fruits are eaten by purple finch and goldfinch (Martin et al. 1951). Old sycamore trees can be good cavity trees (Sullivan 1994), providing good, long-lasting nesting and denning sites for barred owls, wood ducks, raccoons, bats, and other animals.

Commercially, sycamore wood is harvested for use as pulpwood and fuel wood and sometimes as low grade lumber. It has been experimentally planted for the production of biomass for fuel, as it grows fast and coppices well in dense plantations when grown on short rotations (Wells and Schmidtling 1990).

Sycamore is frequently used as a shade tree given its fast growth rate and adaptability. It tolerates air pollution fairly well and is not picky about soil. It does have some drawbacks, however, including its susceptibility to anthracnose that sometimes defoliates trees in late summer. In autumn, the large leaves blow around widely after falling and can be a nuisance. It should not be planted near septic tank drain fields as its roots aggressively invade them.

Clematis (Virgin's Bower)—*Clematis catesbyana* and *C. virginiana*; Leather-Flower—*Clematis glaucophylla*, *C. crispa*, and *C. reticulata*; and Japanese Clematis—*C. terniflora*

Ranunculaceae

These are perennial vines with opposite, compound, deciduous leaves; showy flowers in summer; and showy clusters of fruit in autumn, each fruit displaying feathery structures that stick out in all directions from a fruit cluster. Male and female flowers are borne on separate (dioecious) plants.

All parts of these plants are toxic, containing ranunculin, which is converted to protoanemonin (a blistering agent) when chewed and/or macerated (Colorado State University 2019). The protoanemonin breaks down inside mammals to become anemonin, which is nontoxic, so that the exposure of a person, a deer, or any other mammal to ranunculin is short lived, usually requiring no medical treatment (Colorado State University 2019). The plants can also cause a burning rash on skin if sap gets rubbed there. The blistering agent discourages deer and other mammalian herbivores from eating the foliage.

Virgin's bower is the name for two very similar species of high-climbing vines that occur as uncommon plants in scattered locations from Polk County northward into the southeastern United States, especially in limestone-rich areas such as central Tennessee. *Clematis catesbyana* is by far the species more commonly seen in Florida, occurring in three areas: from Polk County north into Alachua County; in the central panhandle; and in a spot in Duval County (Wunderlin et al. 2020). *Clematis virginiana* is the other species. It is rare in Florida with widely isolated populations found as far south as Hardee County on the central peninsula (Wunderlin et al. 2020). It becomes more common farther north in its range and reaches as far as Canada.

These vines are noticeable when in bloom in midsummer (July–August) with their hanging clusters of white flowers and, when in fruit, with their clusters of fruits with plume-like projections. Large, old stems of these vines, which can be 1 or 2 inches in diameter, are also distinct, having shredded-looking brown bark that reminds me of shredded wheat cereal. When young or cut back, these vines can grow very rapidly with long, slender, green stems that become tan the second year.

Clematis catesbyana is restricted to calcareous hardwood forest areas, both in well-drained, upland hardwood forest and along river floodplains. It occurs in San Felasco Hammock, at Alachua Sink, along Hogtown Creek in Alachua County, and along the Apalachicola River in the panhandle. *Clematis virginiana* diverges somewhat from *Clematis catesbyana* in its habitat preferences, growing best along stream and river floodplains and other moist, fertile places.

The wildlife value of these two clematis species is rather low. The flowers don't seem to be very attractive to pollinators but are sometimes visited by native bees, wasps, butterflies, moths, and flies, and the fruits may be eaten to some extent by songbirds.

The two kinds of virgin's bower, as well as the closely related Japanese clematis, are sometimes used as ornamentals for their pretty flowers. They grow vigorously in landscaped situations, reseed prolifically, and can quickly spread to new areas and become problematic.

Japanese clematis (*Clematis terniflora*), also called "sweet evening clematis," is an invasive exotic vine intentionally introduced into the United States in the nineteenth century for use as an ornamental. It is similar to the preceding two species, *Clematis catesbyana* and *Clematis virginiana*, but its leaves are more leathery with smooth rather than toothed edges. It has gone rampantly wild in many places in eastern North America, including in Gainesville, Florida, especially on stream floodplains such as along Hogtown Creek. It is similar in its ecological characteristics to the preceding two virgin's bower species but spreads even more aggressively. It should never be planted outside of its native range in Asia.

Leather-flower vines, *Clematis glaucophylla*, *Clematis reticulata*, and *Clematis crispa*, are small, slender, uncommon vines that occur in scattered places throughout much of the southeastern United States. They occur in northern

Florida with *C. crispa* and *C. reticulata* ranging south to near Lake Okeechobee and *C. glaucophylla* being restricted to a few spots in northern Florida (Wunderlin et al. 2020). They bloom in May and June with isolated, individual flowers at the ends of long flower stems that bend downward so that the flower is always pointing down.

Whiteleaf leather-flower, *Clematis glaucophylla*, prefers the moist to wet, fertile soil of floodplains, hydric hammocks, and bottomland forests. In Florida, it is rare and found only in Levy County and the central panhandle (Wunderlin 1998). Its leaves are a somewhat waxy white, especially on the underside. The 1-inch-long, urn-shaped flowers on long stalks are hot pink to purplish-red with yellow throats.

Swamp leather-flower (*Clematis crispa*) is perhaps more of a perennial herbaceous vine than a woody vine. Like whiteleaf leather-flower (*C. Glaucophylla*), it also prefers the moist, fertile soil of floodplains and hydric hammocks. It is much more common than *C. Glaucophylla*, occurring in the northern and central Florida peninsula, the Florida Panhandle, and most of the rest of the southeastern United States (Kartesz 2015). The bell-shaped flowers are pale lavender with varying amounts of white showing on the upward-curled ends of the petals.

Netleaf leather-flower (*Clematis reticulata*), like swamp leather-flower (*C. crispa*), is perhaps more of a perennial, herbaceous vine than a woody vine. It is the most common of the leather-flower species, occurring from the central Florida peninsula northward into South Carolina and westward throughout the Florida Panhandle and parts of Georgia, Alabama, Mississippi, and Louisiana into Arkansas and eastern Texas (Kartesz 2015). It grows on the well-drained sandy soil of upland pine forests, including sandhills, and in xeric hammocks. Its urn-shaped flowers are pinkish-lavender with varying amounts of white at the ends of the petals.

The wildlife value of the leather-flower species resides mainly in the benefit of the flowers to hummingbirds and native bees. Bumblebees are frequent visitors to the flowers, and songbirds eat some of the fruits (Hammer 2018). Caterpillars of the mournful thyris moth feed on both wild grape and all the various clematis vines (Metalmark Web and Data 2019).

The leather-flower species are desirable ornamental wildflowers. They are not weedy like the other clematis species and the flowers are unique and beautiful.

Yellowroot (Brook-Feather)—*Xanthorhiza simplicissima*

Ranunculaceae

In Florida, this unique little plant is very rare and listed as endangered. It occurs in the wild from New England southward into the panhandle of Florida and eastern

Texas along shaded stream banks and on moist, shaded soils (Godfrey 1988). It is most widespread and common in West Virginia, Virginia, North Carolina, eastern Tennessee, and Alabama (Kartesz 2015). It has a single, slender stem 1 to 3 feet tall topped with a whorl of deciduous, compound leaves having five leaflets with deeply serrated margins. Its open sprays of small, purplish-brown, star-shaped, five-petaled flowers are produced at the tops of the stems just under the leaves as the leaves are emerging. Seeds are contained within a dry fruit that opens along a single suture in the fall. The autumn foliage, which hangs on into December, turns from yellow to reddish-purple and finally to tan (Nooney 1994).

Yellowroot spreads by sprouting from its widespread, shallow root system, forming flat-topped clonal patches along the well-drained banks of streams (Nooney 1994). It also reproduces by seed.

The wildlife value of yellowroot comes from the flowers, which offer some benefit to insect pollinators, and the seeds which are, perhaps, eaten by birds and other animals.

Yellowroot was used in the distant past as both a dye plant for the yellow dye in its roots and stems and as a medicinal plant.

Yellowroot is valued to some extent in the native nursery trade as a tall, shade-tolerant ground cover that has pretty flowers and colorful fall foliage.

Supplejack (Rattan Vine)—*Berchemia scandens*

Rhamnaceae

This high-climbing, twining vine in the buckthorn family is easily identified by the smooth, black bark on the larger stems (small stems start out green, turning dark brown or gray, then black as they age) and by its simple, alternate, ovate, deciduous leaves with about ten prominent, parallel veins on each side of the blade. The plants are functionally dioecious, with the tiny, greenish flowers being all male or all female on any one plant. The fruits are blue drupes that are about a quarter of an inch in diameter, a bit longer than wide, and often somewhat flattened. Supplejack occurs in appropriate habitats throughout the southeastern United States from southeastern Virginia across into southern Missouri and then south into eastern Texas and the southern tip of the Florida peninsula (but not the Keys) (Kartesz 2015).

Sometimes called "Alabama supplejack," this vine occurs mostly in swamp forests, bottomland forests, and stream floodplains. It also occurs in hydric hammocks and on the lowland edges of pine forests and, less commonly, in hardwood forests on fertile soil on well-drained uplands, especially where limestone is near the surface. It is a calcium-loving plant, and the foliage and fruits are high in calcium content. It is moderately flood and drought tolerant.

It grows in shade when young but prefers sunlight, growing high into the crowns of tall trees. It has no tendrils or thorns to help it climb, but it aggressively wraps around the stems and trunks of upright woody vegetation, often to such an extent that the host stem is deformed by trying to grow around the constricting vine.

Wildlife that eat the ripened fruit, which often persists throughout the winter (Halls 1977), includes raccoon, American robin, northern mockingbird, and brown thrasher (Martin et al. 1951). Wild turkey and gray squirrel also consume the fruit, and I have no doubt that there are additional mammals and birds that eat it as well. (The fruit is not edible by humans, however.) Sprouts and young shoots are browsed by white-tailed deer.

Supplejack is a beneficial nectar and pollen plant for the southeastern region of the United States (Oertel 1967). Beekeepers in eight southern states list it as one of the wild plants useful for honey production (Ayers 2016). Supplejack vines are also used to make furniture and possibly walking sticks. The wood is heavy, hard, moderately strong, but very prone to splitting longitudinally when dried.

Supplejack is not used as an ornamental plant.

New Jersey Tea (Red Root)—*Ceanothus americanus* and Littleleaf Buckbrush (Littleleaf New Jersey Tea, Small Leaf Red Root)—*Ceanothus microphyllus*

Rhamnaceae

New Jersey tea is a small shrub with gray-green, fuzzy foliage topped in late spring to early summer with ball-shaped, fragrant, showy clusters of tiny white flowers. It is rare in Florida, where it occurs in the panhandle and as far south on the peninsula as Highlands County, but it is more common farther north where it grows in scattered locations throughout the entire eastern half of the United States and parts of southeastern Canada (Kartesz 2015).

New Jersey tea has a large, deep root system of reddish roots and a woody base from which thin, mostly herbaceous branches sprout up each year to form a rounded, compact shrub. The branches turn yellow in autumn. It is very drought tolerant and capable of fixing nitrogen. It is neither flood nor shade tolerant and occurs in fire-adapted prairie, oak savanna, and pine savanna in partial shade to full sun on well-drained sandy or rocky soils. In Florida, it was often found in association with longleaf pine, southern red oak, and mockernut hickory in a fire-adapted plant community that is no longer common. Where this community still exists, it is usually not burned frequently enough to maintain New Jersey tea or many of the other less common associated species that require frequent fire.

Wildlife values for New Jersey tea include the foliage, which is browsed by deer and rabbits, the seeds eaten by bobwhite quail and wild turkeys, and the nectar and pollen valuable to a wide range of pollinators including native bees, wasps, flies, beetles, and hairstreak butterflies (Hilty 2019, "New Jersey Tea").

Human use of New Jersey tea occurred during the American Revolutionary War and Civil War periods when colonists used the dried, caffeine-free leaves to make tea. Throughout eastern North American history up to the present time, extracts from the roots have been used medicinally (Lee and Joullie 2018).

New Jersey tea is sometimes planted as an ornamental. It is available from nurseries and should be planted as young potted plants. Transplanting larger, established plants is nearly impossible given New Jersey tea's large, deep root system. It does best in full sun on well-drained soil.

Littleleaf buckbrush is a tiny, shrubby relative of New Jersey tea that grows to about 1 foot tall with up to 2 feet of crown spread. The thin, yellow stems have tiny, deciduous leaves half an inch or less in length and produce small, showy, fragrant, ball-like clusters of tiny white flowers in summer. The flowers are similar to but smaller than those of New Jersey tea. It occurs in sandhill longleaf pine forests in southern Alabama, southern Georgia, in the Florida Panhandle, and on the Florida peninsula south to Highlands County, but it is missing from the counties next to the Suwannee River except for Levy County (Godfrey 1988; Wunderlin et al. 2020).

Littleleaf buckbrush forms part of the ground-cover vegetation in fire-adapted longleaf pine, turkey oak, and wire grass savannas that burn frequently. It is common in sandhill forest areas such as Riverside Island and Salt Springs Island in the Ocala National Forest and also occurs in the Apalachicola National Forest. It is very fire and drought tolerant but not shade or flood tolerant. Like New Jersey tea, it benefits greatly from frequent prescribed burning.

Presumably, this plant is somewhat similar to New Jersey tea in its value to insect pollinators, as a browse plant for deer and rabbits, and as a provider of seeds for bobwhite quail and turkeys.

Littleleaf buckbrush has not been used to any noticeable extent by humans.

Carolina Buckthorn—*Frangula caroliniana* (*Rhamnus caroliniana*)

Rhamnaceae

Carolina buckthorn is a deciduous shrub or small tree 10 to 20 feet tall at maturity with trunks a few inches in diameter. Former national champions recorded in the American Forests publication have been as tall as 46 feet with trunk diameters slightly over 1 foot (Godfrey 1988). The 2- to 6-inch-long leaf blades are shiny dark green with eight to ten prominent parallel veins on each side, running from the midvein to the leaf edge. The small bisexual flowers are inconspicuous, but the one third–inch-diameter edible drupes turn red and finally black as they ripen in autumn. Carolina buckthorn, in spite of its name, has no thorns. It has a somewhat unusual natural distribution, being common in some places, such as the Ozark region of Missouri and Arkansas and the limestone regions of Tennessee, Kentucky, and Alabama, and absent in others. It occurs generally from the Carolinas (but not on the Atlantic coastal plain) to central Texas (Kartesz 2015) and occurs in four isolated spots in Mexico. Its range extends into Florida in a spotty fashion, where it occurs in a few areas in western Alachua and Marion Counties as well as a few spots in Levy, Citrus, Polk, Sumpter, St. Johns, Flagler, Orange, and Sarasota Counties and a few counties adjacent to the Apalachicola River and in the extreme western panhandle (Wunderlin et al. 2020).

Carolina buckthorn is restricted to well-drained soils with abundant available calcium. It is moderately shade tolerant but prefers full sunlight. It is very drought tolerant, slightly flood tolerant, and is quite susceptible to damage by stem borers, which can kill the entire plant.

The wildlife value of Carolina buckthorn is high, largely because of the fruit's desirability for birds and mammals. The fruits (drupes) ripen over an extended period of time in mid- to late summer, with some fruit ripening as late as November 1st. The flowers are visited by bees and other pollinators, and the foliage is a favored browse plant for white-tailed deer.

Carolina buckthorn is used as an ornamental, especially farther north in its range, but it is so particular about soil conditions that it is seldom used in Florida. Common buckthorn, a very invasive relative from Eurasia and North Africa, is a problematic invasive exotic plant in parts of the United States, mostly north and west of North Carolina and Tennessee.

Climbing Buckthorn (Small-Flowered Buckthorn)—*Sageretia minutiflora*

Rhamnaceae

Climbing buckthorn is a rare plant that occurs either as a large shrub with long-reaching branches or as a high-climbing vine. Its tardily deciduous, shiny green leaves are half an inch to 1½ inches long, with two or three prominent veins on each side of the blade, and strong, 1-inch-long, often downward-sloping thorns (that start out as branchlets with a few leaves) along the stems that aid its climbing upward through other vegetation. The long thorns and multiple side branches are this plant's only climbing mechanism as it does not twine or have tendrils. It produces clusters of very small (an eighth of an inch across), very fragrant, white flowers in August and September and purple drupes a quarter of an inch in diameter in autumn that separate into three leathery nutlets by the end of winter (Godfrey 1988). As a high-climbing vine or liana, it has somewhat scaly, tan bark and very hard wood. The thorns gradually die and rot away on old vine stems, which can be up to about 2 inches in diameter. The vine can climb to a height of at least 80 feet into the crowns of forest trees.

Climbing buckthorn occurs along the coast in the Carolinas, Georgia, extreme southern Alabama and Mississippi, and in isolated spots throughout Florida as far south as Collier County on the southwestern edge of the peninsula (Kartesz 2015; Wunderlin et al. 2020). It mostly grows on calcareous soil over limestone or on shell mounds and is most common near the coast. It was common in the higher parts of Gulf Hammock, is common on shell mounds along both coasts, and occurs sparsely in San Felasco Hammock. It often grows in association with live oak, cabbage palm, red-cedar, swamp chestnut oak, Florida maple, sugarberry, red bay, yaupon, and soapberry in upland hardwood forests on calcareous soils.

Climbing buckthorn is shade tolerant to some degree but does best in full sun. In shady situations, it tries to climb into the sun. It is drought tolerant, somewhat salt tolerant, but not very flood tolerant. It spreads by underground runners and by seed, which is distributed by birds and other wildlife.

The wildlife value of this rare plant is probably pretty good. When in bloom in late August and September, it provides flowers for insect pollinators and, in autumn, fruit for wildlife. To date, however, the specifics of these benefits have not been published. I have seen diverse native bees and wasps on the flowers. It is also a browse plant for white-tailed deer.

Climbing buckthorn has no known human uses and is not planted as an ornamental. It is picky about soil and will not grow on poor, sandy soils or poorly drained, acidic, flatwoods soils. Nonetheless, it is tenacious in places where it *will* grow, including in harsh, inner city situations where it tolerates the heat, bright sun, and the impact on the soil of concrete and limerock.

Downy Serviceberry—*Amelanchier arborea*

Rosaceae

Downy serviceberry is a large shrub or small tree, often multistemmed, with gray bark that develops vertical fissures, and simple, deciduous, 1- to 3-inch-long leaves with finely serrated edges. The lower surface of the leaves and the 1-inch-long slender petioles are densely covered with soft, woolly hairs when they first emerge in the spring. In the fall, the leaves provide attractive fall color. The clusters of showy, white, bisexual flowers bloom in early spring just before or as the leaves begin to appear. The flowers are about 1 inch across. The one third–inch-diameter, reddish-purple, apple-like fruits (drupes) containing four to ten tiny seeds ripen in midsummer (Hilty 2019, "Downy Serviceberry"). Downy serviceberry occurs throughout most of the eastern United States, ranging as far south as the Florida Panhandle (Godfrey 1988).

The largest downy serviceberry recorded in Florida is in Leon County and is 37 feet tall with a trunk diameter of about 9½ inches (Florida Champion Trees Database 2020). The largest one in the United States, in 1959, was 50 feet tall with a nearly 3-foot trunk diameter (Godfrey 1988). Downy serviceberry is relatively short lived, rarely getting more than 50 years old (Bernheim Arboretum and Research Forest 2019).

Downy serviceberry grows on well-drained soil in open woodlands, either along streams or in association with longleaf pine, southern red oak, and mockernut hickory in fire-adapted woodlands, often along edges such as the transition to hardwood forests along the upper edges of ravines. It prefers full sunlight but will grow in partial shade. It is drought resistant but not flood tolerant. It resprouts well following fires.

The wildlife value of downy serviceberry is good. Honey bees, native bees, flies, beetles, and other insects visit the blooms in early spring, and caterpillars of various moths and butterflies feed on the foliage including striped hairstreak and red-spotted purple butterflies (Hilty 2019, "Downy Serviceberry"). The fruits are an important wildlife food in early summer, attracting a wide assortment of birds and mammals (Martin et al. 1951).

Human use of downy serviceberry is primarily as a popular ornamental within its native range because of its showy flowers, its fruit, and, especially, the pretty fall color of its foliage. It also provides fruit that is sometimes used for making pies, jams, and preserves (Bernheim Arboretum and Research Forest 2019).

Red Chokeberry—*Aronia arbutifolia (Sorbus arbutifolia)*

Rosaceae

Red chokeberry occurs naturally from Newfoundland and Nova Scotia southward mainly throughout the Atlantic coastal states into the central Florida peninsula and westward through the Florida Panhandle into Arkansas, Louisiana, and eastern Texas (Godfrey 1988; Kartesz 2015). A member of the rose family, it has deciduous, alternate, 1- to 3-inch-long leaves with finely serrated edges and gray, woolly undersides, and produces clusters of white, three eighths–inch-wide, bisexual flowers with pink stamens in spring and clusters of quarter-inch-diameter red, apple-like fruits in the fall that hang on through winter. There are small, purplish-red glands on the tips of the serrations and along the midrib of the upper leaf surface (Godfrey 1988). Red chokeberry grows with multiple stems, each of which is slender and wand-like, reaching heights of 4 to 8 feet. The spreading surface roots put up new stems, thus sometimes forming rather open thickets.

Red chokeberry is found in Florida primarily in pine flatwoods areas either mixed sparingly with the saw palmetto and gallberry shrub layer or, more commonly, along ecotones around wetlands such as cypress domes and stream-side thickets, again mixed with the other shrubs. It is a sun-loving plant that seldom occurs within the shaded confines of swamps and bayheads. It is rarely abundant.

Red chokeberry is tolerant of acidic soils and wet conditions but is not especially flood tolerant. It is fire tolerant, producing vigorous sprouts following a fire, and benefits from periodic fires that keep the habitat open and sunny.

The wildlife value of red chokeberry is moderate. It is a good fruit producer, but the fruits are so acidic that most wildlife species do not eat them right away. Instead, they let the fruit hang on well into winter, at which time other fruit is getting scarce and perhaps the acidity of the chokeberry fruit is diminishing. It is then, in midwinter, that species including cedar waxwings, bluebirds, black bears, and raccoons eat the fruit. The flowers are visited by bees and other insects, and deer will browse on the foliage.

Red chokeberry is well known and often available in the plant nursery trade, with several named varieties offered for sale. It is particularly attractive because of its bright red fall leaf color, but the flowers in spring and the bright red fruits are also attractive.

Hawthorn (Haw)—*Crataegus* spp.

Rosaceae

Hawthorns are large, thorny shrubs or small, thorny trees in the rose family with deciduous, alternate leaves; showy, white, bisexual flowers in spring; and showy red, orange, or yellow fruits in summer or fall. Included here are some of the more common species.

Many hawthorn species have been described, and there is no clear consensus among botanists as to how many species there are or what to call some of the many forms. The cause of much of this confusion links to the fact that hawthorns are able to produce seed by a process called "apomixis" wherein no fertilization takes place, which results in the seedlings being genetically identical to the parent. In this way, populations of genetically identical hawthorn plants are formed that appear to be distinct species but are, in fact, clones. To add to the difficulty, there are hybrids and polyploids within some populations of hawthorns. (Hybrids and polyploids usually look different than the species they are derived from without being different species, which creates visual confusion for anyone trying sort out species diversity.)

Hawthorns tend to inhabit well-drained soil either on uplands or on floodplains that are dry and well drained some of the time. They are not found in swamps, marshes, bayheads, or bogs and only rarely occur in pine flatwoods habitats.

Hawthorns are susceptible to cedar apple rust, which causes yellow- to orange-colored spots on the leaves and fruit and causes galls to form on red-cedar and other junipers, which are the alternate host species for the fungus.

The flowers, which have an unpleasant odor, are visited by solitary bees, honey bees, flies, and other insect pollinators, and the fruits are eaten by a wide range of birds and mammals, although not to as great an extent as one might expect based on availability (Martin et al. 1951). Game birds that make some use of the fruits are wild turkey, bobwhite quail, and wood duck (Halls 1977). The rather dense, thorny crowns are often used as nest sites by birds such as mockingbirds, brown thrashers, cardinals, blue grosbeaks, loggerhead shrikes, eastern kingbirds, and white-eyed vireos. The foliage is browsed to some extent by white-tailed deer (Halls 1977).

The fruits are edible, and people sometimes make preserves, jams, and jellies with them. The wood of most hawthorn species is hard, dense, and strong and has been used for making tool handles. Many hawthorn species also have been used occasionally as ornamentals.

Mayhaw—*Crataegus aestivalis*

Mayhaw is a floodplain specialist that occurs primarily from the Atlantic coastal plain of the Carolinas southwestward into the Gulf coastal plain of Florida, Alabama, and Mississippi, ranging as far south on the Florida peninsula as Levy and Volusia Counties (Kartesz 2015). It has gray bark with often-exposed, reddish inner bark and sometimes has thorns on its trunk. The leaf blades are usually oblong to spatula shaped and may be a dark, glossy green on top; however, the upper leaf surfaces may also be dull and the lower leaf surfaces a gray or rusty red-brown, in both cases because of dense pubescence (Godfrey 1988). Mayhaw blooms well with showy, white to slightly pink flowers in early spring. It produces red fruits about half an inch in diameter that ripen in May. In 1993, a large specimen in Texas measured 43 feet tall with a 10-inch trunk diameter (American Forests 1994).

Mayhaw grows on the floodplains and edges of streams, rivers, ponds, lakes, and bayous. It is flood tolerant and moderately shade tolerant. The frequently abundant fruit production of this hawthorn benefits both wildlife and humans. This is the fruit that is used to make Mayhaw jelly, a popular product in the regions where Mayhaw grows.

Cockspur Hawthorn—*Crataegus crus-galli*

Cockspur hawthorn is a large hawthorn with a broad crown of very thorny branches. Thorns sometimes also occur on the trunk. Its leaves are usually un-lobed and fairly narrow, being broadest toward the tip of the leaf. The form here in Florida is often referred to as variety *pyracanthifolia* because of the leaf shape, which is similar to that of the exotic ornamental pyracantha (firethorn). It occurs naturally throughout most of the eastern United States from Maine into Texas, ranging as far south on the Florida peninsula as Lake County. An exceptionally large one in Virginia measured 40 feet tall with a trunk diameter of 19 inches (American Forests 1994).

Cockspur hawthorn is usually found in upland situations such as open woods, pastures, and fencerows but is also found on moist soil in or adjacent to bottom-land hardwood forests and along streams in pine flatwoods forests (Godfrey 1988).

Beautiful Hawthorn—*Crataegus flava* (*C. pulcherrima*)

Beautiful hawthorn is an uncommon species occurring mostly in Mississippi, Alabama, southwestern Georgia, and the central Florida Panhandle, with isolated populations in Tennessee, eastern Texas, eastern Georgia, and Alachua County, Florida (Kartesz 2015). It is a small tree with brown, furrowed bark; straight twigs; small, quarter-inch-diameter fruits that turn red; evenly lobed leaves with glandular margins and parallel veins; and flowers with 20 stamens (Chafin [2009] 2020). Maximum reported size is 44 feet tall with a trunk 7 inches in diameter (American Forests 1994).

Beautiful hawthorn's habitat is usually upland hardwood forest or mixed pine hardwood forest in ravines and on moist slopes (Chafin [2009] 2020). It is moderately shade tolerant but grows best in full sun. There is a moderately large individual on a slope under a power line in the middle of San Felasco Hammock in Alachua County, Florida.

Parsley Haw—*Crataegus marshallii*

This hawthorn has a distinct appearance. Its deeply lobed, 1-inch-long and wide triangular leaves look like parsley leaves, and its smooth bark, where patches of outer bark flake away on large stems, resembles crape myrtle bark. Parsley haw blooms well with a good show of white to slightly pink flowers in early March followed by small, bright red fruits in fall that hang on into winter. Parsley haw has fewer spines than most hawthorns and is usually slender and tall-growing with an open, irregular crown.

Parsley haw occurs naturally from southeastern Virginia into southeastern Oklahoma and eastern Texas, including the Florida Panhandle and as far south on the peninsula as Polk County (Kartesz 2015). Examples of large specimens include one in Gainesville, Florida, that was 33 feet tall with a 5-inch trunk diameter in 1973 (Ward and Ing 1997) and one in Texas that measured 24 feet tall with an 8-inch trunk diameter in 1993 (American Forests 1994).

The native habitat of parsley haw is the moist, fertile soil of floodplain forests, bottomland forests, stream and river banks, and adjacent slopes (Godfrey 1988). Parsley haw is shade tolerant and moderately flood tolerant but not so flood tolerant that it can survive in swamps or marshes.

A popular ornamental, parsley haw grows well in most landscape situations. Although it usually occurs in the shade of a hardwood forest in nature, it does well in both partial shade and full sun.

Yellow Haw, Summer Hawthorn, Sandhill Hawthorn, Jacksonville Hawthorn—*Crataegus michauxii* (*C. flava*, *C. floridana*, *C. lassa*, *C. quaesita* var. *floridana*, *C. versuta*)

The discussion of this species illustrates the confusion resulting from multiple botanists having differing opinions about the taxonomy of hawthorns. This is a particularly difficult, variable, and perplexing hawthorn or population of hawthorns. One of the latest attempts to sort out this particular group of hawthorns is by Phipps and Dvorsky (2008), who arrived at 22 different species within what they call "*Crataegus* series Lacrimatae," which ranges over the southeastern coastal plain from the Carolinas into Texas. (In the mid-twentieth century, the famed Canadian-born botanist and horticulturist of the southern United States, C. D. Beadle, recognized 54 species within this group.) Phipps and Dvorsky point out that all the hawthorns within this group have several characteristics in common: frequent weeping growth habit; yellow to orange, apple-like fruits; fire adaptation; occurrence on dry upland sites; thick bark that appears partitioned into small blocks; and zigzagging twigs. Godfrey (1988) treats them all as one species here in Florida—yellow hawthorn (*Crataegus flava*)—which includes the form previously known as "Jacksonville hawthorn" (*Crataegus floridana*). The names "yellow haw" and *Crataegus flava* are now being applied to a different species, the "beautiful hawthorn" (Wunderlin et al. 2020); this has produced instances of unclear identification as demonstrated by multiple entries for this plant on the Internet, where some authors use these names for one species and some for the other.

This species or group of species, which is designated here as "yellow haw" (*Crataegus michauxii*) for discussion purposes, is an occasional shrub or small tree on dry, well-drained, sunny uplands. It occurred originally in association with longleaf pine, southern red oak, mockernut hickory, post oak, chinquapin, and wire grass on both sandy and clayhill uplands that burned frequently. Most of the plants that still exist are hanging on in small, leftover parcels and edges now that the bulk of their former habitat has been destroyed. One place where yellow haw might still be found in native habitat is Blackwater River State Forest in the Florida Panhandle.

Yellow haw here in Florida often has a distinctive appearance, being upright but with a crooked, often leaning, trunk and a weeping outer branch habit. It is, however, quite variable. It may or may not have the weeping growth habit, and the 1- to 2-inch-long leaves may take almost any shape from spatula-like to almost round with either entire or toothed margins. There are glands on the leaf margins. Yellow haw occurs both on upland former longleaf pine areas and in scrub on the central Florida peninsula. It is probably the most drought tolerant species of hawthorn in Florida and also probably the most fire tolerant. It is not at all shade tolerant, requiring full sunlight for at least part of the day.

Yellow haw can be rather large. In 1983, one in Levy County measured 30 feet tall with a 15-inch trunk diameter at 4½ feet above the ground (American Forests 1994). It can also be quite small, especially when growing in very xeric habitats such as scrub. Yellow haw is a slow-growing, long-lived plant.

Yellow haw is often an attractive plant with a dense crown and weeping outer branches, and it blooms well in mid-March if growing in full sun. It can make a good ornamental, as it is easily grown, lives a long time, and requires no care in terms of watering, fertilizing, spraying, or pruning. The forms with weeping outer branches would be particularly interesting specimen plants provided they are allowed to assume their natural shape. In general, yellow haw must be grown from seed or purchased as a potted plant from a nursery: it is nearly impossible to transplant because of its deep root system.

One-Flower Hawthorn (Dwarf Hawthorn)—*Crataegus uniflora*

This is an upland hawthorn that usually produces flowers one at a time instead of in clusters, although two or three flowers may form an occasional cluster. It has roundish, serrated leaves 1 to 2 inches long and gray bark that is roughened with shallow ridges on larger trunks. The fruit is green to reddish and from one quarter to half an inch in diameter. Its growth habit is upright and irregular. It occurs from New Jersey and Pennsylvania throughout the southeastern United States into the Florida Panhandle and as far south on the peninsula as Polk County and ranges as far west as Missouri and eastern Texas (Kartesz 2015).

Occurring mostly in upland hardwood forests as an understory shrub or small tree but occasionally found in fire-adapted upland pine forests, this hawthorn is moderately shade and drought tolerant. It is rather small, with the largest one reported being 18 feet tall with a trunk diameter only slightly over 4 inches (American Forests 1994).

Green Haw—*Crataegus viridis*

Green haw is a common, large hawthorn that grows to be a small tree on flood-plains from Virginia as far south as Marion and Volusia Counties on the northern Florida peninsula and westward across the southern United States into southeastern Nebraska, Oklahoma, and Texas (Kartesz 2015). It has flaky gray bark that often reveals a reddish-brown inner bark and usually has branched thorns on its trunk. The leaf blades are 2 to 3 inches long and 1 to 2 inches wide with serrations and, often, lobed margins. It blooms in early spring with attractive white to slightly pinkish flowers in clusters, and it produces ample crops of red-orange, apple-like fruits in the fall that hang on through the winter.

Green haw is one of the largest hawthorns. In 1981, one in West Virginia measured 40 feet tall with a trunk diameter of 20 inches (American Forests 1994). It occurs primarily on the floodplains of streams and rivers and is moderately flood and shade tolerant. Fruit production is often abundant. In the fall of 1973, in San Felasco Hammock in Alachua County, Florida, I found a black bear scat that contained a large amount of the fruit, and on January 3, 2021, Alachua Audubon Society president Debra Segal observed flocks of American robins and cedar waxwings consuming large amounts of green haw fruit at the River Rise Preserve State Park on the Santa Fe River in Alachua County (personal communication).

Southern Crabapple—*Malus angustifolia*

Rosaceae

Southern crabapple, a tall shrub or small tree in the rose family, is native to the southeastern United States and reaches the middle to eastern panhandle region of Florida as far east as Hamilton County (Kartesz 2015). Normally growing about 30 feet tall with a trunk diameter to 10 inches, an unusually large one in Rockville, Maryland, measured 45 feet tall with a 3-foot trunk diameter (American Forests 2018). Southern crabapple is often encountered in the wild as a patch of tall, slender stems originating from the root sprouts of a parent tree that may no longer be present. The alternate, deciduous, 1- to 2-inch-long leaves have serrated edges and are either oblong on spur shoots or nearly oval on long twigs (Godfrey 1988). Clusters of showy, bisexual, sweetly fragrant, light pink flowers about 1 inch wide appear in early spring, followed by the apple- to pear-shaped, yellow-green, 1-inch-diameter fruits. The bark is gray on small trunks and, on older trunks, is roughened by flat-topped ridges between which the reddish-brown inner bark can be seen. The young stems are reddish-brown. The stems and trunks have thorns that initially bear leaves.

Southern crabapple is a sun-loving shrub or small tree of open woodlands on well-drained upland soils containing some clay. It is uncommon in Florida where it is listed as threatened. It is not shade tolerant but is drought tolerant and capable of sprouting back vigorously following fires. It is an alternate host of cedar-apple rust, which can cause reddish spots on the leaves and fruit.

Southern crabapple is moderately beneficial for wildlife. The flowers are visited by honey bees and native bees, such as bumblebees; the fruit is edible by a wide range of mammal and bird species; the foliage and stems are browsed by deer and rabbits; and the moderately dense, thorny crowns make good nesting habitat for songbirds (Hilty 2019, "Wild Crab Apple").

Humans use crabapple primarily as an ornamental tree. It often grows well in landscaped situations, and it puts on a show of pink flowers in spring. It is not especially long lived in Florida but can last several decades. The fruit is sometimes made into jams and jellies, for which it is well adapted because of its acidity and high pectin content.

Ninebark—*Physocarpus opulifolius*

Rosaceae

This is a medium-sized, deciduous, native shrub in the rose family that somewhat resembles spirea (*Spiraea*) and occurs throughout much of eastern North America, extending as far south as Jackson and Calhoun Counties in the Florida Panhandle (Godfrey 1988). It is rare in Florida and nearby states, being more common farther north, and is listed as endangered in Florida. It has triangular leaves that can be 3 to 5 inches long and 2 to 4 inches wide with serrated edges and often two lobes near the leaf base. The brown bark on older stems peels open, revealing tan younger bark. The white to slightly pinkish, bisexual flowers occur in dense, round clusters on the ends of twigs in late spring, and the reddish-green fruit clusters ripen in autumn. The branches of young plants are upright, but older specimen plants have long, arching branches. The flower clusters on plants in sunny, open situations can be showy.

Ninebark occurs along stream banks and on adjacent moist slopes in open woodlands, where it reproduces primarily by seed. It is most common and most often used as an ornamental in the northern part of its range.

The wildlife value of ninebark resides mainly in its benefit to insect pollinators such as honey bees, native bees, wasps, flies, and butterflies (Hilty 2019, "Ninebark"). It is browsed to some extent by white-tailed deer.

Ninebark is a well-known ornamental with many named varieties. It is used as a landscape plant primarily in the northern United States and southern Canada, where it is a tough, adaptable shrub requiring little care.

Black Cherry—*Prunus serotina*

Rosaceae

Black cherry (wild cherry) is a common tree in the eastern United States including the Florida Panhandle and as far south on the Florida peninsula as Desoto County (Wunderlin et al. 2020). It also occurs as a native tree in Mexico, Honduras, and South America and as an invasive exotic tree in Europe, Australia, and New Zealand. It is often seen as a young tree of medium size, from 30 to 60 feet tall with 6- to 18-inch trunk diameters. However, at maturity, it is often 100 feet tall with a straight trunk 2 feet or more in diameter. The largest one in Florida was recorded in 1981 in Sugarfoot Hammock on the west side of Gainesville with a trunk diameter of 5 feet (192 inches in circumference at 4½ feet above ground). A second big one nearby had a 4½-foot trunk diameter. Young trees are often quite pretty, with silvery-gray, birch-like bark and shiny dark green, alternate, deciduous leaves that move easily in the slightest breeze. The 3- to 4-inch-long by 1½-inch-wide leaf blades have serrated edges and a strong, distinctive odor when crushed. A pair of nectaries (nectar-secreting glands) occurs on the top surface of the petiole near where it joins the leaf blade. Small, bisexual, white flowers appear as racemes in March on stalks that grow from the previous season's twigs. The cherries turn from green to red to black as they ripen in June and July. The bark on older trunks divides into square scales and becomes dark reddish-brown or gray to nearly black, but small diameter parts of the upper trunk and branches retain the silvery bark.

Wild cherry grows best on well-drained sandy soil of moderate fertility. Its natural habitat is upland hardwood forest, either mesic or xeric, in association with laurel oak, live and sand live oaks, pignut hickory, sweetgum, persimmon, magnolia, and red bay. However, it is not very common in these forests. It is much more abundant in fencerows, power line rights-of-way, abandoned groves, and cutover upland pine plantations where Hercules'-club, laurel oak, and cherry-laurel are common associates. Black cherry commonly invades upland pine habitats if they do not burn frequently.

Black cherry is an early successional species colonizing new habitats rapidly. It does this by producing abundant fruits that are eaten by birds, which then widely distribute the hard seeds (pits) as they pass back out of their digestive tracts. Therefore, most black cherry reproduction occurs in places where birds perch. In addition, the cherries are eaten and distributed by mammals such as deer and raccoons, often after the fruit has fallen. Many of the seeds germinate the first year, but many also remain dormant, some for as long as twenty years.

Black cherry seedlings can withstand considerable shade for the first year or two, but this shade tolerance decreases as the trees grow larger, with mature trees needing full sunlight. On appropriate soil, black cherry grows fast. In young pine plantations it easily outgrows slash or loblolly pine, often growing up through and eventually overtopping the pines, even after getting a later start than they do. Black cherry will also sprout vigorously from its base if the top is cut or killed, but it seldom produces root sprouts. In its natural hardwood forest habitats, black cherry reproduces by colonizing disturbed areas and small openings in the forest such as treefall gaps. The ideal width of such openings for obtaining black cherry reproduction is about 100 feet (Fowells 1965).

Black cherry is one of the first trees to leaf out in the spring, beginning in mid-February in Gainesville, Florida. The new growth can withstand cold and frost better than most other species. However, much of the new growth is sometimes eaten by eastern tent caterpillars, which seem to prefer this tree to any other plant. This does not seem to hurt the tree very much, for cherry trees can continue to grow rapidly even when they are partially defoliated every spring. A more serious problem is their susceptibility to borers and bark beetles that attack the trunk and roots, killing some trees and damaging others. Because of this, black cherry trees often die before reaching maturity. On the other hand, they occasionally live in good health for over 200 years.

Black cherry is quite drought tolerant and wind-firm in spite of its shallow root system, which is often dominated by one exceptionally large surface root. Conversely, black cherry will not withstand flooding or even prolonged soil saturation and is easily damaged by the slightest amount of fill dirt placed near its trunk. It is also easily damaged by fire.

The wildlife value of black cherry is high. The leaves are regularly eaten by caterpillars, thus providing food for birds and producing several kinds of moths and butterflies, including the eastern tent caterpillar moth, another 26 species of moths, the eastern tiger swallowtail, the red-spotted purple butterfly, and the coral hairstreak butterfly (Hilty 2019, "Wild Black Cherry"); and the leaves, twigs, and bark of seedlings are a favorite food for deer, rabbits, and cotton rats. There are also several kinds of beetles that feed on the foliage. This feeding by mammals and insects occurs in spite of the fact, first, that the wilted foliage contains cyanic acid and is therefore at that stage poisonous to domestic livestock (Fowells 1965), and second, that the little nectaries at the juncture of leaf blades and petioles support tree-dwelling ants which help protect the leaves from some leaf-eating insects (Hilty 2019, "Wild Black Cherry"). The flowers provide pollen and nectar for honey bees and a variety of native bees and wasps, and the abundant fruit provides a reliable and important food supply in early summer for a variety of birds and mammals. Some trees produce fruit earlier than others, with the first fruit usually ripening about June 1st, the peak fruit production occurring from mid-June to mid-July, and with some fruit still available until August 1st.

Wild cherry has been identified as the fourth most valuable woody plant for wildlife in the southeastern United States (after oaks, pines, and blackberries), primarily because of its fruit production (Martin et al. 1951). (This ranking is an average for the entire population of each species or genus of tree. For instance, a single cherry tree may sometimes not be as valuable for wildlife as a single red mulberry tree, but because red mulberry is not nearly as abundant, the cherry population as a whole is ranked higher.) Some of this high ranking pertains to wildlife use in the northern part of the United States. For instance, while wild cherries are highly valuable for cedar waxwings in the north, they ripen in Florida well after this bird has left for its northern breeding areas. Even so, there is abundant wildlife use of the fruit in Florida by birds such as bobwhite quail, wild turkey, cardinal, mockingbird, crow, blue jay, eastern kingbird, great-crested flycatcher, northern flicker, red-bellied woodpecker, red-headed woodpecker, pileated woodpecker, summer tanager, brown thrasher, and eastern towhee, plus some use by mammals such as black bear, gray fox, raccoon, opossum, squirrels, and mice (Martin et al. 1951).

Black cherry wood has a smooth texture, is of medium weight and strength, and the heartwood is rot resistant, even when the tree has died and the wood comes in contact with the soil. Thus, dead cherry tree snags will sometimes remain standing for decades. Even though rot resistant, old cherry trees are sometimes hollow.

From the southern Appalachians north into New England, black cherry is one of the most valuable of hardwood timber trees, second only to black walnut. The beautiful, durable, reddish-brown heartwood is highly valued for making furniture, paneling, and many other items. A sufficient amount of large, good quality cherry timber is not available in Florida to support this type of market, although individual trees are sometimes selected for making paneling or handmade items. It is a favorite wood for many local artisans who make furniture, bowls, salad forks, and a host of other items. It makes good firewood, burning slowly and producing an abundance of hot coals. Heartwood snags can be cut into fence posts.

Considering the many fine qualities of black cherry, one might think it would be valued as a landscape tree, but, in fact, the opposite is true. It is almost never planted intentionally. This is partly due to its reputation as a weed tree that comes up in fencerows and other places where it is often unwanted. However, it does have some genuine serious drawbacks as a landscape plant. Perhaps the greatest one is that it is often infested with eastern tent caterpillars in early spring. Although this doesn't really hurt the tree, it looks bad, and the caterpillars migrate from the tree to other plants after growing big and healthy on the cherry tree. A second drawback is that black cherry does not grow nearly as well in manicured, watered, and fertilized lawns as it does in the wild. Excessive watering, in particular, is bad for it, as is any sort of reshaping of the contour of the soil. Also, direct sunlight on the trunk due to the removal of nearby trees or the removal of its own lower limbs will often damage the trunk by encouraging boring insects to attack it. Nearby construction or reconstruction of streets, sidewalks, or buildings within the tree's root zone is also damaging, as is soil compaction caused by lawn mowing with heavy equipment, by the parking of vehicles within its root zone, and by other activities involving heavy machinery. Whatever the cause, the resulting unhealthy tree will have a thin crown, dead branches, and a generally straggly appearance, and it will often die. On the other hand, in rural situations where it grows in low density subdivisions and even in urban situations where it has come up after all the land contouring and construction activity is done, black cherry is often a beautiful, healthy, durable, and desirable tree.

American Plum—*Prunus americana*

Rosaceae

American plum is distinctly different from the other wild plums in Florida. The deciduous leaves are larger, broader, thicker, and have an elongated tip, and its 1-inch-diameter red fruits are larger, mature later, and are better to eat. The scaly bark on large stems is quite distinctive, somewhat resembling river birch bark. American plum is more likely to be single trunked and, on average, is larger, more robust, and taller than Chickasaw or flatwoods plum. Like these other plum species, it is quite showy when in bloom in the early spring.

American plum grows on fertile, calcareous soils in upland hardwood forests and was somewhat common in parts of Gulf Hammock before the destruction of this forest in the late twentieth century. It also occurred as a rare tree in upland hardwood forests around Gainesville such as San Felasco Hammock. Unfortunately, it seems to be disappearing from these locations. It is a bit more common in the central panhandle of Florida and in scattered locations to the north into the rest of the eastern United States. It may still be found in the hardwood forests on the bluffs along the Apalachicola River south of Chattahoochee, in the live oak hammocks along the Aucilla River, and perhaps in scattered locations in between. The second largest one reported in Florida, now long dead, was on the east side of Wilmot Botanical Garden on the University of Florida campus in Gainesville. The largest one grew on the side of the bluff at Flat Creek Landing beside the Apalachicola River. In 1993, it was 48 feet tall, 3 feet 3 inches in circumference at 4½ feet above ground (1 foot in diameter), and had an average crown spread of 36 feet (Ward and Ing 1997). The University of Florida tree was shorter but otherwise of comparable size.

American plum is considerably more shade tolerant than the other plum species native to Florida although this tolerance is only moderate. Able to live 50 years or more, it seems more durable and longer lived than the other native plums.

The wildlife value of all the native plums is probably similar. The bisexual flowers come early, in late February to early March, and attract a wide variety of bees and other insects, and the fruit of American plum, which ripens later than that of the other wild plums, is eaten by a variety of mammals and birds. A black bear scat I found in San Felasco Hammock in 1973 was full of American plum seeds. In the southeastern United States, Martin et al. (1951) identify gray fox as the main consumer of wild plum fruit, while Miller and Miller (1999) identify white-tailed deer, black bear, gray fox, raccoon, and Virginia opossum as eating the fruit.

For humans, the main value of American plum is the landscape potential of this small tree. Although rarely planted, it is better suited than the other wild plums for planting on good soil in North Florida. It is a bright and beautiful addition to any landscape when covered with white blossoms in late February and early March. Its foliage is attractive, its fruit is edible, and it is well suited for planting under power lines, near buildings, and in other places where a small tree is desired. Unfortunately, little to no attention has been paid to this species here. A selection of plants that produce good fruit and provide a good show of spring flowers would result in a landscape plant of high value.

Chickasaw Plum—*Prunus angustifolia*

Rosaceae

This wild plum is quite common in scattered rural parts of North Florida, often having escaped from old homesites. Its ancestors were among the earliest horticultural plantings in the state, having been cultivated by both early European settlers and the Native Americans who predated them. It is a large shrub to small tree, commonly reaching 10 to 20 feet tall with trunk diameters of 3 to 6 inches. The largest one ever reported in Florida was in Gainesville. It was 29 feet tall and had a trunk diameter at breast height of slightly over 1 foot. The deciduous leaves of this plum average about 2 inches in length and are narrower than those of the other wild plum species and somewhat folded upward along the midrib. The bisexual white flowers are produced abundantly in late February to early March. The fruit, which matures from late May to late June, ranges in color from yellow to orange to red as it ripens.

Chickasaw plum grows around old homesites, in old fields, along fencerows, and occasionally in upland pine forests and pine plantations. It often grows in clusters of plants that come from root sprouts. It is not shade or flood tolerant, and it does not usually live much longer than 30 years.

Though the fruit is eaten by gray fox (Martin et al. 1951) and by white-tailed deer, raccoon, opossum, and black bear (Miller and Miller 1999), Chickasaw plum does not seem to be a preferred wildlife food item. The foliage is often eaten by eastern tent caterpillars, and the twigs are browsed by deer and rabbits. The low, dense crown provides a good nesting place for birds such as mockingbirds, shrikes, and kingbirds.

Chickasaw plum starts blooming in early March before the leaves begin to grow. This is about the same as the bloom time for flatwoods plum, slightly earlier than for American plum, and much earlier than for Ichetucknee plum. Chickasaw plum's bloom is distributed along its branches in a linear pattern. The fruit is tastier than that of flatwoods plum and makes good jelly. There are cultivars of this plum still used for landscaping with larger, sweeter fruit than the average wild plant. This plum is very adaptable and hardy when provided with sufficient sunlight. It is an excellent small tree to plant under power lines or in other sunny locations where small size and a bright show of early blossoms is desired, although it spreads by root suckers much more readily than either flatwoods or American plum.

Flatwoods Plum (Hog Plum)—*Prunus umbellata*

Rosaceae

This is the common wild plum of northern Florida. It occurs naturally from the Carolinas southwestward into eastern Texas and southward into the Florida Panhandle as far south as Desoto County on the Florida peninsula (Wunderlin et al. 2020). It is a large shrub to small tree, commonly getting 10 to 20 feet tall with trunk diameters of 3 to 6 inches. The largest one ever reported was in Colclough Pond Audubon Sanctuary in Gainesville. In 1974 it was 33 feet tall and had a trunk diameter at breast height of slightly over 1 foot. The deciduous leaves of this plum average about 2 inches in length. The white bisexual flowers bloom in late February to early March, and the plums, which mature in June, range in color from yellow to orange to red to dark purple as they ripen.

Flatwoods plum grows on the edges of hardwood forests, in old fields, along fencerows, in xeric hammocks, and occasionally in upland pine forests. It is not shade or flood tolerant, and it often does not live much past 30 years, although professional botanist David Hall has two healthy individuals in his yard that are over 40 years old (personal communication).

The fruit of this plum is not avidly sought by wildlife but is eaten by gray fox

(Martin et al. 1951) and by white-tailed deer, raccoon, opossum, and black bear (Miller and Miller 1999). The foliage is often eaten by eastern tent caterpillars, and the twigs are browsed by deer and rabbits. The low, dense crown provides a good nesting place for birds such as mockingbirds, shrikes, and kingbirds.

Flatwoods plum blooms very early in the spring before its leaves begin to grow, usually starting in mid- to late February. It blooms slightly earlier than Chickasaw or American plum and much earlier than Ichetucknee plum. Compared to the other plum species of Florida, its bloom is finer, denser, and more evenly distributed throughout the crown of the tree. It is the prettiest of the plums both in blossom and in fruit and is adaptable and hardy when provided with sufficient sunlight. It is an excellent small tree to plant under power lines or in other sunny locations where small size and a bright show of early blossoms is desired.

Ichetucknee Plum—*Prunus* (undescribed species or variety)

Rosaceae

This small shrub resembles flatwoods plum but is much smaller, only growing to between 2 and 5 feet tall with main stems only up to about 1 inch in diameter. Its leaves are also smaller (about 1 to 1¼ inches long) with much smaller serrations that lack visible glands on the serration tips. The fruit ripens in late July, is dark purple to nearly black when ripe, is long stalked (similar to flatwoods plum [*Prunus umbellata*] but unlike scrub plum [*Prunus geniculata*]), has reddish flesh, and averages slightly over a half inch in diameter. Ichetucknee plum blooms in mid-March, often two weeks to one month later than either flatwoods plum or Chickasaw plum and much later than scrub plum. The fruit also ripens later. The length of Ichetucknee plum's flower stalk is similar to that of flatwoods plum. The young growing twigs and the leaf petioles are often red.

This plant grew on both sides of U.S. Highway 27 on both sides of the Ichetucknee River within a half mile of U.S. 27's crossing of the Ichetucknee River. The largest remaining population is in Ichetucknee Springs State Park east of the river and north of the highway.

Ichetucknee plum grows in a fire-adapted community containing longleaf pine, southern red oak, mockernut hickory, sand post oak, bluejack oak, turkey oak, chinquapin, woolly pawpaw, sparkleberry, deerberry, shiny blueberry, wire grass, splitbeard bluestem, butterfly pea, partridge pea, pineland wild indigo, silver croton, and many other herbaceous species. Where it grows next to the highway at Ichetucknee Springs State Park, slash pine occupies the overstory of this former slash pine plantation, but the plant community is being managed with prescribed burning and is in good condition.

This plum forms large clones by sprouting from runners. It is well adapted to fire, resprouting vigorously following burns. It is not shade tolerant and probably not at all tolerant of flooding.

The wildlife value of this little plum is probably similar to that of the other Florida plums. Additionally, because it forms dense thickets in the otherwise rather open ground cover of the sandhill habitat where it grows, Ichetucknee plum may offer the benefit of providing cover for rabbits, bobwhite quail, and towhees.

Ichetucknee plum has no current landscape value. Although easy to transplant, it seems to be less adaptable to different soil types than the other Florida plum species. I planted two specimens in my yard in Gainesville, and although they lived for about 10 years, they refused to grow any larger and eventually died. Flatwoods, Chickasaw, and American plum all grow vigorously in my yard.

Carolina Laurel Cherry (Cherry-Laurel)—*Prunus caroliniana*

Rosaceae

"Cherry-laurel," the most commonly used name for this species, is nearly identical to the name "cherry laurel" used for a different species of cherry tree, *Prunus laurocerasus*, that is native to Eurasia and commonly planted as an ornamental in the United States and many parts of Europe. Therefore, in order to be clear which species is being described here, "Carolina laurel cherry" will be the name used for *Prunus caroliniana* in this account.

Carolina laurel cherry is a common, weedy, small tree occurring from the Carolinas across the southeastern coastal plain into eastern Texas (Kartesz 2015). It is not common in native, undisturbed forests but is common to abundant in urban, suburban, and rural untended or disturbed sites such as property boundaries, pine plantations, and fencerows, reaching as far south on the Florida peninsula as Lake Okeechobee (Wunderlin et al. 2020). It has dark, shiny, evergreen, alternate leaf blades 3 to 4 inches long and 1 to 1½ inches wide. The leaves sometimes have entire margins but usually have serrations on the edges that resemble small spines. The undersides of the leaves often have glandular spots at the base on either side of the midrib (professional botanist David Hall, personal communication). According to University of Florida botanist Walter Judd, for some people, the crushed leaves have a strong odor reminiscent of maraschino cherries (personal communication). Carolina laurel cherry produces drooping, linear clusters of showy white, bisexual flowers in early spring and black, half-inch-long, inedible fruits (drupes) in late summer that hang on through the winter. The bark is smooth and dark brown to black on small trunks but becomes roughened with age and size to resemble black cherry bark on exceptionally large trunks.

The size of Carolina laurel cherry is variable. Often considered a large shrub because it is frequently used as a hedge plant, if left alone, it usually attains the stature of a small- to medium-sized tree. It often matures, stagnates, and eventually dies after getting 40 to 50 feet tall with a trunk up to about 1 foot in diameter. However, in exceptional cases, it becomes much larger. A large one in Lakeland, Florida, measured 47 feet tall with a trunk diameter of 3⅓ feet in 1987 (American Forests 1994). The average life span of Carolina laurel cherry seems to be about 50 years.

The difficulty in attempting to know the original range or habitat of Carolina laurel cherry is perhaps best described by Robert Godfrey: "The Carolina laurel cherry has been extensively used as an ornamental or for hedges and screens. It naturalizes so freely that it is scarcely, if at all, possible to know what its original range or habitat may have been, but perhaps maritime communities" (1988, 573). Carolina laurel cherry is moderately shade, drought, and salt tolerant but not flood tolerant. It reproduces abundantly by seed and also to some extent by root suckers. The fruits are eaten in winter and early spring by birds such as robins and cedar waxwings, which then distribute the seeds widely. There are patches of Carolina laurel cherry in treefall gaps in San Felasco Hammock northwest of Gainesville, Florida. There are also large numbers of Carolina laurel cherry seedlings and saplings in pine plantations and in urban and suburban neighborhoods throughout northern Florida, where it is often a problematic weed.

The wildlife value of Carolina laurel cherry is fairly good. Its fruits are eaten in late winter by robins, cedar waxwings, and other birds. The foliage is sometimes consumed by caterpillars—although not to my knowledge by the eastern tent caterpillar—and the flowers are visited by bees and other insect pollinators.

Carolina laurel cherry is sometimes used as an ornamental for hedges and screens. It is easy to grow and produces a dense barrier of dark green foliage. The main drawback for such use is that the plant spreads aggressively by seed, coming up wherever there is a planting bed, hedge, or untended spot. It also spreads to some extent by root suckers. The foliage contains hydrocyanic acid, which is highly toxic if ingested.

Swamp Rose—*Rosa palustris*

Rosaceae

Swamp rose is the only native rose species commonly found in the wild in Florida. (Carolina rose [*Rosa carolina*], also a native, has been found a few times in three panhandle counties and in Alachua and Clay Counties, and there are various exotic rose species that occasionally escape cultivation.) Swamp rose occurs in wetlands throughout most of the eastern United States including the Florida Panhandle, and it ranges as far south as Polk County on the Florida peninsula (Kartesz 2015). It is an upright, deciduous shrub growing 2 to 6 feet tall, with compound leaves having five to seven leaflets and strong, sharp thorns (prickles) on the stems. The sweetly fragrant, pink, bisexual, midsummer flowers are 1½ to 3 inches across and are followed by rose hips (fruits) a third of an inch in diameter that become red when ripe.

Swamp rose is a sun-loving shrub commonly found in marshes and the marshy edges of streams, ponds, lakes, and swamps. It is also found in very open, marshy spots within swamps and bayheads. Swamp rose is neither shade nor drought tolerant, but it is very flood tolerant, being able to grow in seasonally flooded areas of shallow water.

Swamp rose reproduces both by seed and by rhizomes and sometimes forms thickets.

Swamp rose provides pollen (but not nectar) to bumblebees and other insect pollinators, and the rose hips are eaten by various birds and mammals (Hilty 2019, "Swamp Rose"). A shrub thicket of swamp rose can provide cover for birds and small mammals and a nesting structure for songbirds (Martin et al. 1951).

Swamp rose is sometimes planted as an ornamental and is available from some plant nurseries. It is better adapted to wet ground than other rose species, blooms well in midsummer with attractive flowers, and is relatively trouble free compared to hybrid roses.

Blackberry—*Rubus pensilvanicus (R. argutus)* and *R. cuneifolius* and Dewberry—*Rubus flagellaris* and *R. trivialis*

Rosaceae

This group of abundant, thorny shrubs is well known, although differentiating the individual species can sometimes be difficult because of hybridization and polyploidy. The two blackberry species (*Rubus pensilvanicus* [*R. argutus*] and *R. cuneifolius*) are upright shrubs, whereas the two dewberry species (*Rubus flagellaris* and *R. trivialis*) trail along the ground with some upright stems. They all have palmately compound leaves with three, four, or five leaflets; prickles along the stems; white flowers in spring on year-old stems; and fruits, commonly called "berries," that turn from red to black as they ripen in late spring or early summer. They reproduce abundantly from seed and spread profusely by underground runners. Blackberries and dewberries are sun-loving plants that are not especially shade tolerant. They are fast-growing pioneer plants that rapidly colonize open areas. Though they are moderately drought tolerant, fruit production can be greatly reduced by drought. They respond to fire by resprouting vigorously.

Blackberries and dewberries are among the most important native plant species for wildlife. The fruit ranks at the very top of the list of summer foods for wildlife (Martin et al. 1951), being eaten by a very wide range of bird and mammal species including wild turkey, bobwhite quail, cardinal, towhee, catbird, yellow-breasted chat, blue jay, orchard oriole, summer tanager, brown thrasher, mockingbird, raccoon, black bear, striped skunk, gray fox, squirrels, and mice. The flowers are important for insect pollinators, especially honey bees and native bees, the foliage and stems are important browse for white-tailed deer and cottontail rabbits (Martin et al. 1951), and the thickets of brambles are important cover for cottontail rabbits, towhees, and many other wildlife species. Of the five wild turkey nests I have found, three were in sand blackberry thickets.

The blackberries and dewberries are important plants for humans as well, both as fruit producers in the wild and as cultivated fruit producers. Sand blackberry, in particular, produces abundant fruit that people eat plain or use to make pies and pastries, jams and jellies.

Sand Blackberry—*Rubus cuneifolius*

Sand blackberry is the most common species and the most important one for wildlife and humans. Its leaf shape and structure are more variable than those of the other species, with simple leaves and various forms of compound leaves often mixed together. It is widespread throughout Florida and the southeastern United States (Kartesz 2015). Sand blackberry occurs most abundantly in well-drained pine forest areas, pine plantations, pastures, and old fields repurposed from former upland pine forests. It is highly adaptable, however, occurring in many other places including prairies, pine flatwoods forests, and along the edges of wetlands. Sand blackberry leaves are white to grayish underneath, its stems are very thorny, and it grows in extensive waist-high patches (thickets). Produced abundantly in June, its fruits are tasty and edible by both humans and wildlife.

Of the four species here in northern Florida, sand blackberry provides the best wildlife habitat and supplies the most fruit. Besides the wildlife values already mentioned for all four species, sand blackberry thickets are excellent habitat for a variety of insect and spider species. While blackberry picking, one must often shake a stink bug, grasshopper, or spider off the fruit before placing it in a bucket. Paper wasps often nest in dense blackberry patches, and it is wise to wear snake-proof boots or leggings because a diamondback rattlesnake might be lurking there in hopes of making a meal of one of the rabbits, cotton rats, mice, or birds that gather in these thickets when the fruit is ripe.

Highbush Blackberry (Sawtooth Blackberry)—*Rubus pensilvanicus* (*R. argutus*)

Like sand blackberry, highbush blackberry is also common, mostly in and on the edges of wetlands but also scattered about in a wide variety of habitats. It occurs throughout much of the eastern United States including the Florida Panhandle and the Florida peninsula as far south as Glades and Charlotte Counties on the west side of Lake Okeechobee (Wunderlin et al. 2020). It is much more flood and wet-soil tolerant than the other *Rubus* species and sometimes dominates the very wet centers of seepage bogs. Compared to the other *Rubus* species, highbush blackberry also grows much taller (sometimes getting over head high), is more wickedly thorny, and has larger leaves and stems. The leaves are not whitish underneath, and though its fruit is less desirable for humans to eat, it *is* edible. Highbush blackberry is not as prolific a fruit producer as sand blackberry, nor does it form low thickets the way sand blackberry does. The taxonomy of highbush blackberry is still open to various interpretations, with the following Latin names (binomials) sometimes applied: *Rubus argutus*, *Rubus betulifolius*, and *Rubus pensilvanicus*.

Southern Dewberry—*Rubus trivialis*

Southern dewberry is moderately common on roadsides, in fencerows, on the edges of and in openings within upland hardwood forests, and in various other places throughout most of Florida and the rest of the southeastern United States (Kartesz 2015). It has prominent reddish hairs in addition to the thorns on its stems and usually grows in vine-like fashion along the ground. It blooms and fruits earlier than the blackberry species, and the fruit, though very tasty, is not produced in as great abundance as that of sand blackberry. Dewberry fruits ripen beginning in early April in Gainesville, Florida.

Northern Dewberry—*Rubus flagellaris*

Northern dewberry is an occasional species in the Florida Panhandle (Wunderlin 1998), becoming more common farther north in its range. It does not have the red hairs of southern dewberry, is not as plentiful a fruit producer here in Florida (Godfrey 1988), and drops its leaves earlier than the other species of *Rubus* in our area.

Partridgeberry—*Mitchella repens*

Rubiaceae

This is a small, perennial, creeping, ground-cover vine with half-inch-long op-posite, roundish, evergreen leaves (with a yellow stripe along the midrib) spaced apart along slender stems that trail along on the ground surface and root at the nodes. The small, bisexual, white to slightly pinkish flowers have four petals that are covered on top with hairs. These flowers occur as pairs joined at the base in late spring and summer, with the resulting red fruit having two dimples on its surface left from the two flowers that produced it. The edible, one third–inch-diameter red fruit ripens in late summer and often persists through the winter. Partridgeberry occurs on stream banks and in hardwood or mixed hardwood and conifer forests throughout the eastern United States from southern Canada south to the central Florida peninsula and west into eastern Texas (Godfrey 1988).

Partridgeberry is a shade-tolerant and moderately drought-tolerant ground-cover plant that is most common in Florida on well-drained soil in upland hardwood or mixed pine–hardwood forests. It is so low growing that it is easily smothered by leaf litter, so it tends to occur scattered about in spots where leaf litter does not accumulate such as on stream slopes and the tops of hummocks or in xeric hammocks where leaf litter is less abundant. Partridgeberry produces clonal patches as the growing stems take root. It also reproduces by seed. It is not fire tolerant, occurring in upland pine forests only if they no longer burn frequently.

The wildlife value of partridgeberry is low to moderate. The flowers are pollinated by native species of bumblebees (Hicks et al. 1985), and the fruit is eaten to a limited extent by wild turkeys, bobwhite quail, and white-footed mice (Martin et al. 1951). Because consumption of partridgeberry fruit by white-footed mice has been observed in regions north of Florida, it may be, by extension, that here in Florida related native woodland mice, such as the cotton mouse and the golden mouse, perhaps also eat partridgeberry fruit. Miller and Miller (1999) note gray squirrel and raccoon as additional species that eat the fruit, and white-tailed deer may browse the foliage to some extent.

Partridgeberry is occasionally used as an ornamental ground cover and is sometimes gathered for making decorations. As the world of nature is in rapid retreat everywhere, the gathering of a slow-growing native plant such as partridgeberry to make temporary decorations should be discouraged.

Buttonbush—*Cephalanthus occidentalis*

Rubiaceae

Buttonbush is a deciduous wetland shrub or small tree with opposite or whorled leaves 2 to 5 inches long and 1 to 3 inches wide. In midsummer, it produces round, white, ball-like flower clusters about an inch in diameter. It often grows 5 to 10 feet tall with a trunk a few inches in diameter. The bark on older stems is dark brown and rough with vertical ridges. Buttonbush occurs naturally throughout Florida, except in the Florida Keys, and throughout most of the eastern United States (Wunderlin et al. 2020). It also grows in the Sacramento Valley in California. Maximum size is represented by one in Alachua County, Florida, that measures 28 feet tall with a 1½-foot trunk diameter (Florida Champion Trees Database 2020).

Buttonbush is exceptionally flood tolerant, often growing in the deeper-water parts of freshwater swamps where the flooding is deep and prolonged enough to cause the forest canopy to be somewhat open. Buttonbush is also common in shallow water around the margins of lakes, ponds, and streams. In habitats subject to prolonged flooding, its most common tree associates are bald-cypress and swamp tupelo. When growing in association with red maple and water-elm, buttonbush grows farther out into deep water than water-elm and much farther out into deep water than red maple. It is shade tolerant although it can also grow in full sun.

Buttonbush is not a preferred browse plant and is toxic to cattle, but it produces seeds (technically fruits) that are eaten by songbirds and by ducks such as mallards and wood ducks (Miller and Miller 1999). Green-winged teal, blue-winged teal, gadwall, Florida duck, and ring-necked duck also eat the seeds to some extent (Martin et al. 1951). The bisexual flowers are attractive to butterflies and bees. Stands of buttonbush in open water are sometimes used for roosting or nesting by wading birds.

Buttonbush is not used commercially in any way, even for landscaping, although it could be usefully planted in retention basins or wetland reclamation areas where it will attract pollinators. It could also make an attractive, wildlife-friendly specimen plant for wet areas in native plant gardens.

Snowberry (Milkberry)—*Chiococca alba*

Rubiaceae

Snowberry is an evergreen tropical shrub with smooth, leathery, opposite leaves 2 to 4 inches long. Its bell-shaped, bisexual flowers, about a third of an inch long, start out white but quickly turn yellow and hang downward in panicles. The bright white, rounded fruits, about a quarter inch long, each contain two dark brown seeds (pits). Snowberry grows in South America, Central America, the Galapagos, Mexico, the southern tip of Texas, the Caribbean, and on the Florida peninsula along both coasts as far north as Dixie and Duval Counties (Wunderlin et al. 2020).

Snowberry is found in nature as a sprawling shrub or vine that often crawls into and on top of other shrubs. It occurs occasionally along the coast in live oak hammocks, on shell mounds, and in shrubby areas among and behind the dunes. It is very salt tolerant. Freezing temperatures can kill it back to the ground.

Snowberry is insect pollinated and reproduces by seed. To date, I find no published information on its use by wildlife.

Snowberry is sometimes used as an ornamental because of its dark evergreen leaves, white to yellow flowers, bright white fruits, salt tolerance, and adaptability to growing on trellises.

Pinckneya (Fever-Tree)—*Pinckneya bracteata* (*P. pubens*)

Rubiaceae

Pinckneya is an extremely rare, small, wetland tree with opposite, deciduous leaves 2 to 8 inches in length. The tree becomes showy when in bloom in May because of the large pink (or cream or white) sepals that occur next to the relatively inconspicuous tubular, bisexual flowers. It sets seed in capsules in September. Pinckneya occurs in seepage wetlands in a few spots in South Carolina, southern Georgia, the eastern half of the Florida Panhandle, and Clay and Marion Counties on the northern end of the Florida peninsula (Kartesz 2015).

As a shrub or small tree, pinckneya gets up to 20 feet tall with an open crown of few branches and a trunk 1 or 2 inches in diameter. The largest one reported as of 1994 had a trunk diameter of 4 inches and a height of 32 feet (American Forests 1994).

Pinckneya occurs in seepage wetlands that are continuously wet but do not flood. It needs full sunlight to grow well, only occurring in the most extreme of these seepage wetlands where there is no closed canopy of larger trees. Black titi is a frequent associate.

The individual stems of pinckneya are short lived, but the plant will sprout back from the root collar, which results in many plants being multistemmed. Pinckneya spreads by seed and is not weedy. To date there is little information on the wildlife value of this plant. The flowers are insect pollinated and, of the fruit, Gilman and Watson note that it "does not attract wildlife" (1993, "*Pinckneya pubens*"). Indications that the inner bark is "extremely bitter" may account for the dearth of information on pinckneya's wildlife value.

Human use of pinckneya is mostly historic. In the eighteenth century, the bitter bark was used to make a drink for treating fevers, particularly those associated with malaria, hence the common name "fever-tree" (Michaux 1810–13). It is now used to a very limited extent as an ornamental. However, pinckneya is so specialized in its need for a constant water supply that it is very difficult to grow except in pots that are watered daily.

Wild Coffee (Shiny-Leaved Wild Coffee)—*Psychotria nervosa*

Rubiaceae

Wild coffee is a small evergreen shrub that grows from 1 to 6 feet tall. Its opposite leaves, 3 to 6 inches long by 1 to 2 inches wide, are shiny dark green on top with a distinctive puckered appearance caused by impressed veins. The lower leaf surface is not shiny and has hairs along the raised veins. Clusters of small white flowers are produced at the ends of the upper twigs in summer, followed by clusters of bright red, round, one third–inch-wide drupes which, together with the evergreen foliage, create the visual impression of a coffee plant (*Coffea*). Wild coffee occurs mostly near the coast of the Florida peninsula from Duval and Levy Counties southward; it also grows inland in widely scattered, small populations from Alachua County southward (Wunderlin et al. 2020).

Wild coffee is an uncommon to rare plant in Florida, occurring mostly on shell mounds and in limestone outcrop hardwood forests and growing in the shade of cabbage palms, live oaks, and other trees. It is cold sensitive, does not grow well in full sun, and does not withstand flooding. It is shade tolerant and seems to do best in the shade of a closed-canopy forest. Although often growing near the coast on shell mounds, it is only moderately salt tolerant. Wild coffee periodically freezes to the ground in northern Florida and is therefore usually smaller there than it is farther south.

The flowers are a nectar source for butterflies, including the atala and the great southern white in South Florida, and the fruits are attractive to birds such as the northern mockingbird (Hutchinson 2013).

The fruits are marginally edible by people but are seldom consumed. The seeds are not a suitable substitute for coffee beans or for making a coffee-like beverage. Wild coffee is occasionally used as an ornamental in South Florida.

Firebush (Scarletbush)—*Hamelia patens*

Rubiaceae

This tropical, evergreen shrub has simple, opposite (or more often whorled) leaves and red-orange tubular flowers. The leaves are thin, have smooth margins, and vary from less than 2 inches long to 6 inches long and from 1 to 2 inches wide. New twigs, leaf stems, and the central vein and margin of the leaves are usually reddish. The bisexual flowers, which are produced continuously in terminal clusters from June until frost, are from one half to 1 inch in length. The berries start out green, turning orange, then red, then black. Just under half an inch long and about a quarter of an inch in diameter, they contain numerous tiny seeds, each slightly smaller than the head of a pin.

Firebush reproduces primarily by seed, distributed by the birds that eat the fruit. However, it can also spread slowly by root suckers, forming small clonal patches.

Firebush is native to coastal hammocks and Indian mounds on the Florida peninsula from Clearwater and Vero Beach southward and also occurs inland on the edges of and in openings in hardwood forests in the interior, where it gets as far north as Marion County (Christman [1999] 2003). It also occurs in Mexico, Central America, and South America. Firebush occurs in northern Florida as an increasingly popular ornamental. It prefers a well-drained sunny spot with good soil underlain by limestone but is adaptable. Where soils lack limestone, firebush will grow best if crushed shell or limestone is mixed deeply into the soil before planting. It is moderately drought tolerant once fully established and moderately salt tolerant.

Firebush is especially valuable as an ornamental for use in butterfly and hummingbird gardens. When in bloom, it is more attractive to zebra longwing butterflies, gold rim swallowtail butterflies, and ruby-throated hummingbirds than any other plant I have grown. Other butterflies also come to it, as do bumblebees. The berries are occasionally eaten by birds. Experienced Gainesville gardeners Ruth and Rayna Wallbrunn report cardinals, mockingbirds, and hermit thrushes eating the fruits, and I have seen cardinals, mockingbirds, and, especially, catbirds eating them. The wildlife value of this small shrub is high, mainly because of its flowers.

Firebush is virtually trouble free as an ornamental. Once established, it requires no water, fertilizer, pesticides, or trimming. It freezes to the ground almost every winter, but springs back the next year more vigorous than ever. (I do trim away the dead stems once they have been killed by frost.) It does not spread rapidly or become a nuisance and blooms continuously from early summer until first frost. The leaves, flowers, and fruits are all beautiful, and the butterflies and hummingbirds this plant attracts and supports add to the overall appeal of a landscape.

Skunkvine—*Paederia foetida*

Rubiaceae

Skunkvine is a vine species of Asiatic origin introduced into Hernando County, Florida, some time prior to 1897 by the United States Department of Agriculture for testing the plant's potential use as a fiber crop (Flores 2003). It escaped into the wild and has been a highly problematic invasive exotic pest ever since. It is particularly abundant on the central and north central Florida peninsula and is spreading rapidly into new areas. It is listed as a Category I invasive exotic pest plant in Florida (FLEPPC 2013).

Skunkvine is a thin-stemmed, rapidly growing, evergreen vine with opposite, soft, smooth-margined leaves that are rounded to heart shaped at the base, long pointed at the tip, and about 1 to 3 inches long. The vines often form a dense, dark green mass of vegetation on top of other vegetation. The young stems are green, but older stems are tan in color. On the ground, the stems continuously take root at the nodes as they grow. Having no thorns, spines, or tendrils to help them climb, the vines constantly spiral clockwise as they grow upward around the stems of other plants or around other structures. One key to identifying this plant is to walk on or otherwise crush it, releasing the skunk-like odor that gives the plant its name. The small but attractive bisexual flowers are grayish-pink with a red center. Skunkvine produces small, round, fleshy, shiny brown fruits from midsummer to late fall, each of which contains two black seeds.

Unfortunately, skunkvine is highly adaptable and aggressive, able to establish itself and grow vigorously in almost any kind of forest from densely shaded hardwood forests to sunny pine forests. It can also grow in swamps and marshes (Flores 2003). The aboveground part of the plant is not cold tolerant, but the root systems survive hard freezes to send up vigorous sprouts each spring. Skunkvine grows so vigorously and covers other plants so thickly—including ground-cover plants, shrubs, and trees—that it can kill most of them by depriving them of sunlight. It can also grow in lawns, surviving regular mowing. It can be killed by manual removal and by applying herbicides such as glyphosate (Roundup) and triclopyr (Garlon), but complete control is nearly impossible, necessitating repeated treatments for many years (Flores 2003). I have been battling skunkvine in my yard and neighborhood for many years and am gradually losing the battle.

Skunkvine has invaded the Withlacoochee State forest near Brooksville, Colclough Pond Sanctuary in Gainesville, and Silver Springs State Park east of Ocala, to name a few places. I know of no place of any size where it has been successfully eliminated. There are currently efforts based in both Florida and Hawaii to find insects in the native range of this plant in Asia that will help control this pest (Flores 2003).

The wildlife value of this plant is hugely negative as it smothers and kills native vegetation that supports wildlife. On one hand, the flowers are visited by bees. On the other, though the fruits are eaten to some extent by mockingbirds, they unfortunately spread the seeds to new areas. Skunkvine, which roots easily, can also spread when discarded vine segments are left on the ground in brush and waste piles. The vines spread rapidly by producing long ground runners which root and branch at each joint as they spread into new areas.

Trifoliate Orange (Hardy Orange)—*Citrus trifoliata* (*Poncirus trifoliata*)

Rutaceae

Trifoliate orange is a very thorny large shrub or small tree in the citrus family with deciduous, trifoliate leaves about 2 inches long that are aromatic when crushed. It has green twigs and thorns, orange blossom–like flowers, and a citrus fruit about 2 inches in diameter that turns yellow when ripe and then hangs on into winter. It is an invasive exotic species from China and Korea.

Scattered about widely in the southeastern United States, trifoliate orange is especially problematic in Texas but is also common in Florida where it is a troublesome invasive exotic in places such as San Felasco Hammock Preserve State Park. It is very cold tolerant, being able to withstand temperatures down to minus 10 degrees Fahrenheit. It is also quite drought tolerant, more tolerant of the various citrus diseases now plaguing the citrus industry than most other citrus species, and is in general quite hardy, thus earning trifoliate orange its other common name, "hardy orange."

The flowers of trifoliate orange attract honey bees and native bees, and it is one of the host plants of the giant swallowtail butterfly. The dense, thorny crown makes good nesting habitat for songbirds.

Trifoliate orange is sometimes used as an ornamental, mostly well north of Florida where the novelty of growing a citrus tree in a cold climate prompts some people to plant it. It will grow and survive as far north as New York City. It has stiff, strong branches and twigs armed with large, stiff, and very sharp thorns, which have supported its occasional use as a barrier hedge. The fruit has been used to a limited extent to make preserves, but, unprocessed, it is nearly inedible because of its offensively bitter and sour taste.

The main use of trifoliate orange in Florida has been as a rootstock for grafted citrus trees. Orange, satsuma, tangerine, lemon, lime, kumquat, and grapefruit trees grafted onto trifoliate orange rootstock are somewhat dwarfed but, compared with trees grafted onto other rootstocks, slightly cold hardier, more resistant to some citrus diseases, and produce sweeter fruit. Nonetheless, fruit production is somewhat reduced on trees grafted onto trifoliate orange rootstock, and there have been problems with citrus decline associated with this rootstock as well. Hybrid rootstocks where one of the parents is trifoliate orange have been developed and used with considerable success.

Sour Orange (Seville Orange, Bitter Orange)—*Citrus* x *aurantium*

Rutaceae

Sour orange looks like a typical orange tree. It is evergreen with glossy, dark green leaves 3 to 5 inches long, green twigs, and sharp thorns, and it produces typical-looking orange blossoms and fruit. It escaped into the wild in Florida over 400 years ago, having been brought to St. Augustine from Spain, and can be found scattered about in native forests as an occasional tree throughout most of Florida and into southern Georgia.

Sour orange trees are similar to sweet orange trees but are more upright with narrower crowns, and, while the fruit looks similar, the taste is sour and bitter (although not nearly as disagreeable as the taste of trifoliate orange). It is often found in moist, fertile hardwood forests as an uncommon understory tree. It is usually between 10 and 30 feet tall with a 4- to 6-inch trunk diameter. It is more tolerant of wet soil conditions than are sweet orange trees. Sour orange is moderately shade tolerant although it does best in full sun. It is also moderately drought and salt tolerant but not fire tolerant. Sour orange is often found on shell mounds along the coasts on either side of the Florida peninsula.

Sour orange is slightly more cold tolerant than other kinds of citrus, with the exception of trifoliate citrus, which can survive much colder temperatures. Other types of citrus have escaped into the wild, such as tangerines on the central Florida peninsula and Key limes in extreme South Florida and the Keys. All of these citrus types are susceptible to the various disease problems that plague the citrus industry in Florida, the most devastating of which is currently "citrus greening," caused by the bacterium *Candidatus liberibacter asiaticus*. This disease appears to be in the process of killing all the citrus trees in Florida, including the invasive ones such as sour orange. It may be that bioengineering efforts will produce citrus resistant to this disease, but the citrus trees now in groves and in people's yards, plus all wild citrus trees, will likely be gone in a few years.

The wildlife value of sour orange is fairly low. The flowers are attractive to honey bees and native bees, and the fruit is eaten to a limited extent by some animals. The foliage is a larval food plant for the giant swallowtail butterfly.

Sour orange has occasionally been cultivated for its beauty, blossoms, and fruit. The fruit is traditionally used in various Spanish recipes and to make marmalade and sour orange pie. Its main use since the beginning of the citrus industry in Florida has been as a rootstock for growing other types of citrus that are grafted onto it.

Wafer-Ash (Hoptree)—*Ptelea trifoliata*

Rutaceae

Wafer-ash is a deciduous shrub or small tree with alternate, glabrous, bright green, trifoliate leaves with 2-inch-long petioles and 3-inch-long leaflets. The bark is smooth and gray; the inconspicuous white, bisexual flowers bloom in terminal clusters in March and April; and the wafer-like fruits ripen in September in Gainesville, Florida. Normally a very small single-trunked or multistemmed tree, one in my yard in Gainesville got to be 21 feet tall with a trunk 4 inches in diameter (Ward and Ing 1997). The national champion in 1994 from Hartford, Connecticut, was 34 feet tall with a trunk diameter of 17 inches (American Forests 1994).

Wafer-ash grows in scattered locations throughout much of the eastern United States, ranging into the Florida Panhandle and as far south on the Florida peninsula as Polk County (Wunderlin et al. 2020). It occurs as a rare plant in Florida, mostly as an understory shrub or tree in upland hardwood forests on limestone outcrops. It is drought tolerant, but not salt or fire tolerant, although it will sprout back from its root collar following a fire. Wafer-ash is moderately shade tolerant but grows best in full sun.

The bitter foliage is not browsed by deer, but it is one of the larval food plants for the giant swallowtail butterfly. The flowers supply nectar for native bees, wasps, and flies.

Wafer-ash is sometimes planted as an ornamental, especially in butterfly gardens, mainly north of Florida.

Hercules'-Club (Southern Prickly-Ash, Toothache Tree) —*Zanthoxylum clava-herculis*

Rutaceae

Hercules'-club is a small- to medium-sized deciduous tree with compound leaves that usually possess thorn-like prickles along the leaf stem (rachis). There are also strong, sharp, thorn-like prickles along the branches and on the trunk that gradually form a cork pad under the base, thus raising each prickle up on a pyramidal pedestal of cork. On old trunks, the prickles may be missing, leaving only the pyramidal cork structures. In any case, the cork projections on mature trunks give the trees a distinctive appearance. Hercules'-club produces small, pale yellow-green, bisexual flowers in terminal clusters in March and April, followed by clusters of small, round fruits in September that contain black seeds.

Hercules'-club is often 20 to 50 feet tall with a trunk diameter of 4 to 10 inches. A large one from Texas measured 51 feet tall with a trunk diameter of 16 inches, and a large one from Florida measured 65 feet tall with a trunk diameter of 11 inches (American Forests 1994). Hercules'-club is a rather short-lived tree, rarely living more than 50 years.

There are two close relatives of Hercules'-club that rarely and barely reach northern Florida, one from farther north and one from farther south. The northerly relative, northern prickly-ash (*Zanthoxylum americana*), typically ranges from Alabama through the Mid-Atlantic states and parts of the Midwest all the way up into Canada, but it has also been found in three small spots in northern Florida. The southerly relative, lime prickly-ash (*Zanthoxylum fagara*), is native to Texas, Mexico, the Caribbean, Central and South America, and also to the Florida peninsula from Citrus, Marion, and Volusia Counties southward (Wunderlin et al. 2020). Both are shrubs or very small trees.

Some people confuse Hercules'-club with devil's walkingstick (*Aralia spinosa*). These two species are quite different. Although both are spiny, the prickles of Hercules'-club have stout bases and are arranged at random along the stem, whereas the prickles of devil's walkingstick are longer and more slender and are arranged both in rings around the stem *and* at random. Also, the leaves of the two species differ significantly from each other: the leaves of Hercules'-club are moderate in size and once pinnately compound with 9 to 15 leaflets, whereas the leaves of devil's walkingstick are huge and twice or thrice compound with as many as 60 leaflets.

Hercules'-club is native near the coast in the Carolinas and Georgia, throughout most of Florida, and westward into Louisiana, Arkansas, and eastern Texas (Kartesz 2015). It is rather weedy, often growing along fencerows in areas of pasture, pine plantation, or rural subdivisions on well-drained uplands and is capable of spreading from fencerows into adjacent lands by seed and root suckers. It also occurs along the edges of forested areas and in treefall gaps within upland hardwood forests. Hercules'-club requires full sunlight to do well, tolerates poor, dry soil, and is quite drought tolerant. It is moderately salt tolerant, often occurring near the coast. It is not amenable to shade, flood, or fire and is susceptible to severe damage from occasional hard freezes that occur in mid- to late spring after the tree has started growing.

Hercules'-club flowers provide nectar for honey bees, native bees, and flies. The fruits are a low-quality food eaten to some extent by birds that distribute the seeds widely, which results in the trees often coming up under power lines and along fences. The bark of seedlings and sprouts is sometimes chewed by rodents. The foliage is a larval host for the giant swallowtail butterfly.

Hercules'-club is rarely planted except as a curiosity because of the strong, thorn-like prickles that can cause serious injury. The lightweight, soft, weak wood is not at all rot resistant, and the roots are thick and weak, breaking easily if bent. Hercules'-club has no commercial value but has been used medically. All parts of the plant contain a chemical that produces tingling and numbing in the mouth if chewed. Hercules'-club bark was used in the past to alleviate toothaches. Care should be taken with this remedy, however, because if the saliva thus induced is swallowed, it will also numb the throat and can cause choking.

Eastern Cottonwood—*Populus deltoides*

Salicaceae

Eastern cottonwood is a large, deciduous tree with moderately large, triangular, glabrous leaves with serrated edges and petioles (leaf stems) that are somewhat vertically flattened close to the leaf blade. The leaves are 2 to 6 inches long and 2 to 4 inches wide, and the petioles are 2 to 4 inches long. The light brown bark on large trunks is deeply furrowed vertically. Cottonwood can grow to be over 100 feet tall with trunk diameters exceeding 4 feet and large, spreading crowns of massive branches. The largest reported trunk diameter, measured in 2013 on an 88-foot-tall tree in Gage, Nebraska, is nearly 12 feet (American Forests 2018). Cottonwood is a common tree in most of the eastern and central United States, but it is not common in Florida, either where it occurs as a native tree on the floodplains of the panhandle's major rivers or as a planted tree elsewhere in the state.

Eastern cottonwood is an early pioneer species on new silt bars in river floodplains, usually in association with river birch, black willow, sycamore, and silver maple. As the forest on such sites matures, other tree species arrive by seed, including bald-cypress, sweetgum, sugarberry, American elm, water-elm, swamp tupelo, green ash, pumpkin ash, water hickory, overcup oak, and red maple. Throughout the development of the forest, cottonwood remains as an emergent tree, with the top of its crown extending above the average height of the rest of the canopy.

When planted on moist, fertile sites in the Mississippi River basin, cottonwood is the fastest-growing large tree in the country. It can grow 10 to 12 feet in height per year for ten years, reaching over 100 feet tall with a 2-foot trunk diameter in one decade (Cooper 1990). One reason cottonwood and other species in the willow family can grow so fast, even on newly deposited sand and silt bars that may be low in available nitrogen, is that they have nitrogen fixing bacteria in the intercellular spaces of their stems (Doty et al. 2009). Even so, eastern cottonwood is very site sensitive and grows more slowly on less ideal sites. On dry upland sites, it may not survive at all unless its roots reach the water table, a stream, a drain field, or some other permanent water source. Cottonwood seldom lives more than 100 years.

Cottonwood is extremely intolerant of shade, requiring full sun to survive. It is only moderately tolerant of flooding and can only withstand drought if a permanent water source remains within reach of its roots. It is not fire tolerant.

The partially flattened leaf petiole of eastern cottonwood is an adaptation to living in full sunlight. The petiole allows the leaves to hang downward with the edge of the leaf pointing upward, and it allows the leaves to sway in the slightest breeze. The result is that sunlight strikes the leaves more fully in the early morning and late afternoon than it does at midday, and the constant motion of the leaves allows sunlight to penetrate through the leaves in intermittent bursts that permit the inner and lower leaves to receive more light than they would otherwise.

Cottonwood is a wind-pollinated, dioecious species that reproduces by seed; each seed has a tail attachment of cottony fibers that facilitates wind dispersal. The cottony fluff with its tiny seeds is often produced in abundance, resulting in wind-blown rafts and piles that look like cotton, hence the name "cottonwood." If the seeds happen to land on cottonwood's optimum habitat of new, moist silt bars next to a river, a new stand of trees may be the result. The tree can also reproduce by stump sprouts and can be planted by simply inserting a cut twig into the ground during the winter.

In Florida, where cottonwood is rare, it is seldom bothered by insects and diseases. This points to the advantages of not overplanting any one kind of tree. Farther north and west, where cottonwood is much more common, numerous insects and diseases attack the tree, including the cottonwood leaf beetle, the larvae of several kinds of borers, several kinds of leaf hoppers, several canker diseases, the viceroy butterfly caterpillar, the poplar tentmaker, and the cottonwood leaf curl mite (Morris et al. 1975).

The wildlife value of cottonwood is moderately low until it matures and begins to decline, at which time it can be a valuable den, nest, and roost tree. Seedlings and sprouts are browsed by deer and rabbits, and beavers cut saplings and pole-sized trees for food and dam construction (Cooper 1990).

Cottonwood trees are harvested from natural stands and also grown in plantations for their wood, which is used as core stock in furniture manufacture and for high quality pulpwood (Cooper 1990).

Eastern cottonwood makes the quickest possible shade tree if planted in full sun and provided with plenty of water and fertilizer. It can be attractive with its large, shiny, bright green leaves that tremble in the breeze, and it does not become a nuisance by producing seedlings or multiple root sprouts. Cottonwood has two main drawbacks as a landscape tree. First, its wood is neither very strong nor rot resistant, which results in its susceptibility to storm damage.

Second, its large, deciduous leaves, when widely scattered by autumn winds, make lawns temporarily messy.

Swamp Cottonwood—*Populus heterophylla*

Salicaceae

Swamp cottonwood is an uncommon large tree closely related to eastern cottonwood. It occurs in the coastal plain of the Carolinas along the Ohio and Mississippi River drainages in the middle of the United States and in widely scattered populations elsewhere, including along the Apalachicola and Ochlockonee Rivers in the Florida Panhandle (Johnson 1990). It resembles eastern cottonwood, but the leaves are more rounded and the petiole (leaf stem) is not flattened. At maturity, swamp cottonwood is often 100 feet tall with a large trunk. The largest one, reported in 1991 from Ohio, was 140 feet tall with a trunk diameter of almost 9 feet (American Forests 1994).

Swamp cottonwood grows in river floodplain swamps and other wet floodplain habitats and is more flood tolerant than eastern cottonwood. It commonly grows in association with bald-cypress, swamp tupelo, water tupelo, black willow, green ash, pumpkin ash, pop ash, American elm, water-elm, buttonbush, swamp privet, possum haw, water hickory, red maple, sycamore, sugarberry, water locust, persimmon, and overcup oak (Johnson 1990). It is intolerant of shade, though perhaps not quite as intolerant as eastern cottonwood. Its place in the forest is at the top as an emergent tree that reaches above the average height of the other canopy trees.

The wildlife value of swamp cottonwood is similar to that of eastern cottonwood. It has low value for wildlife until it becomes a large tree with cavities, at which time an individual can be quite useful as a den and nest tree.

Swamp cottonwood is not abundant enough in Florida to be economically important. The wood is mixed with eastern cottonwood and sold as cottonwood to be used for box veneer and pulpwood. Cottonwood makes excellent pulpwood, vital in the manufacture of high-grade book and magazine paper (Johnson 1990). Swamp cottonwood has never been planted for timber production.

Swamp cottonwood is not used as a landscape tree.

Black Willow—*Salix nigra*

Salicaceae

Black willow is the largest of our willows. In Florida, it grows 50 to 100 feet tall with a trunk diameter of 1 to 2 feet. Its maximum potential size, with heights up to

140 feet and trunk diameters up to 4 feet, has been attained by trees in the Mississippi valley (Pitcher and McKnight 1990). The foliage is similar to that of coastal plain willow, with leaves that are long and narrow, especially toward the tip. The undersides of black willow leaves tend to be grayish-green to olive, whereas those of coastal plain willow are usually a frosty whitish shade. As with all willows, male and female flowers are borne on separate catkins on the same (monoecious) tree. The small seeds have silky hairs at one end to facilitate wind dispersal and are produced in capsules. The bark of black willow is deeply ridged on old trunks and varies in color from brown to nearly black.

Black willow is common throughout most of the eastern United States, reaching into the Florida Panhandle where it often forms hybrids with coastal plain willow in the central panhandle (Nelson 1994). It extends east in Florida to the Suwannee River (Godfrey 1988) and perhaps farther east along the upper Santa Fe River at least to Worthington Springs (see the species account on Santa Fe Willow that follows). Black willow grows in wetlands, primarily in the floodplains of rivers and streams, where it is a pioneer on newly formed sand bars and silt bars. It reproduces both by seed and by the rooting of broken twigs and branches. Common associates include coastal plain willow, river birch, cottonwood, red maple, water-elm, sweetgum, and sycamore.

Black willow is very flood tolerant. It is a short-lived tree that is not fire, shade, or drought tolerant, and the wood is not rot resistant. It is a fast grower, however, able to attain average heights of 49 feet at 10 years of age in planted stands in Mississippi (Pitcher and McKnight 1990), and it presumably has nitrogen fixing bacteria in intercellular spaces within its twigs as do other species of willow (Doty et al. 2009).

The wildlife value of black willow is low to moderate in Florida. The stems and foliage are a host for the larvae of the viceroy butterfly (Glassberg et al. 2000) and provide food for a large number of other insects such as moth larvae, wood-boring beetles, long-horned beetles, leaf beetles, leaf miner flies, weevils, flea beetles, gall flies, plant bugs, stink bugs, leaf hoppers, spittlebugs, saw flies, and tree hoppers (Hilty 2019, "Black Willow"). The catkins are pollinated by many kinds of flies and bees. Deer browse the foliage, and beavers eat the inner bark and twigs, although this use is less significant in Florida than it is farther north (Martin et al. 1951).

Black willow wood is lightweight, straight grained, easily worked, takes stains and paints well, and is relatively free of splinters, making it ideal for specialty uses such as the manufacture of artificial limbs (historically crucial before the invention of plastics) and for the making of crates and boxes; it is also used for pulpwood (Pitcher and McKnight 1990). Black willow is not abundant enough in Florida to be commercially important, but it is valuable in its role of stabilizing soil on stream banks, sand bars, and floodplains.

Black willow is rarely used as an ornamental in Florida. It is not well adapted to the well-drained soils of most developed landscapes but could be planted in sunny places on the edges of water features or wetlands.

Santa Fe Willow—*Salix* (undescribed population)

Salicaceae

A population of willow trees on the upper Santa Fe River between O'Leno State Park and the Santa Fe Swamp on the northern border of Alachua County is similar to black willow and coastal plain willow but visually distinct from both. The fast-growing new tips of foliage on the upper and outer crown of the tree give a distinctly reddish cast to the tree's crown during the growing season. It is a medium-sized tree, larger, taller, and more upright than coastal plain willow. The long-tapering, narrow leaves are narrower and more gradually tapering on the basal end and have longer petioles than either black willow or coastal plain willow but are otherwise similar to those of black willow, except that the new leaves are often white underneath. As with other willows, male and female flowers are borne in separate catkins. The gray bark on the trunks and larger branches is strongly ridged. Santa Fe willow has not been described botanically, perhaps with good reason. Between 400 and 500 species of willows have been described worldwide (Nelson 1994), making the task of ensuring that this one is really a new species or variety problematic. It is most likely a local variant of black willow.

Santa Fe willow is restricted to low, frequently flooded parts of the floodplain of the upper Santa Fe River, where it is abundant in some places, such as where Highway 121 crosses the river at Worthington Springs. (It probably also grows along the lower parts of the New River and Olustee Creek, which are tributaries of the Santa Fe River.) It prefers newly formed sand and silt bars adjacent to the water. It is a fast-growing pioneer tree that often grows in association with river birch. Like black willow, it is quite flood tolerant but not at all shade tolerant, and its lightweight wood is not rot resistant.

The wildlife and potential commercial and landscape values of Santa Fe willow are presumably similar to those of black willow. Santa Fe willow looks attractive where it grows in the wild, suggesting that it might be worth trying out as an ornamental on moist ground next to water. Willows are easily propagated by taking cuttings and rooting them. The drawbacks to using willows as ornamentals in Florida include their short life span, their being prone to storm damage, and their not being drought tolerant when planted on uplands away from a water source.

Coastal Plain Willow—*Salix caroliniana*

Salicaceae

Coastal plain willow, sometimes called "Carolina willow," is the common willow of the Florida peninsula, growing in wet areas throughout the state except for parts of the western panhandle and growing in many places elsewhere in the southeastern United States. In the central panhandle, this willow and black willow occur together and often hybridize (Nelson 1994). Coastal plain willow is a small, often shrubby tree, usually not much more than 30 feet tall, with a maximum potential height of 70 feet and trunk diameter of 1 foot. It has slender twigs and branches and long, narrow leaves that are whitish underneath. Male and female flowers are borne in separate catkins on the same (monoecious) tree. The female flowers produce capsules containing tiny seeds that have silky hairs at one end to facilitate wind dispersal. The variable bark is brownish and becomes roughened with flat-topped ridges on older trunks.

This fast-growing, short-lived, shrubby pioneer tree is water dependent. It is neither fire nor shade tolerant but is very tolerant of flooding. It grows in marshes, swamps, creek and river floodplains, drainage ditches, and on pond and lake edges and will invade and take over fire-dependent marshes and wet prairies that no longer burn. It invades many man-made wetlands such as water retention areas, phosphate mine slime ponds, and the edges of canals and ditches. Its ability to thrive in such situations may be based on its nitrogen-fixing capacity. Like other willow species (Doty et al. 2009), coastal plain willow has nitrogen-fixing bacteria between the cells of its stems. These bacteria allow the plants to convert unusable nitrogen from the atmosphere and soil into forms they can absorb, which supports and enhances their development.

Coastal plain willow is not common in the acidic wetlands of the pine flatwoods forests of northern Florida, perhaps because of the fire histories of these marshes and prairies or because of the acidity of the soil. Common associates include waxmyrtle, buttonbush, salt bush, red maple, sweetgum, pop ash, pumpkin ash, green ash, swamp tupelo, and bald-cypress.

The wildlife value of coastal plain willow is moderate. The same diverse array of insect types that feed on black willow also feed on coastal plain willow, although some of the species may be different. These insects, in turn, provide food for birds and other wildlife. The flowers on the catkins are pollinated by flies and bees. Deer feed on the foliage to some extent, and beavers eat the twigs and bark, but the wildlife value of willows for mammals is more important farther north than it is here in Florida (Martin et al. 1951).

The wildlife impact of an invasion of willow and other woody plants into an herbaceous marsh can be negative, damaging the open marsh habitat used by cranes, herons, egrets, ibis, gallinules, grebes, ducks, rails, and other marsh-dependent wildlife. Payne's Prairie provides an example of this. Its invasion by willows began around 1950 with a few coastal plain willows in the delta area of Sweetwater Branch. The footprint of these trees subsequently and continually expanded until they became a dominant force, turning large areas of the marsh and prairie into shrub thickets. In consequence, the managers of Payne's Prairie State Park have used herbicides to successfully combat the invasion and restore marsh habitat.

On the other hand, an isolated stand of willows in a deep marsh or on an island in a pond or lake can be very valuable by providing the structure for a breeding rookery of herons, egrets, white ibis, anhingas, and boat-tailed grackles. On Payne's Prairie, the endangered snail kite builds nests in coastal plain willows that are surrounded by marsh habitat.

Coastal plain willow is not used to any significant degree commercially or for landscaping.

Florida Willow—*Salix floridana*

Salicaceae

Florida willow is a very rare small tree that is listed as endangered by the state of Florida and is being studied for possible federal listing. I've seen two of these trees along Mormon Branch in the Ocala National Forest and perhaps five on a small tributary of Deep Creek about three miles south of the town of Interlachen, Florida. It reportedly occurs in a total of perhaps ten general locations from southwestern Georgia southward into the Florida Panhandle and the Florida peninsula as far south as Orange County. The Florida Natural Areas Inventory (2019) reports 22 specific spots of occurrence in Florida.

Florida willow grows in permanently wet seepage areas along spring runs. Its physical appearance is distinct from that of the other willows in our area. The leaves are rather large and broad, not narrow and long tapering toward the base like those of black and coastal plain willows, and Florida willow stems and buds are thicker than those of the other willows. Like other willows, the male and female flowers are borne in separate catkins on the same (monoecious) plant. The largest Florida willow tree I've seen is, or was, at Mormon Branch. (I haven't been to the site in many years.) It was perhaps 20 feet tall with a trunk diameter of about 6 inches.

Florida willow is exceedingly water dependent, always growing in or adjacent to permanently flowing water. (It will grow well in a well-drained pot that is watered regularly, and it is easy to propagate from cuttings.) Florida willow is associated with moisture-loving trees such as Atlantic white cedar and swamp tupelo. It is not especially shade tolerant, however, and does best in open spots dominated by herbaceous plants. Where I have seen it, the also-rare largeleaf Grass-of-Parnassus (*Parnassia grandifolia*), listed in Florida as endangered, is one of the plant species growing in the same location.

Florida willow is so rare and small and restricted in its requirements that it has no known wildlife, commercial, or landscape values. It is of scientific interest because of its rarity.

Prairie Willow—*Salix humilis* and Missouri River Willow—*Salix eriocephala*

Salicaceae

These two willow shrubs typically occur far north of Florida but have a few small, relict populations in the central Florida Panhandle in the case of Missouri river willow and in the eastern Florida Panhandle in the case of prairie willow (Godfrey 1988). Both are deciduous shrubs with leaves neither as long nor as narrow as those of black willow and coastal plain willow. The pubescence on the leaves' undersides causes them to look white or gray. Both species, as is true of all willows, are monoecious, having separate male and female catkins on the same plant. Both occur in wet places, although farther north in the midwestern states prairie willow also occurs on well-drained, dry uplands.

Both prairie willow and Missouri River willow are so rare in our area that neither is important as wildlife habitat or for human uses. They are primarily interesting as representatives of the large number of shrubby willow species abundant in the far north, where they *are* very important for wildlife. Like other willows, they provide food for the caterpillars of moths and the viceroy butterfly and for many other insects such as leaf beetles, leaf hoppers, saw flies, plant bugs, gall flies, stink bugs, and aphids. The catkins are pollinated by flies and small, native bees, and the foliage is browsed by white-tailed deer (Hilty 2019, "Prairie Willow").

Florida Maple—*Acer saccharum* subsp. *floridanum* (*A. barbatum*) (including a note on Chalk Maple [*A. leucoderme*])

Sapindaceae

Florida maple is a southern subspecies of the sugar maple. It is a medium-sized tree, usually 60 to 80 feet tall and 1-foot in trunk diameter, with deciduous leaves shaped like the maple leaf on the Canadian flag. The leaves of all maples are arranged opposite each other, two by two on the stem. Red maple leaves have serrated edges in addition to their lobes, whereas Florida maple leaves are not serrated on the edge. Florida maple produces inconspicuous, separate male and bisexual but functionally female flowers on the same tree (Godfrey 1988) followed by paired winged fruits that are wind distributed. Cross pollination is perhaps achieved primarily by wind, but insects *do* visit the flowers. Florida maple bark is fairly smooth and light gray or tan in color.

The native range of Florida maple encompasses scattered populations on appropriate soils on the Piedmont and coastal plain from southeastern Virginia to Arkansas and extreme east Texas, extending south into Florida in widely scattered spots in the panhandle and on the peninsula, both along the Gulf Coast south to Tampa (Jones 1990) and inland in widely scattered, isolated populations as far south as Polk County (Wunderlin et al. 2020). There is an isolated population of the closely related and ecologically similar chalk maple (*Acer saccharum* subsp. *leucoderme*) along the Apalachicola and Chipola Rivers in the Florida Panhandle (Godfrey 1988).

Florida maple is unusual in that many of the inner crown leaves, after turning a light to dark tan in late November or early December, will remain on the tree all winter. This probably serves the function of providing the thinly barked trunk with some protection from the bright winter sun.

The maximum size attained by Florida maple here in Florida is about 80 feet in height with a trunk diameter of 2 feet. As a mature tree, it usually has a well-balanced, rounded crown of dense foliage about 50 feet in diameter.

Florida maple requires a soil with at least a moderate supply of calcium, fertility, and moisture. It can withstand some flooding, although not nearly as much as red maple. It usually occurs in hardwood forests growing on limestone outcrops where it has a high number of tree associates including spruce pine, red bay, sugarberry, winged elm, white ash, southern magnolia, sweetgum, pignut hickory, live oak, swamp chestnut oak, bluff oak, Shumard oak, laurel oak, basswood, persimmon, hornbeam, hophornbeam, and soapberry. White oak and beech are additional associates in the panhandle. In the Gulf coastal hammocks, cabbage palm, live oak, red-cedar, swamp chestnut oak, Shumard oak, sweetgum, and sugarberry are common associates. Florida maple also grows on stream slopes and river floodplains. It is common in San Felasco Hammock, Peacock Slough, Ichetucknee Springs State Park, Florida Caverns State Park, Torreya State Park, and in some of what remains of Gulf Hammock and other coastal hammocks.

Florida maple is especially shade tolerant and normally grows as an understory or subcanopy tree in native hardwood forests. Among the trees that are part of the upper canopy of Florida's forests, only southern magnolia and American beech come close to being as shade tolerant. Florida maple reproduces prolifically by seed, and the seedlings and saplings can grow in the full shade of the hardwood forest. Even so, deer, rabbits, and insects feed on the leaves and twigs sufficiently to keep this tree from completely taking over the forest. Florida maple has thin bark and is not at all fire tolerant, but it can withstand minor bouts of flooding, drought, and exposure to mildly brackish water. Compared to red maple, it is much more wind-firm, drought resistant, and shade tolerant. However, it is often damaged by gray squirrels feeding on its bark. Indeed, it is much more susceptible to bark damage by squirrels than any other tree in Florida, which could be one of the factors behind Florida maple's being an uncommon tree. It appears to be a rather short-lived tree with a maximum life span of perhaps 100 years.

The wildlife value of Florida maple is moderately high. It produces a crop of paired, winged fruits (seeds) each year, which reach full size in May and then remain on the tree in an unripe condition from early September to early October, when gray squirrels begin feeding on them. If they are not all eaten, they remain on the tree until completely ripe in the middle of October when they begin to fall. Once they fall, bobwhite quail and other birds and mammals also feed on them (Martin et al. 1951). Florida maple bark is an important food for both gray squirrels and flying squirrels. The twigs and foliage are browsed by deer and rabbits. The flowers are visited by bees that collect pollen and perhaps some nectar. Finally, old Florida maple trees are often hollow, making good den trees.

The wood of Florida maple is hard and strong, making it suitable for the manufacture of high quality flooring and furniture. In Florida, however, it is not abundant enough to support a market for high quality lumber. Florida maple also makes good firewood.

Florida maple is a beautiful, medium-sized shade tree that is best used in landscapes where gray squirrels are not overly abundant. It is easy to grow and has been planted to some extent as an ornamental tree, often with disappointing results where squirrels eat the bark to such an extent that the tree is damaged. Because of its amenability to shade, Florida maple is particularly well adapted to growing under old, declining trees to provide a replacement for the day the older trees die.

Red Maple—*Acer rubrum*

Sapindaceae

Red maple is a medium- to large-sized tree. It is one of the best-known and most abundant tree species in the eastern United States, ranging from Newfoundland in eastern Canada all the way down to South Florida and as far west as Minnesota

and eastern Texas (Walters and Yawney 1990). Red maple is quite variable genetically, with considerable variation in morphological and adaptive characteristics
over its large area of distribution (Walters and Yawney 1990). It has opposite, thin,
deciduous leaves 2 to 5 inches long that have serrated edges and usually three
somewhat indistinct lobes of variable shape. (Farther north, red maple leaves
sometimes have five lobes.) The leaf stalk (petiole) is often red; the fall color of
the leaves is quite variable, ranging from pale yellow to a glorious bright red; and
the winged fruits produced in late winter may be greenish, brown, or pink but are
more often a bright scarlet at maturity. Male and female flowers typically occur
on separate (dioecious) trees, but some flowers are bisexual. Pollination is accomplished both by wind and by insects (Barta 1985).

The size of mature trees is variable. When growing in landscaped situations
on well-drained land, red maple's height reaches about 60 feet with a 1-foot trunk
diameter. In the wild on moist to wet, fertile soil it commonly attains a height
of 100 feet and a trunk diameter of 2 feet, with exceptional trees attaining trunk
diameters of as much as 4 feet. The largest one reported in Florida is 115 feet tall
with a 4½-foot trunk diameter (Florida Champion Trees Database 2020). The
largest one reported nationally, from New Jersey, has a trunk diameter of 7 feet
and a height of 75 feet (American Forests 2018).

In the wild in Florida, red maple is largely restricted to swamps and stream
floodplains. In these situations, some of its most common tree associates are bald-
cypress, swamp tupelo, Florida elm, pumpkin ash, green ash, water oak, laurel
oak, sweetgum, and, on the Florida peninsula south of Gainesville, cabbage palm.
Farther north, red maple is often both an upland and a wetland tree. Both in Florida and farther north, red maple's abundantly produced, wind-dispersed, winged
fruits enable it to invade open areas that are being allowed to reforest naturally. It
is also able to invade pine flatwoods forests that are no longer subject to frequent
fire.

Red maple is a rapidly growing, early successional tree that is short lived and
has lightweight, weak, rot-prone wood. It is neither tolerant of fire nor dense
shade. On the plus side, red maple adapts to a wide range of soil types and conditions (Walters and Yawney 1990). It is more flood tolerant than laurel oak, water
oak, Florida elm, or sweetgum, but not as flood tolerant as bald-cypress, cabbage
palm, swamp tupelo, water-elm, buttonbush, pop ash, or pumpkin ash. It is not
especially salt tolerant, being less so than bald-cypress, pumpkin ash, and swamp
tupelo. The root system is normally shallow and extensive, and because the wood
is weak, red maple is easily damaged by high winds.

One aspect of red maple ecology remains unsettled: that is, whether the flowers are wind or insect pollinated. It seems that the flowers of most maples, including red maple, are wind pollinated at least some of the time, whereas some
species of maples are exclusively wind pollinated. However, it is also true that

bees and other insects visit both the male and the female flowers of red maple. In the northern part of its range, red maple blooms so early and the weather is often so cold and windy, that insects are sometimes prevented from visiting the flowers. However, in Florida, bees and other insects visit red maple flowers abundantly on warm, sunny days. The proportion of pollination done by wind compared to that done by insects remains undocumented, but research by Barta (1985) indicates that at least some cross-pollination of red maple is performed by insects.

The wildlife value of red maple is moderate. The fruits are eaten to some extent by bobwhite quail and squirrels (Martin et al. 1951) (they are also edible by humans), and the foliage, twigs, and bark provide food for insects and for mammalian browsers such as deer and rabbits. Beavers also use red maple stems and twigs for food. The very early blooming of the flowers, which occurs mainly in January in Florida, well before most other flowers appear, is a benefit to bees and other insect pollinators. A study conducted in Maryland found that red maple flowers provided abundant nectar and pollen for honey bees and a wide variety of native bees, wasps, and flies, some of which were effective cross-pollinators of the flowers (Barta 1985).

When forested wetlands are logged, red maple is primarily used for pulpwood. Lightweight, weak, and rot prone, its wood is nonetheless sometimes processed for low quality lumber and firewood and occasionally used in the manufacture of higher quality products such as clothes pins, furniture, and musical instruments.

Red maple is often grown and sold in nurseries as a landscape tree. It is popular and can be quite pretty with its light gray trunk and thin leaves that turn color in the fall. The particular varieties of female trees that produce winged, scarlet fruits in February are especially attractive. (Some red maple trees produce male flowers exclusively.) Ornamental use of red maple has several drawbacks, however: it is not well adapted to upland soils in Florida; it is not resistant to drought; it is easily damaged by wind and ice storms; and its leaves are toxic to horses when wilted or newly fallen (Alward et al. 2006). It is also notoriously short lived: though it can live somewhat longer in the wild on appropriate soils, in landscaped situations it often lives no longer than about 50 years.

Silver Maple—*Acer saccharinum*

Sapindaceae

Silver maple is a common tree throughout most of its native range, which covers much of the eastern United States and adjacent southern Canada, extending south into the Florida Panhandle along the Apalachicola River (Gabriel 1990). In native wetland and floodplain forests farther north, silver maple is a medium-sized tree that gets up to 100 feet tall with a 2- to 3-foot trunk diameter. It is usually smaller in Florida or when planted on upland sites as an ornamental. Uncommon in Florida, silver maple used to be planted as an ornamental as far south as Orlando, usually with poor results. It has opposite, deeply lobed leaves with toothed margins, and their gray to silvery undersides have given the tree its name. The male and female, mostly wind-pollinated flowers occur on separate (dioecious) trees. The bark is light gray in color and gets shaggy in old age.

Silver maple is found in Florida mainly along the Apalachicola and Choctawhatchee river floodplains and on adjacent banks and bluffs (Godfrey 1988). It is a rapidly growing, weedy tree that often has poor form because of weak, brittle wood that results in broken branches and multiple forks of the main leaders. While fairly adaptable to varying soil conditions and to flooding, silver maple is not shade, fire, or drought tolerant.

Silver maple has moderately good wildlife value. Its winged fruits are eaten to some extent by squirrels and birds; the winter buds and bark are an important food source for squirrels farther north; beavers eat the bark; deer browse the bark, twigs, and foliage of young plants; and old trees often provide denning cavities used by various species of wildlife, including wood ducks (Gabriel 1990).

Farther north, the soft, weak, rot-prone wood is sometimes sold as soft maple lumber, but it is typically used as pulpwood. In Florida, silver maple is not common enough to be valuable commercially.

In the 1950s and 1960s, silver maple was often planted as an ornamental in Florida and, more frequently, throughout much of the eastern United States. It is now much less popular because of the poor results of these plantings. Silver maple often has poor form, breaks up easily in ice and wind storms, and has extensive, shallow roots that often find their way into water lines, sewer lines, and drain fields and cause cracks and heaving in sidewalks and foundations as well as blocked drainage tiles. Moreover, it grows rather poorly and is short lived when planted on upland sites in Florida. For these reasons, it is now seldom planted here and is used less often than previously in its more northerly ranges.

Boxelder—*Acer negundo*

Sapindaceae

Boxelder is a medium-sized deciduous tree in the maple genus with pinnately compound, opposite leaves. The stems, green at first, turn tan after a year or two. The leaf shape is reminiscent of poison ivy, except that the leaves usually have five leaflets instead of poison ivy's three. Male and female, wind-pollinated flowers occur on separate (dioecious) trees. The paired, winged, wind-distributed fruits are similar to other maple fruits. Boxelder occurs in the Great Plains provinces in Canada and across most of the United States south of Canada (Overton 1990). Its native range extends into Florida along the Gulf Coast from the Apalachicola River southward to Tampa Bay and inland to Gainesville and Ocala.

Boxelder is a weedy early pioneer on floodplains and fertile uplands that have abundant calcium in the soil. In Florida, it is common in a few places, such as the Hogtown Creek floodplain in Gainesville, and absent or rare in most other places. Tree associates include red maple, sweetgum, water oak, laurel oak, sugarberry, and coastal plain willow.

Fast growing and short lived, boxelder grows 40 to 60 feet tall with 1- to 2-foot trunk diameters and rarely lives more than 60 years. It is tolerant of flooding, drought, and alkaline soils. This drought and high pH tolerance, combined with its rapid growth and ability to withstand cold temperatures, made it a favorite tree for planting in windbreaks on the Great Plains.

The wildlife value of boxelder is moderate. The fruits are eaten by squirrels and some birds, and insects feed on various parts of the tree, in turn providing food for birds and other animals. Because boxelder breaks up easily in storms and the wood is not rot resistant, old boxelder trees often provide cavities for nesting and denning.

Boxelder has no commercial value in Florida and is not a preferred landscape tree.

Red Buckeye—*Aesculus pavia*

Sapindaceae

Red buckeye is a shrub or small tree with deciduous, palmately compound, opposite leaves. The leaves are distinctive, usually with five leaflets all joined together at a central point. Red buckeye leaves emerge early in spring, usually early February, and are quite frost resistant. A hard freeze will wilt the new leaves, but they recover with no damage. The spikes of showy tubular red flowers come out early as well, reaching full bloom in early March. The fruit, which ripens in September, contains one to three shiny brown poisonous seeds about 1 inch in diameter encased in a thin, leathery hull.

Red buckeye occurs from Orange County, Florida, northward into southeastern North Carolina, westward through the Florida Panhandle into eastern Texas, and north through Arkansas into the southern tip of Illinois (Godfrey 1988). Red buckeye is slow growing, attaining small tree–size after a couple decades' growth. The largest ones I've seen are beside the Waccasassa River in Levy County where some are 30 to 40 feet tall with 1-foot trunk diameters. The current Florida champion, measuring 31 feet tall with a trunk diameter of about 10 inches, grows at

Dudley Farm State Park in western Alachua County (Florida Champion Trees Database 2020). (Larger ones have been reported well north of red buckeye's native range, but these large individuals may be the result of hybridization between red buckeye and one of the larger-growing species of buckeye.)

Red buckeye is usually found on moist, fertile soil in hardwood forests on stream floodplains, limestone outcrops, ravines, and fertile uplands. Mildly flood tolerant, fairly drought tolerant, and very shade tolerant, red buckeye does best in partial shade to full sun but will grow, flower, and make fertile seed in the full shade of a hardwood forest canopy. Red buckeye does not do well on infertile, sandy soils or on highly acidic soils.

The wildlife value of red buckeye rests primarily on its early production of flowers, which are visited by ruby-throated hummingbirds when they first arrive in the spring (Miller and Miller 1999). The plant and its seeds have little other wildlife value because of their high toxicity.

Red buckeye is a popular ornamental with its showy red flowers, small size, and good shade tolerance. It is often selected for planting in situations where a small, shade-tolerant flowering tree is desired. Its attractiveness to ruby-throated hummingbirds is key to its popularity among wildlife gardeners. One drawback to red buckeye as an ornamental plant is the toxicity of its seeds, which are attractive but dangerously poisonous.

Soapberry (Wingleaf Soapberry, Florida Soapberry)—*Sapindus saponaria (S. marginatus)*

Sapindaceae

Soapberry is a small- to medium-sized deciduous tree with alternate, pinnately compound, light green leaves 12 to 18 inches long that turn bright yellow in late autumn. Its eight to eighteen lanceolate, smooth-margined leaflets, 2 to 6 inches long, turn white after falling to the ground in December. Soapberry's smooth gray bark becomes scaly on old trees. The tree produces terminal clusters of inconspicuous yellow-green, functionally unisexual flowers that are followed by a round "nut" (fruit) about 1 inch in diameter containing a single seed. This hard, black seed is poisonous.

Soapberry is most common in Mexico, Texas, Oklahoma, and nearby states but occurs on the Florida peninsula and in a few spots in the Florida Panhandle on shell mounds and on limestone outcrop soil intermixed with various other hardwood trees. Some botanists have split the Florida population of this tree into two species, one with a winged rachis (the stalk part of the compound leaf) and one without the wings; some botanists consider the western population of this tree to be *Sapindus drummondii* or *Sapindus saponaria* var. *drummondii*. Soapberry is also native to the Big Island of Hawaii, making it the only tree native to both Florida and Hawaii.

Soapberry is often an understory tree or a tree that grows on the edge of or in a gap in a forest. It prefers partial to full sun, and in good light conditions it can reach heights of 70 feet. The largest one reported from Florida was 72 feet tall with a trunk diameter of 28 inches (Ward and Ing 1997). It was growing near Alachua Sink on the north edge of Payne's Prairie in an area of extreme sinkhole activity and abundant outcroppings of limestone and chert in association with live oak, sugarberry, sweetgum, pignut hickory, silver buckthorn, red bay, and yaupon. Only slightly shade tolerant, soapberry is very drought tolerant but not especially amenable to flood or fire, although the western population often grows in riparian areas. It seems to have few insect or disease problems.

Soapberry has very little wildlife value other than providing nectar for bees and other pollinators.

Because it can tolerate drought, salt, high soil pH, and urban conditions, soapberry is occasionally used as an ornamental tree. Its moderate size and bright yellow fall leaf color can be attractive features in landscape situations. Soapberry's growth rate is slow to moderate, however. Historically, Native Americans and early European settlers made use of soapberry's fruit pulp to make a soap for washing clothes.

Taiwanese Goldenrain Tree (Goldenrain Tree, Flamegold)— *Koelreuteria elegans* subsp. *formosana* (*K. formosana*)

Sapindaceae

This medium-sized deciduous tree from Taiwan is not native to the Americas but, unfortunately, is commonly planted here as an ornamental. It has twice pinnately compound leaves and produces clusters of bright yellow flowers in early autumn followed by pinkish-red papery capsules containing six small, round, black seeds. It is closely related to another species of goldenrain tree, *Koelreuteria paniculata*, with singly compound leaves, that is commonly planted north of Florida throughout most of the United States and is also an invasive exotic pest. Both of these species are commonly called "goldenrain tree." *Koelreuteria elegans* subsp. *formosana* is listed as an invasive exotic pest species in Florida (FLEPPC 2013) and several other southern states and is highly invasive in urban and suburban areas as well as in adjacent native forests. It commonly grows to about 40 feet tall with a trunk diameter of 1 foot. The largest one reported in Florida, in Gainesville, had a trunk diameter of 3 feet and a height of 47 feet in 1991 (Ward and Ing 1997).

Taiwanese goldenrain tree grows well on almost any soil, and it reproduces prolifically from seed. The capsules come apart in three sections, and each section has two seeds attached to the papery capsule wall, which serves as a sail to facilitate wind dispersal. The seedlings are very competitive, enabling this tree to invade landscaped areas, stream corridors, abandoned lots, and native upland hardwood forests. It is already taking over the understory and midstory in some places along creek slopes in Gainesville and probably in other cities. Because this tree provides almost no value to wildlife, its presence bodes ill for the future health of our remaining upland hardwood forest areas and wildlife corridors.

The only wildlife value of Taiwanese goldenrain tree appears to be that of the leaves, stems, flowers, and seeds as food for some insects. I have never seen birds feeding either on the seeds or on the few insects that feed to a limited extent on the tree. One insect, a red-orange and slate-gray scentless plant bug known as the "red-shouldered bug" (*Jadera haematoloma*), is sometimes abundant beneath goldenrain trees, feeding on the seeds but doing no damage to the tree or anything else (Mead 1985). These bugs were often copious beneath my next door neighbor's goldenrain tree, but I have never observed birds or any other animals eating them.

In the past, the ornamental value of this tree was considered high by landscape architects and nurserymen because of the fall color of its flowers and capsules. Consequently, it was grown and sold in vast numbers and is now a very common yard tree, often forming monocultures that push out native trees and shrubs. This is unfortunate, because Taiwanese goldenrain trees are a real threat to wildlife, native plants, and the ecology of native forest areas in proximity to them. They are also a nuisance in landscaped areas: the seedlings come up so prolifically every year that much work must be done to weed them out to prevent their taking over the entire landscape. This tree is usually no longer available in local nurseries because of its invasiveness, but unfortunately it is still available and promoted on the Internet.

Southern Buckthorns—*Sideroxylon* spp. (*Bumelia* spp.)

Sapotaceae

"Buckthorn," "bully," and "bumelia" are common names used for these species. (Over time, the Latin [scientific] name of this genus has changed back and forth between *Sideroxylon* and *Bumelia*.)

Southern buckthorns are shrubs or small trees with straight, sharp thorns, some located at the ends of stiff twigs, and with simple, alternate leaves with entire margins. On older twigs, leaves are often produced on short spur shoots and are thus grouped in clusters along the stem. The twigs of several of these species often have irregular swellings caused by gall wasps. Another insect specializing on this plant genus is the eastern bumelia borer, a brightly colored and large longhorn beetle whose larvae bore into the stems and roots, sometimes killing the plants. Very small, white, bisexual flowers are produced in dense clusters in the leaf axles along the stems in late May and early June. The round to oblong, shiny, tasty, edible, black fruits, one-quarter to three-quarters of an inch in diameter, begin to ripen during the first half of September and remain available until mid-October.

The wildlife value of southern buckthorns is high in the few places where they are common enough to have a significant impact. The sweet fruits are edible by humans, other mammals, and birds. Cardinals eat the flesh without swallowing the single hard seed, whereas brown thrashers, mockingbirds, and catbirds swallow them whole, thereby presumably aiding in the distribution of the seeds. There is little information in the literature on the wildlife benefits of these plants, but most species that eat the fruit of other plants probably also eat buckthorn fruits. The flowers attract large swarms of many kinds of bees and wasps. The foliage is browsed by white-tailed deer.

Southern buckthorns are not often used in landscaping, but saffron plum in South Florida and along the coast of the central Florida peninsula, and tough buckthorn and silver buckthorn in the central and northern Florida peninsula have great potential for such use. Big-fruited buckthorn is sold in nurseries in Georgia. They are all shrubs or small trees that fit easily under power lines and in other places where a shrub or small tree is needed, and they all produce sweet, tasty, edible fruits. The main drawback to their use is that they have sharp thorns. They are easy to grow from seed, grow well in pots, and transplant easily. They are drought tolerant and require no special care.

Southern buckthorns are never very common, and some species are among the rarest plants in Florida.

Silver Buckthorn—*Sideroxylon alachuense*

Silver buckthorn is one of the rarest and most beautiful of the southern buckthorns. Because it is so rare and seems to be declining in the wild, it is officially listed by the state of Florida as endangered. Its leaves are glossy dark green on top with a shiny, silvery pubescence below. Its stems are light gray, smooth, hairless, and relatively thick and stiff. While it can grow to about 30 feet tall with a trunk diameter of 5 inches or more, it is usually smaller. It occurs in the limestone outcrop hammock (calcareous southern upland hardwood forest) at Alachua Sink on the north side of Payne's Prairie; on an isolated hilltop in calcareous mesic hammock (hardwood forest) in Silver Springs State Park under a forest of swamp chestnut oak and pignut hickory; and at five other small spots, one in Lake County, one in Nassau County, one near Orlando, and one in Hamilton County. It also occurs on an island in the Okefenokee Swamp in Georgia (Florida State University botanist Loran Anderson, personal communication). The known wild population encompasses fewer than 100 plants. There are also a few plants in cultivation, including in my yard in Gainesville.

Saffron Plum—*Sideroxylon celastrinum*

Saffron plum is well known in the Florida Keys for its fruit. Its northernmost population occurs on the north end of Seahorse Key, offshore from Cedar Key in Levy County. This buckthorn is primarily a coastal-hammock, shell-mound, and salt-flat plant of the southern half of the Florida peninsula and the Florida Keys. It also grows in southern Texas, Mexico, Central America, northern South America, the Bahamas, and Cuba. It is the least cold hardy and most salt tolerant of the southern buckthorns. It has small, narrow leaves without raised veins and produces numerous, stiff, spine-tipped twigs. It grows to be a small tree with a maximum height of 30 feet and a trunk diameter of about 1 foot. Saffron plum's delicious fruits are frequently eaten by people and are, in my opinion, one of the best-flavored of Florida's native fruits. The fruits are also very attractive to wildlife and are reportedly a favorite food of both white-tailed deer and the endangered white-crowned pigeon of the Florida Keys.

Gum Bumelia—*Sideroxylon lanuginosum*

Gum bumelia is widely scattered about in dry, sandy hammocks, mixed pine–hardwood forests, and high pine forests (sandhills and clayhills) in northern Florida, western central and southern Georgia, and southern Alabama. It is the most common of the southern buckthorns. When they first emerge, the leaves of gum bumelia are dully pubescent above and below, and the leaves' undersides remain this way through the summer. The upper surface of the leaf remains dull even after the pubescence wears away. The twigs are also pubescent during their first summer. The tree's maximum size is about 40 feet tall with a 1-foot trunk diameter.

Buckthorn Bully—*Sideroxylon lycioides*

Buckthorn bully occurs in hardwood forests on floodplains and the adjacent bluffs and natural levees in the central Florida Panhandle area and adjacent parts of Alabama and Georgia; it also occurs in scattered, isolated populations as far south on the Florida peninsula as Lake and Orange Counties (Wunderlin et al. 2020). It is considered rare in Florida (Clewell 1985) and appears on the Florida endangered species list. Buckthorn bully is able to withstand more flooding and wetter soils than the other buckthorns, perhaps with the exception of Everglades bully. Two tiny populations of this buckthorn abide on muck soil in forests of Atlantic white cedar, cabbage palm, tupelo, and bay trees in Putnam and Marion Counties. Its leaves, twigs, and flowers are hairless except when first emerging from the bud. Its maximum size is about 40 feet tall with a trunk diameter of 1½ feet.

Big-Fruited Buckthorn—*Sideroxylon macrocarpum*

Big-fruited buckthorn (Ohoopee bumelia) is not known to occur in Florida but is included in this discussion to give a more complete picture of this genus. It occurs on sandy soil in 13 counties in south-central Georgia in the Ohoopee Dunes area in the vicinity of the upper Ohoopee River. It is similar in size and ecology to rufous buckthorn, being a small, fire-adapted shrub rarely more than 1 or 2 feet tall. Big-fruited buckthorn spreads by underground stems to form patches in the ground cover of sandhill forests populated by longleaf pine, turkey oak, and sand post oak. The leathery leaves are rounded at the tips and tapering at the bases, about three-quarters of an inch to 1½ inches long, with blond- to rusty-colored hairs on the lower surfaces. When ripe, the fruit is black, round, and about three-quarters of an inch in diameter. This species is sold in local nurseries as "Ohoopee bumelia" (Chafin [2008] 2020, "*Sideroxylon macrocarpum*").

Smooth Buckthorn—*Sideroxylon reclinatum*

Smooth buckthorn (Florida bully) is a small buckthorn with small, slender, thin, hairless leaves and relatively small fruits. The leaves often display some white, cottony pubescence below when young (Godfrey 1988). It is widespread in hardwood forests on and next to floodplains and in mesic to hydric hammocks on the northern and central Florida peninsula. Smooth buckthorn rarely grows taller than 10 feet or has stems thicker than 1 or 2 inches in diameter. A subspecies restricted to the subtropical rockland and marl prairie habitats at the south end of the Florida peninsula, Everglades bully, *Sideroxylon reclinata* subsp. *austrofloridense* (Corogin and Judd 2014), is listed as threatened by the federal government and as endangered by the state of Florida.

Rufous Buckthorn—*Sideroxylon rufohirtum*

Rufous buckthorn is a fire-adapted species occurring on sandy soil in sandhill and scrub habitats from Suwannee and Columbia Counties southward into Hillsboro and Manatee Counties on the Florida peninsula (Godfrey 1988; Wunderlin et al. 2020). It spreads underground, forming clonal thickets from 6 inches to 2 feet tall. It is especially susceptible to the stem galls that specialize on this genus. The fruits of this species are rather large and round: about half an inch in diameter, they contain more of the sweet green flesh and are more edible than the fruits of the other buckthorn species with the exception of saffron plum and possibly big-fruited

(Ohoopee) buckthorn. Although previously considered a variety of smooth buck-thorn (*S. reclinatum*) by Wunderlin (1998), rufous buckthorn is distinct from that species in leaf, fruit, growth form, ecology, and habitat characteristics and is now considered a separate species (Corogin and Judd 2014). It seems to be most similar to big-fruited buckthorn.

Tough Buckthorn—*Sideroxylon tenax*

Tough buckthorn (tough bumelia) inhabits deep sands in scrub, sandhill, and xeric hammock forests and coastal dunes and hammocks on the northern and central Florida peninsula. Its leaves are shiny on top but vary considerably from one population to the next in size, shape, and underside pubescence. The most common form in coastal habitats has oblong to spatulate leaves, the undersides of which are densely covered with shiny golden hairs. Populations in Florida scrub habitats such as the sand pine scrub in the Ocala National Forest have smaller, less elongated leaves with dark brown to dark gold, dull to shiny pubescence under-neath. There is a newly described species of native bee, *Colletes francesea*, which has only recently been observed in association with tough bumelia in scrub habi-tat on the Lake Wales Ridge on the southern Florida peninsula (Deyrup and Dey-rup 2011). The most attractive plants of this species are among the most beautiful native Florida trees, but there are also many individuals that are thin and scrubby looking.

Thorne's Buckthorn—*Sideroxylon thornei*

Thorne's buckthorn (Georgia bully) is a plant I have not seen. Godfrey (1988) describes it as a rather small and spindly shrub with leaves that are shiny above and woolly pubescent below. It was known to him from only one spot in Florida in Jackson County but grows in several counties in southwestern Georgia, occur-ring beside open water in hardwood forests that flood occasionally. Wunderlin et al. (2020) list it as growing in six counties in the western Florida Panhandle. It appears on the Florida endangered species list.

Note: Even though it is a Georgia endemic not known to occur in Florida, big-fruited buck-thorn is included in this discussion as is Everglades bully. Both are presented in order to provide a more complete view of this interesting genus.

Florida Anise—*Illicium floridanum*

Schisandraceae

Florida anise grows to be a large shrub with a maximum height of about 30 feet. It is densely vegetated with 3- to 6-inch-long, dark green, smooth, slightly leathery, evergreen leaves that have a whitish midrib. The leaves are aromatic, with a spicy odor reminiscent of grapefruit and fish. It occurs in wetlands primarily in the western Florida Panhandle, central and southern Alabama, southern Mississippi, and extreme southeastern Louisiana (Kartesz 2015). The foul-smelling, maroon flowers with their numerous, narrow, radially displayed petals are 1 to 2 inches wide. The star-shaped, 1-inch-wide fruit ejects a small, hard seed for up to several meters' distance as each segment opens explosively.

Florida anise is exceptionally wetland dependent, having no drought resistance yet little tolerance for flooding. It primarily occurs in steephead ravines, along the streams that originate in these ravines, and in seepage bogs—all places where water is never lacking but where flooding is either nonexistent or very temporary. Florida anise is very shade tolerant and highly competitive on sites where it is well adapted. Although capable of coming up from seed, it reproduces primarily by root suckers or rhizomes, often producing extensive clonal thickets along the stream bottoms it inhabits (Chafin [2007] 2020).

There are no reports of wildlife use of this plant (Miller and Miller 1999), although it could provide escape cover for large mammals and provide nectar for pollinating insects. The flowers are primarily pollinated by flies, which are attracted by the foul smell of the flowers and obtain nectar secreted from the base of the stamens (Chafin [2007] 2020).

Florida anise is a handsome evergreen shrub with attractive maroon flowers that is rarely planted as an ornamental. The main difficulty is that it requires a constant supply of water. It does best on acidic sandy soils with a high organic matter content.

This plant is toxic and should not be used for flavoring food (Christman [1997] 2003, "*Illicium floridanum*").

Yellow Anise—*Illicium parviflorum*

Schisandraceae

Yellow anise is a medium-sized, multistemmed shrub with 3- to 6-inch-long, smooth, leathery, pleasantly aromatic, evergreen leaves and green twigs that turn tan the second year. Stem height is rarely more than 20 feet, and trunk diameters rarely exceed 2 inches. The yellow, insect-pollinated flowers are small and inconspicuous, and the star-shaped fruit opens explosively when ripe to expel the small, hard seeds. The leaves produce a delightful aroma reminiscent of anisette or licorice when crushed.

Yellow anise has a small, relict wild population, occurring in only seven counties in the northern and central Florida peninsula from Marion and Volusia to Polk and Osceola Counties (Wunderlin et al. 2020) and, then, only in rather small, isolated spots. The largest populations known to me are along Salt Springs Run and along Mormon Branch, both in the Ocala National Forest in Marion County. It grows mixed with needle palm, swamp azalea, pipestem, and fetterbush on low, wet ground along Mormon Branch and to the southeast under a bottomland hardwood forest canopy of swamp laurel oak, water oak, red maple, cabbage palm, swamp red bay, swamp tupelo, pumpkin ash, loblolly bay, sweetbay, and Florida elm. Next to Mormon Branch, it grows under Atlantic white cedar as well, where it is accompanied by a very small, relict population of Florida willow and larger relict populations of climbing fetterbush (*Pieris phillyreifolia*) and largeleaf Grass-of-Parnassus (*Parnassia grandifolia*). This unique, relict plant population contains thousands of yellow anise plants and thousands of needle palms over an irregularly shaped area covering about one square mile of permanently moist to wet organic soil that never floods.

In the wild, yellow anise grows where there is a constant and reliable water source. It is neither drought tolerant nor especially flood tolerant and is not fire adapted. Though very shade tolerant, it can also grow in full sunlight. It does not seem to have problems with insects or diseases.

There is no published information on the wildlife value of yellow anise, but the habitat encompassing yellow anise thickets and its associates along and to the southeast of Mormon Branch is inhabited by black bears, which no doubt benefit from the dense cover these plants provide.

Yellow anise is a popular landscape plant often used for hedges or as a foundation plant next to buildings. Though much more adaptable and drought hardy than Florida anise, it still needs supplementary watering during droughts. Otherwise, it is care free, needing only occasional trimming if desired. It looks best if trimming is done with hand clippers to avoid shearing the large leaves in half. The sweet-smelling leaves might tempt one to add them to food for flavoring, but this is not a good idea. All parts of yellow anise are reportedly toxic (Christman 2003).

Bay Starvine—*Schisandra glabra*

Schisandraceae

This high-climbing, twining vine has alternate, deciduous leaves 1 to 5 inches long by one-half to 3 inches wide with pointed leaf tips and a few scattered teeth on the margins. It also forms mats of vegetation on the ground. Bay starvine is uncommon, occurring in scattered spots in the central Florida Panhandle and from there northward and westward into Georgia, Alabama, Mississippi, Louisiana, and Arkansas with a few spots farther north (Kartesz 2015). It is listed as a threatened species by both the states of Florida and Georgia. This vine can be mistaken for climbing hemp vine, but the latter has opposite leaves and attaches to tree trunks by means of aerial roots (Godfrey 1988). Bay starvine produces small red flowers on long, drooping stalks and drooping clusters of half-inch-long red berries.

Bay starvine is a shade tolerant vine and ground-cover plant occurring in moist, fertile hardwood forests on lower slopes and stream terraces. It is often associated with American beech.

There is little information on the wildlife value of this vine. The flowers are likely insect pollinated and the berries are likely eaten by birds.

Corkwood—*Leitneria floridana*

Simaroubaceae

Corkwood is a colony-forming shrub or very small tree with lightweight wood and deciduous, elliptical, often slightly drooping leaves clustered toward the ends of the stems. The leaves are 3 to 6 inches long and smooth margined, with a slight silky hairiness on the undersides. Corkwood spreads widely by sprouting up from surface roots or underground stems, and the individual plants often present a single upright stem 1 or 2 inches in diameter and 5 to 10 feet tall with very few branches. It occasionally gets larger, however. One growing in the Waccasassa Bay Preserve State Park south of Cedar Key, Florida, was perhaps 20 feet tall with a trunk diameter of about 5 inches and a well-developed crown of branches. It was cut down in the 1980s to supply someone with a sample of the wood. The current national champion, in Jefferson County, Florida, is 17 feet tall with a trunk diameter of 3 inches (American Forests 2018).

Corkwood occurs in about ten geographically isolated populations. The largest of these occurs in the Big Bend area of the Gulf Coast of Florida from Waccasassa Bay to Apalachicola Bay and along the Apalachicola River floodplain. Corkwood also occurs in a sizable population located in northeastern Arkansas and southeastern Missouri and in populations on the Gulf Coast of Texas south and east of Houston. There is a small population on the Atlantic coast of Georgia, another population (perhaps extirpated) in southwest Georgia, one on the west boundary of Alabama, and several scattered about in central and southeastern Arkansas and south-central and east-central Missouri (Sharma et al. 2008).

Corkwood is confined to wetlands, growing in coastal marshes and woodlands, tidal estuarine shores, and swampy woodlands and prairies (Sharma et al. 2008). On the Gulf Coast of Florida, it occurs in marshes, swamps, and hydric hammocks, growing on 6 to 10 inches of soil over limestone bedrock (Sharma et al. 2008). These areas are occasionally flooded by brackish water during tropical storms. In some of the marshes, corkwood grows in full sun all day long. In swamps and hydric hammocks, it grows in moderate shade. Corkwood grows in both brackish and freshwater marshes and often in standing water. It is obviously highly amenable to flooding and brackish conditions.

Corkwood is a rare plant that survives in only a few places. Hydrologic alteration is one of the main threats to its limited populations. Such alterations come about either as a consequence of human activities such as draining or diking or as the result of sea-level rise brought about by global warming (Sharma et al. 2008). Another threat to corkwood is posed by the invasive exotic Chinese tallow tree, which can overwhelm and crowd out entire corkwood populations (Sharma et al. 2008).

Corkwood produces male and female catkins on separate (dioecious) plants, which results in each clonal patch being either male or female. Since corkwood is wind pollinated, separate clones may become so isolated that pollination occurs only infrequently. Corkwood females produce inch-long, oblong, green to yellowish drupes that each contain one seed.

Legend has it that corkwood's very low density wood was once used to make fishing net floats (Godfrey 1988).

Christmasberry—*Lycium carolinianum*

Solanaceae

Christmasberry is a salt tolerant, thorny, evergreen shrub with succulent leaves, purplish flowers, and red berries that occurs along both coasts in Florida, ranging north into Georgia and west into Texas (Godfrey 1988). It can grow to about 8 feet tall with buff colored stems. The half- to 1-inch-long leaves are linear and lack petioles. The flowers have a white center and four lobes that can be blue, purple, pale lavender, or nearly white. The half-inch-long berries, which ripen in late fall or early winter, are a lustrous orange-red to bright red.

Christmasberry grows on the edges of saltwater flats and salt marshes in moist places. It is exceptionally salt tolerant, growing well in places intermittently flooded by salt water. Neither drought nor shade tolerant, Christmasberry nearly always does best in full sunlight.

The flowers are attractive to butterflies and moths, and the fruits are eaten by songbirds, raccoons, and perhaps other species.

This species is a member of the nightshade family and is potentially toxic to both humans and livestock. Eating the bright red berries is ill advised.

Christmasberry is an exceptionally colorful plant, and when its purple, confetti-like flowers contrast with its bright red berries and pear green leaves, the effect is vivid and cheerful. It is no surprise, then, that it is sold in nurseries as an ornamental. Although doing best in coastal landscapes, it can also be cultivated in upland situations (professional botanist David Hall, personal communication).

Bladdernut—*Staphylea trifolia*

Staphyleaceae

Bladdernut is a deciduous, 10- to 20-foot-tall shrub with opposite, dark green, trifoliate leaves. It occurs in much of the eastern United States with scattered populations as far south as the central Florida Panhandle where it occurs in a few spots along the Chattahoochee and upper Apalachicola Rivers (Godfrey 1988). Its finely serrated leaflets are mostly 2 to 3 inches long and 1 inch wide, tapering to a point at the end. The bark on small branches is smooth with longitudinal streaks of black and light gray. It produces dangling clusters of greenish-white, bell-shaped flowers in the spring and bladder-like, 1- to 2-inch-long, papery capsules in late summer that persist into winter.

Bladdernut grows beneath hardwood forest canopies on moist, fertile soil along streams and on stream floodplains. It can withstand temporary flooding and is shade tolerant. It sometimes produces thickets by root suckering.

The flowers attract honey bees, bumblebees, halicid bees, andrenid bees, syrphid flies, dance flies, and the giant bee fly; thrips and the bladdernut plant bug (red plant bug) suck juices from the plant, at least where it grows in the Midwest (Hilty 2019, "Bladdernut").

Bladdernut is sometimes sold as an ornamental shrub.

Two-Winged Silverbell—*Halesia diptera* and Carolina (Little) Silverbell—*Halesia carolina*

Styracaceae

These two similar species are small, deciduous, slender-trunked, flowering trees that occur as uncommon floodplain trees or shrubs mainly in the Florida Panhandle and northward into Alabama and beyond. Although typically rather small, the largest two-winged silverbell reported in Florida was 55 feet tall with a trunk diameter of over 1 foot (Godfrey 1988), and a Carolina silverbell in the mountains of Tennessee measured 118 feet tall with a trunk diameter of over 2½ feet (American Forests 2019). Silverbells have two buds at each leaf junction instead of one as in most plants. The flowers bloom in March and are white and showy, hanging down in small bunches from the leaf axils of last year's twigs. The fruits, which also hang down from the twigs, are dry and winged, containing one to three seeds.

Two-winged silverbell is rare on the Florida peninsula, occurring mostly in the Florida Panhandle, Alabama, Mississippi, Louisiana, and East Texas. Its fruit has two thin opposite wings extending from and running the length of its sides.

Carolina silverbell occurs in Gulf Hammock and along the Suwannee River floodplain, in the Florida Panhandle, and then ranges north into Alabama, Arkansas, Georgia, the Carolinas, Tennessee, and West Virginia, with a few isolated populations even farther north. Its fruits have four wings. In Florida, these two species occur mostly in hardwood forests on river floodplains and on slopes near streams. I have seen them on the natural levies beside major rivers. They are shade tolerant but prefer some sunlight.

The main wildlife value of the silverbells seems be to that of their flowers for pollinators, mainly honey bees and bumblebees.

Both of these species are used to some extent as ornamentals for their attractive blossoms and winged fruits. Because of their shade tolerance and freedom from disease and insect problems, they offer an attractive alternative in places where a crape myrtle or a flowering dogwood tree would not do well.

American Snowbell—*Styrax americanus* and Big-Leaf Snowbell—*Styrax grandifolius*

Styracaceae

The very similar American and big-leaf snowbells are shrubs or small trees with broad, simple, alternate, deciduous leaves with entire or slightly toothed margins. Each leaf axil has two buds, and the trees produce hanging, showy white flowers in late spring and hanging, roundish, capsule fruits in autumn. Both species occur throughout most of the southeastern United States. Big-leaf snowbell ranges south into Florida only in the panhandle and in one small location in Duval County in the northeastern corner of the Florida peninsula; American snowbell occurs in scattered locations as far south as Charlotte County on the central Florida peninsula (Wunderlin et al. 2020). Both are small trees, with the largest American snowbell reportedly measuring 21 feet tall with a trunk diameter of nearly 4 inches and the largest big-leaf snowbell measuring 17 feet tall with a trunk diameter of 6½ inches (American Forests 2019).

These little flowering trees are uncommon in Florida and are mostly found in the understory on moist, fertile, calcareous soil in hardwood forests on stream floodplains, on well-drained soil near streams, or on limestone outcrops. I have seen American snowbell in Gulf Hammock. They are shade tolerant but able to grow in full sunlight. They are also moderately flood tolerant.

The main wildlife benefit of these trees seems to be in providing food for honey bees, bumblebees, and other insects that visit the flowers.

Both species of snowbell are used on occasion as ornamentals for their showy white spring flowers.

Sweetleaf (Horse Sugar)—*Symplocos tinctoria*

Symplocaceae

Sweetleaf is a large shrub or small tree. Resembling those of red bay in size and appearance, its alternate, semievergreen leaves are nonetheless slightly thicker, not shiny, and have a coating of short hairs on their undersides. The leaves are not aromatic, but have a slightly sweet taste when bitten. The flowers, which appear before or as the new leaves begin to emerge in March, are yellowish-white clumps or balls arrayed along the previous season's stems with profuse stamens sticking out beyond the rest of the flower. A small, green, oblong fruit (drupe) ripens in September. The largest sweetleaf currently reported, from Virginia, is 48 feet tall with a trunk diameter of nearly 1½ feet (American Forests 2019).

An oddity of sweetleaf's appearance involves the fungal gall (*Exobasidium symploci*) that often forms on a few of the newly growing leaves in spring. This gall results in large, fleshy growths that sometimes resemble human ears.

Sweetleaf occurs from southeastern Virginia south and west into southeastern Texas, reaching into the Florida Panhandle and into the peninsula as far south as Putnam, Alachua, and Levy Counties (Kartesz 2015) with one population recorded as far south as Hillsboro County (Wunderlin et al. 2020). Sweetleaf is typically sporadic in its distribution but is sometimes abundant, as it is in an upland forest area populated with longleaf pine, southern red oak, and mockernut hickory on well-drained sandy soil in the center of San Felasco Hammock in Alachua County. It is often found on the ecotone between upland longleaf pine forests and either upland hardwood forest or some sort of wetland or floodplain.

Sweetleaf is moderately shade tolerant but does best in full sun. It is drought tolerant but not especially flood tolerant. It responds to fire by dying to the ground and then resprouting vigorously. Sweetleaf reproduces both by seed and by root suckers, often forming clonal patches.

The wildlife value of sweetleaf is low to moderate. It is the host plant for the caterpillars of the king's hairstreak butterfly (Glassberg et al. 2000); the flowers are attractive to bumblebees, honey bees, and butterflies; and the clonal thickets provide cover for wildlife. The foliage is browsed by livestock but is not particularly attractive to deer. The value of the fruit as wildlife food has not been documented.

Although potentially useful as a hedge plant, sweetleaf is rarely planted as an ornamental as it has proven very difficult to propagate (professional botanist David Hall, personal communication). It was formerly used to produce a yellow dye.

Loblolly Bay—*Gordonia lasianthus*

Theaceae

Loblolly bay is a medium-sized tree in the tea family with alternate, thick, dark shiny green, evergreen leaves 3 to 6 inches long that have slightly scalloped (shallowly toothed) edges. Throughout the year in the crown of a loblolly bay tree a few individual leaves will turn bright red prior to falling, giving the appearance of a few male cardinals perching in the tree. Because individual leaves live about three years, this tree does not drop many leaves at any one time. Showy white flowers 3 inches in diameter are borne singly in a scattered array throughout the tree's crown in June, July, and August, standing out brightly against its dark green foliage. The growth form of the tree is conical with a single straight trunk until the tree is quite large, at which time it develops a large crown supported by a few large branches. On trunks less than 1 foot in diameter, the bark is light gray and smoothly furrowed; eventually, however, when the trunks attain diameters over 2 feet, the bark becomes dark reddish-brown, thick, and deeply furrowed and ridged.

The largest loblolly bay ever reported was in a bayhead on Hughes Island in the Ocala National Forest. In 2004, it was 96 feet tall with a trunk diameter of 4 feet 4 inches at 4½ feet above ground (American Forests 2004). It has since died. I have seen perhaps a dozen other large loblolly bay trees, but the vast majority of this species' population is made up of younger, smaller trees, most with trunk diameters of less than 1 foot.

Loblolly bay occurs in wetlands and areas adjacent to wetlands on the coastal plain of North Carolina southward to Lake Okeechobee on the Florida peninsula and westward into the Florida Panhandle and the southern tips of Alabama and Mississippi (Gresham and Lipscomb 1990). It is quite habitat specific, needing a constant and reliable source of acidic water near the ground surface with no threat of prolonged flooding. Although very well adapted to a persistently high water table at, or even slightly above, some of the land surface, loblolly bay does not tolerate flooding above that level nor does it tolerate dry, well-drained soil conditions.

Loblolly bay is not considered fire tolerant, but the larger trees will often withstand moderate ground fires. One thing that makes loblolly bay susceptible to fire damage is its shallow root system. Fire damage to trees often involves the root system as well as the visible aboveground parts of a tree. Loblolly bay is moderately shade tolerant but does best with ample sunlight. There are no reported insect or disease problems associated with loblolly bay, but I have seen evidence of insect borer damage on the trunks of some trees.

Loblolly bay produces seeds in abundance from capsules every year. It reproduces and spreads both by seed and by root sprouts. As fire has become less frequent in many pine flatwoods forest areas, loblolly bay has spread from the edges of wetlands into adjacent pine flatwoods forests.

Common tree associates of loblolly bay include pond-cypress, slash pine, loblolly pine, pond pine, sweetbay, swamp red bay, swamp tupelo, swamp laurel oak, sweetgum, water oak, dahoon, and red maple. In a few places, Atlantic white cedar and cabbage palm are also present. Common shrub associates include gallberry, large gallberry, fetterbush, swamp doghobble, and white titi. In the Florida Panhandle, black titi is another associate and a strong competitor where the two occur together.

The wildlife value of loblolly bay is moderate. Deer browse the foliage, especially stump sprouts, and bees and butterflies visit the flowers.

The commercial value of loblolly bay is low. It is harvested for hardwood pulp and potentially for fuel wood.

Loblolly bay is one of the most beautiful of our native trees and yet is used only rarely as a landscape tree because it is not nearly adaptable enough to be successfully used in most landscape situations. When planted on well-drained uplands, it either fails completely or needs frequent watering with soft, acidic water. Fertilizer is often harmful to this species. When planted on wet sites, it usually dies the first time the weather gets really rainy and the site floods. If one has a spot in their landscape that never floods but where water naturally and reliably seeps to the surface, then that is a place to try this beautiful tree.

Silky Camellia (Virginia Stewartia)—*Stewartia malacodendron*

Theaceae

Silky camellia is an uncommon large shrub or small tree in the tea family with 2- to 4-inch-long, thin, deciduous leaves and, in April or May, showy, 3-inch-wide, white flowers with a center of purple stamens. The fruit is a woody capsule. On each major branch, the secondary branches, leaves, and flowers are often oriented in one horizontal plane. Silky camellia is found in scattered, isolated populations from southeastern Virginia through the Carolinas, Georgia, Alabama, southern Mississippi, and Louisiana, with the Florida population, which is listed as endangered, confined to the panhandle west of the Ochlockonee River and with small populations in southern Arkansas and extreme eastern Texas (Kartesz 2015; Godfrey 1988).

Silky camellia is an understory species occurring in upland hardwood forests in ravines, on slopes, and along streams. It is shade tolerant but not drought or fire tolerant. When not in bloom, this little tree is rarely noticed. It is usually 3 to 20 feet tall with a small trunk; the maximum size is represented by one recorded in 2019 in North Carolina that is 19 feet tall with an 8-inch-diameter trunk (American Forests 2019).

The main wildlife value of silky camellia is the value of its flowers for bees and butterflies.

Silky camellia is available from some native plant nurseries and is sometimes planted as a specimen landscape plant. It can be difficult to grow, requiring rich soil, partial shade, and supplemental watering during droughts.

Eastern Leatherwood—*Dirca palustris*

Thymelaeaceae

Leatherwood is a deciduous, wide-branching shrub 3 to 7 feet tall with distinctive twigs (with very noticeable nodes) and roundish, alternate, 2- to 4-inch-long leaves. It is much more common near the Canadian border in New England and the Lake States than in the southeastern United States, but it does range into the Carolinas and Alabama, and it occurs in three counties in the Apalachicola River drainage basin in the middle of the Florida Panhandle (Kartesz 2015). Among leatherwood's notable features is the exceptional flexibility of its twigs and branches. Another is that the bark, when stripped from the branches, is exceedingly strong. Leatherwood produces small, one-third–inch-long, drooping, yellow flowers in early spring before the leaves are fully grown.

Leatherwood is quite shade tolerant, occurring as an understory shrub in moist, fertile upland hardwood forests. The flowers are self-fertile, so many of the plants in an isolated population may be clones of one another.

Leatherwood has little wildlife value. The flowers are visited to some extent by small native bees.

Leatherwood is used occasionally for landscaping farther north. It has bright yellow fall foliage.

Water-Elm—*Planera aquatica*

Ulmaceae

This uncommon deciduous tree really doesn't have a common name. It is often called "planer tree," "planera," or "water-elm" by the few people who know it.

When not flowering or fruiting, water-elm is recognizable with some difficulty by its growth form, habitat, leaf size and shape, and bark color and texture. The tree most similar in appearance to water-elm is Florida elm. Water-elm, in comparison, has leaves that are smaller, somewhat broader at the base, less rough to the touch, and quite variable in size, even on the same twig. The bark of water-elm has irregular plates that are gray-brown on the old surfaces and a lighter orange-brown on newer surfaces between the plates.

Water-elm is a low-growing, stout, crooked, medium-sized tree that often has more than one trunk. The largest one reported to the Florida champion tree program, growing on Suwannee River Water Management District lands about a half mile northwest of the town of Wannee, Florida, was 66 feet tall with a trunk diameter of 5 feet and a crown spread of 55 feet when measured on February 18, 2000. The former champion in San Felasco Hammock northwest of Gainesville, Florida, has a height of 41 feet, a trunk 3½ feet in diameter, and a crown spread of 60 feet. Both of these champion trees have much larger trunk diameters than is typical for this species. More typical dimensions for water-elm would be a height of 30 to 50 feet with a trunk diameter of 6 inches to 2 feet.

Water-elm occurs in scattered populations from Levy and Alachua Counties on the northern Florida peninsula (Wunderlin et al. 2020) northward and westward throughout most of the southeastern coastal plain and up the Mississippi River floodplain to the southern tip of Illinois (Godfrey 1988). Scattered populations of water-elm grow on the floodplains of the Apalachicola, Aucilla, Suwannee, and Santa Fe Rivers. Concentrated populations, sometimes in pure stands, grow in karst depressions called "prairies" in San Felasco Hammock northwest of Gainesville and on Hogtown Prairie on the west side of Gainesville.

Water-elm, as the common name implies, grows in places that are frequently flooded. It is adapted to sites of moderate to high fertility that flood deeply, remain flooded for many months, but also dry out periodically to well below the soil surface. It also grows on river floodplains where it is often restricted to river banks or to sites that are flooded more deeply than most. Its most common associate is pop ash. Pop ash and water-elm sometimes grow together in stands that contain no other tree species. Other common associates include bald-cypress, coastal plain willow, sweetgum, red maple, overcup oak, green hawthorn, buttonbush, and Florida elm.

Obviously, water-elm withstands flooding well, although not as well as bald-cypress. It is only moderately shade tolerant, which probably explains why it is restricted to severely flooded sites, where most of the large canopy trees cannot survive, and to river banks where it gets light by being on the edge next to open water. It only grows on sites that rarely, if ever, burn. Its bark is thin, so its ability to withstand fire is probably low. Water-elm seems to live a long time, but there are no data on this. Old trees are usually hollow.

Interestingly, water-elm provides an ideal habitat for epiphytic plants (in much the same way that pond apple does in South Florida). This is probably the result of a combination of qualities including its irregular, plated bark, the advanced age and slow growth of many of the trees, the many large horizontal branches that define its growth habit, and the humid, sunny habitats where it usually grows. The plants that most abundantly festoon the branches in San Felasco Hammock Preserve State Park are Bartram's airplant, greenfly orchid, and resurrection fern. Ball moss and Spanish moss also occur in somewhat reduced abundance or on smaller branches as do various lichens, leafy liverworts, and true mosses.

The wildlife value of this tree is high in comparison to the amount of space it occupies. It provides a low, complex forest structure with its growth form combined with the epiphytic plant growth that is usually quite distinctive. Stands of water-elm are sometimes used by nesting colonies of herons and egrets as in the northwestern corner of Sanchez Prairie in San Felasco Hammock. Such stands also provide good foraging habitat for migrating warblers and vireos. Botanist and field biologist Michael Drummond and I, along with several other folks, observed a flock of about 50 goldfinches feasting on water-elm flowers at Planera Pond in San Felasco Hammock on February 21, 2004. The roundish little fruits with a profusion of small projections on their surfaces are probably food for wildlife such as gray squirrels and wild turkeys, although I have no data on this. These fruits, which, including the green leafy covering, are about a half inch in diameter, ripen in late March. White-tail deer browse the foliage within their reach. The often-hollow trunks provide nesting and denning cavities for an assortment of creatures.

Water-elm has rarely been used as an ornamental. However, it is easy to grow from seed (which should be planted immediately after collection) and probably from cuttings. It could be ideal for planting in retention basins in situations where trees are desired.

Florida Elm (American Elm)—*Ulmus americana* var. *floridana*

Ulmaceae

Florida elm is a medium- to large-size deciduous tree. It usually has the classic vase-shaped crown of the American elm (*Ulmus americana* var. *americana*), and its leaves are similar to those of American elm as well: dark green, asymmetrical, and more than twice as large as the leaves of winged elm and cedar elm. Florida elm is a smaller tree than American elm, however, and its bark, quite variable in appearance, often has roundish plates or scales instead of the deep, interlacing, longitudinal ridges of American elm. Seedling Florida elm leaves are much smaller and more symmetrical than those of sapling and mature trees. Florida elm produces inconspicuous, bisexual, wind-pollinated flowers in early February and winged seeds ("samaras") in late February. Because it leafs out in early March, ahead of most of its associates, it is easy to spot in a diverse hardwood forest at that time of year. One unique feature of mature Florida elm trees is the frequent occurrence of large, narrow, vertical buttresses where the main roots leave the base of the trunk. These buttresses and the accompanying very strong and numerous surface roots visible on stream banks are better developed on Florida elm than on any other Florida tree.

The natural range of Florida elm extends throughout the Florida Panhandle and south on the peninsula into Palm Beach County on the east coast and into Lee County in the Big Cypress Swamp on the west coast (Wunderlin et al. 2020). Florida elm also reaches north of Florida into the coastal plain of North Carolina. In the vicinity of Raleigh, North Carolina, the physical appearance of Florida elm, especially the character of the bark, is quite distinct from that of American elm.

In the wild, Florida elm is largely restricted to lowland hammocks, bottomland hardwood forests, stream floodplains that flood occasionally, the shallow edges of swamps and other wetlands, and to stream banks. Its common associates include swamp laurel oak, water oak, live oak, sweetgum, red maple, sugarberry, persimmon, green ash, pumpkin ash, pop ash, water-elm, water hickory, loblolly pine, and bald-cypress. Florida elm is scattered along the banks and floodplains of most of our rivers and many smaller streams. It is also common in hydric hammocks such as Gulf Hammock in Levy County and Twelve Mile Swamp in St. Johns County. The two largest Florida elms I have seen were located in a hydric hammock area of San Felasco Hammock State Preserve Park and on the floodplain of the Silver River in Silver Springs State Park. Both were about 95 feet tall with trunk diameters slightly over 4 feet at 4½ feet above ground and had crown spreads of about 60 feet.

Florida elm prefers fertile soil of near neutral pH with a high calcium content. It can withstand more flooding than the other elms in our area, but not as much flooding as green ash, pumpkin ash, pop ash, water-elm, red maple, swamp tupelo, cabbage palm, or bald-cypress. Florida elm is not particularly tolerant of shade, fire, or drought. Because of its powerful root system, it is less likely to be uprooted by strong winds than many other kinds of trees on similar soils but is susceptible to trunk and branch breakage. It is probably more susceptible to the Dutch elm disease than the other elms of our area, but, so far, this disease has not reached Florida. Florida elm does not grow as tall as winged elm and cedar elm, nor is it as shade tolerant or as able to compete with other hardwood trees. It is not especially long lived, probably rarely living more than 100 years. It produces abundant winged fruits in the early spring. Though many seedlings come up each spring, most of them die from drought or dense shade or are eaten before they can grow tall enough to be above the browsing level of deer and rabbits. Once well established, sapling elms are largely free of serious pests until they reach maturity when they may begin to suffer from heart rot (internal decay of the heartwood). Florida elm supports many kinds of insects that feed on the foliage, but these rarely do much defoliation or damage, with the exception of a small, leaf-feeding beetle observed in Gainesville in 2017 and 2018.

The wildlife value of Florida elm is moderate. The winged fruits, available mid-February to early March, are larger than those of the other elms and are eagerly consumed by gray squirrels. They are an important food item for wild turkey, wood duck, and purple finch (probably house finch in Florida) (Martin et al. 1951). As with the other elms, deer and rabbits browse on the leaves and twigs, seeming to prefer them to most other browse plants. Florida elm is sometimes used for food by beavers, and the trunk is used by yellow-bellied sapsuckers who drill rings of holes in the bark that produce sap and attract insects. The many kinds of insects that feed on the foliage provide some food for birds. Elm trees rarely make good cavity trees for wildlife, although Florida elm is somewhat better in this regard than the other elms.

The wood of Florida elm is used to some extent for pulpwood and for making crates. It is also suitable for the manufacture of flooring, furniture, and hockey sticks. It makes good firewood though it is difficult to split because of its interlocking grain.

Florida elm is not often used for landscaping. This is rather odd, given its strong similarity to the American elm that was once the most popular shade tree among all landscape trees in the northern half of the eastern United States until the Dutch elm disease was introduced and killed the vast majority of these magnificent trees. Florida elm makes a fine shade tree. One special landscape use for which Florida elm is unsurpassed is in controlling erosion on stream banks. Its massive, intertwining roots that grip the banks of creeks and streams far surpass those of any other tree in Florida in preventing bank erosion.

Winged Elm—*Ulmus alata*

Ulmaceae

Winged elm is a medium- to large-size deciduous tree native to the southeastern United States from Virginia into Missouri and eastern Texas (Snow 1990) and ranging throughout northern Florida and as far south on the Florida peninsula as Pasco County (Wunderlin et al. 2020). It is usually a tall, slender tree with a graceful appearance and sometimes an almost weeping habit at the edges of the crown. The largest one reported in Florida, at Torreya State Park in 1993, was 126 feet tall with a trunk diameter slightly over 3½ feet and an average crown spread of 59 feet (Ward and Ing 1997).

Winged elm's small, narrow leaves, typically about 2 inches long, are more than twice as long as they are wide. Vigorously growing twigs of young trees often have corky projections or "wings" that give this tree its name. These wings are not always present, however, and the twigs of the rather similar cedar elm are even more likely to have wings. The easiest way to tell these two elms apart is by looking at the leaves, which on cedar elm are smaller, blunter, and less than twice as long as they are wide. Another difference between the two species is that winged elm usually has straight to gracefully-arched, flexible twigs and branches, whereas cedar elm has stiffer and more crooked branches. The clearest way to distinguish winged elm from cedar elm is in the timing of flowering: winged elm produces its bisexual, wind-pollinated flowers in early spring, whereas cedar elm produces its flowers and fruit in the fall.

Winged elm grows best on moist, well-drained, fertile upland soil. It occurs naturally as an uncommon tree in association with other hardwoods in fertile upland hardwood forests and slope forests. In these situations, it is always tall and slender, and, at maturity, is usually an emergent tree, sticking up above the general canopy. Its most common associates include laurel oak, water oak, sweetgum, pignut hickory, white ash, swamp chestnut oak, sugarberry, hornbeam, hophornbeam, spruce pine, and magnolia.

Winged elm is moderately shade tolerant but clearly does best in full sunlight where it can grow rapidly. It is quite adaptable, growing well on both sandy and clay soils and able to withstand both drought and occasional flooding of short duration. It is firmly rooted with both shallow and deep roots that are slender but strong. However, winged elm can be damaged or blown over by high winds more easily than many other tree species because of its tall, slender form. When surface roots are cut, they resprout vigorously, although winged elm does not often reproduce this way. It produces an abundance of small winged fruits in February that are scattered by the wind.

The fruits of winged elm are eaten by wood ducks, wild turkeys, and purple finches (Martin et al. 1951). I have seen cardinals, house finches, and goldfinches eat them in Gainesville, Florida. Gray squirrels often feed extensively on them from mid-February to early March. Deer browse on the leaves and twigs that are within reach (Short et al. 1975), seeming to prefer them to most other browse plants, and rabbits eat many seedlings. Indeed, given the adaptability, shade tolerance, rapid growth, and high reproductive capacity of winged elm, it would probably be an abundant tree were it not for the browsing of deer and rabbits. Elms rarely become cavity trees. The overall wildlife value of winged elm is moderate.

The wood of winged elm is hard and strong and has interlocking grain, making it difficult to split. It makes fine firewood and is used commercially for making floors, furniture, crates, and high-quality hockey sticks (Snow 1990).

Winged elm has become a popular shade tree in northern Florida in recent years because of its adaptability, ease of nursery production, ease of transplanting, and graceful appearance. Its leaves are small enough to filter down into lawns, making it unnecessary to rake them. It is a fine choice for a tall, graceful shade tree but perhaps should not be planted near houses or other buildings given how easily it can be broken or toppled by high winds. It adapts well to parking lots, although in this harsh environment its growth is somewhat stunted.

Cedar Elm—*Ulmus crassifolia*

Ulmaceae

Cedar elm is a tall-growing, slender elm that often has a single straight trunk with stiff, crooked, horizontal lower branches and a relatively narrow crown. However, in the open, it produces a large, rounded crown. The oval, 1- to 2-inch-long, deciduous leaves are the smallest of the native elm species although not as narrow or as pointed at the tips as winged elm leaves. The upper surface of cedar elm leaves feels rough like sandpaper, whereas the surface of winged elm leaves feels smooth. The bark on mature trees is composed of narrow vertical ridges that curve out from the trunk to some extent, giving it a slightly shaggy appearance. The small leaves and unusual shape of the crown and branches often give cedar elm a distinctive, airy look, as if it belonged in a Chinese or Japanese painting.

The main native range of cedar elm is the eastern half of Texas, southern Arkansas, northern Louisiana, and northwestern Mississippi; throughout this range, it occurs both on river floodplains and on well-drained uplands (Stransky and Bierschenk 1990). A small, widely disjunct population occurs in northern Florida where it grows on clay or alluvial soils over limestone or on clay soils that contain shells. As a mature tree, cedar elm is often taller than its hardwood associates. It occurs in or next to the upper floodplains of the Suwannee, Ocklawaha, and Silver Rivers and in Gulf Hammock. Its most common associates along the rivers are water oak, laurel oak, swamp chestnut oak, overcup oak, water hickory, sugarberry, winged elm, Florida elm, sweetgum, and American hornbeam. Its most common associates in the Gulf coastal hammocks are live oak, cabbage palm, and southern red-cedar.

Cedar elm occurs naturally in several areas along the Suwannee River, including a large stand along Holton Creek in the Holton Creek Wildlife Management Area in Hamilton County and a few trees along the dry run to Lime Sink at Suwannee River State Park in Suwannee County, at Troy Spring in Lafayette County, and in the Suwannee River floodplain in Dixie County. It also occurs in hardwood forest along a narrow belt about one mile inland of the Gulf Coast from the north edge of Gulf Hammock, just south of State Road 24 in Levy County, south through the center of Gulf Hammock into Hernando County. Finally, it occurs in hardwood forest on the slopes at the edge of the floodplain of the Silver River in Silver Springs State Park in Marion County and to some extent on the edge of the Ocklawaha River floodplain south of its confluence with the Silver River. The largest cedar elm in Florida (and national co-champion) was in Silver Springs State Park until it blew down in March 1993 in what was called "The Storm of the Century." It was 118 feet tall, had a trunk diameter of about 2¾ feet, and a crown spread of about 66 feet. Measured in the same location in 1994, the replacement state champion measured 107 feet tall with a trunk diameter of 3 feet and a crown spread of 69 feet. Several other cedar elms in this vicinity are 120 feet tall with trunk diameters of 2 feet.

The ecology of cedar elm is similar to that of winged elm, except that its soil requirements are much narrower. It requires a calcareous soil with some clay in it. It tolerates more shade and also more salt than the other elm species that grow in Florida. While it grows more slowly than winged elm in most situations, cedar elm grows rapidly in ideal situations of full sunlight, good soil drainage, high moisture availability, neutral pH, and high fertility. Cedar elm seems to live a long time, perhaps 200 years or more, but it is not very wind-firm. It flowers in late August and early September and produces abundant fruit in October (winged elm and Florida elm produce fruit in February).

Cedar elm's wildlife value is similar to that of winged elm. Some insects feed on the leaves in the spring, and gray squirrels eat some of the fruit in October. Wild turkeys and wood ducks also eat significant quantities of the fruit (Martin et al. 1951). The purple finch reportedly feeds heavily on elm seeds farther north (Martin et al. 1951), so they probably feed on them here to some extent as well. The house finch, which has become common in Florida, also feeds on elm seeds. Cedar elm seeds may not be as valuable to wildlife as winged elm and Florida elm seeds, however, because they are not produced when mast from other trees is normally in short supply. Nonetheless, I have seen squirrels eating the fruit on a number of occasions. Cedar elm rarely provides cavities. The twigs and foliage are excellent browse for deer and rabbits. Although elm trees are primarily wind pollinated, I have watched native and honey bees visiting the flowers of cedar elm in late August and early September.

The wood of cedar elm is hard and strong and has the interlocking grain that, like that of winged elm, makes it hard to split. Also like winged elm, cedar elm wood makes excellent firewood and is used in the commercial manufacture of flooring, furniture, crates, and hockey sticks (Snow 1990).

Cedar elm was unknown in Florida until 1970. It is not yet commonly used for landscaping in this state, but it obviously has landscape potential, particularly as a specimen or shade tree. Although found on moist sites in the Florida wild, cedar elm is quite drought tolerant. It is widely used as a shade tree in Austin and Dallas, Texas, and is one of the few trees that is well adapted for planting in parking lot medians and the spaces between sidewalks and streets. Many years ago, Noel Lake, former head landscape designer for the University of Florida main campus, planted two cedar elms in the parking lot of the Campus USA Credit Union at the northwest corner of the intersection of southwest 34th Street and southwest 20th Avenue in Gainesville, Florida. To date, both are doing well. One advantage cedar elm has over most other shade trees is that its leaves are small enough to filter down through the grass, making it unnecessary to rake them. It is very easy to grow and transplant and is quite adaptable to most landscaped situations. Its tolerance of high pH and of some salt in the soil enable this elm to withstand the harsh conditions of cities better than most other shade trees.

Slippery Elm (Red Elm)—*Ulmus rubra*

Ulmaceae

Slippery elm is a common, medium- to large-sized tree in much of the eastern part of the United States from New England into southern Minnesota and from the Carolinas into East Texas, but it is uncommon south of Kentucky and barely reaches into the northern part of the central Florida Panhandle (Cooley and Van Sambeek 1990). It is similar in most ways to American elm, but the leaves are not quite as asymmetrical and have long, tapering tips. Also, the leaves are sand-paper rough on top and soft underneath. The bark is thick and rough with gray ridges and reddish-brown color between the ridges. Slippery elm gets its name from the mucilaginous nature of the reddish inner bark. The largest one currently reported, from Kentucky, is 90 feet tall with a trunk diameter of 7½ feet (American Forests 2019).

Slippery elm is an uncommon tree along the Apalachicola River, occurring on slopes along the river and its east-side tributaries on moist, fertile soil with some clay content. It prefers soils that have high calcium content. It also occurs on floodplains and on dry, rocky uplands farther north. Slippery elm is moderately shade tolerant (Cooley and Van Sambeek 1990). Common tree associates in Florida include pignut hickory, white ash, sweetgum, winged elm, laurel oak, and basswood.

The wildlife value of slippery elm is similar to that of the other elms. Its fruits ripen in the spring and are eaten by gray squirrels, wild turkeys, and various other mammals and birds. The foliage is eaten by insects, deer, and rabbits.

Slippery elm timber is used in the commercial manufacture of the same products for which American elm timber is used, including furniture, paneling, packing crates, and hardwood pulpwood.

Overshadowed as it was in the past by American elm, slippery elm has seldom been used as an ornamental tree, and now it is somewhat susceptible to the Dutch elm disease. In the Florida Panhandle, outside the current range of the Dutch elm disease, it would be an interesting specimen tree to plant on appropriate soil.

Florida Lantana—*Lantana depressa* var. *floridana*; Common Lantana—*L. strigocamara* (incorrectly *L. camara*); and Trailing Lantana—*L. montevidensis*

Verbanaceae

In northern Florida, there are two invasive exotic species of lantana and one rare, endangered, native species. The most common, largest, and most invasive is *Lantana strigocamara*, usually just called "lantana" but also known as "West Indian lantana" (incorrectly), "common lantana," and "shrub verbena." It is a medium-sized shrub with multiple stiff, somewhat prickly stems, a strong root system, and opposite, broad, aromatic leaves from 1 to 4 inches in length with coarsely serrated margins and rough surfaces. The bisexual flowers, that bloom any time of year, occur in paired clusters with multiple flowers per cluster. These flowers change color as they age, especially after being pollinated. This causes the central and youngest flower clusters to be yellow and the older flowers on the outer edges to change color to pink or orange or red. These pretty, color-changing flower heads have made lantana a popular ornamental worldwide, especially in the past. As a result of its popularity, this species has escaped cultivation and become a problem invasive exotic in many tropical and subtropical parts of the world, including Florida where lantana is classified as a Category I invasive plant (FLEPPC 2013).

The other common species of lantana in northern Florida is trailing lantana. It has slender, nonprickly stems that trail along the ground or climb into and on top of other vegetation. Its opposite leaves, up to 1 inch long, are roundish and not as rough to the touch as those of common lantana. Trailing lantana's flowers also occur in clusters, but they are always bluish with white centers, and they do not change color. This species is also a popular ornamental that has been distributed around much of the world.

Both of these species are most common in cities, villages, and subdivisions, often occurring in waste places such as vacant lots, road edges, and fencerows. They produce fruits that turn a metallic purple when ripe and are then scattered widely by the mammals and birds that eat them. All parts of both species, except for the ripe fruit, are toxic to mammals including livestock and humans.

The native species, Florida lantana, inhabits coastal strands and coastal scrub along the Atlantic coast of the Florida peninsula as far north as St. Johns County (St. Augustine). It is a very rare mat-forming shrub with paired clusters of yellow flowers. Florida lantana is closely related to common lantana, and one of the threats to its survival as a species is hybridization with common lantana (retired University of Florida botanist Walter Judd, personal communication).

The wildlife value of all of these species resides primarily in the benefit of their flowers for pollinators, especially butterflies. The fruits are eaten by songbirds. There is a native moth species known as the "lantana moth" (*Diastema tigris*) whose larvae feed on the foliage of common lantana. This moth has been introduced as a biological control agent into places such as Hawaii, Australia, Micronesia, and various countries in Africa where common lantana is an invasive pest.

The main use of lantana is ornamental, with the exotic species planted extensively and, until very recently, the native species rarely used. Because native Florida lantana has become increasingly available from many native plant nurseries, it is showing up more often now in Florida gardens. People continue to plant both exotic species, however, even though common lantana has long been identified as invasive, and, regrettably, both are widely available from nurseries.

Elderberry (Elder)—*Sambucus nigra* subsp. *canadensis* (*S. canadensis*)

Viburnaceae, Adroxaceae, Caprifoliaceae

Elderberry is a common shrub found throughout Florida and most of the rest of the United States (Kartesz 2015). It commonly grows 5 to 10 feet tall with stems 1 to 2 inches in diameter. The largest one reported in Florida, in Gainesville, was 20 feet tall and had a trunk diameter of 11 inches. Mature plants often produce new stems from the base. These grow very rapidly, producing straight stems about three-quarters of an inch in diameter and containing a large, soft, white pith. Elderberry has opposite, pinnately to twice pinnately compound leaves and produces large heads of white, bisexual flowers followed by large heads of black berries (drupes). Although flower and fruit production occurs over a broad time period throughout summer and fall, peak flowering is in May and peak fruit production in June.

Elderberry's main requirements are water and abundant sunlight. It grows best on wet sites on soils that have either some clay or considerable organic matter. It sometimes forms pure stands in the shallow areas of freshwater marshes, on the edges of lakes, ponds, and marshes, and in the wet centers of some bayheads and seepage bogs. It can be common in sunny openings in swamps. It also invades moist to wet upland fields, fencerows, ditches, roadsides, clearings, and other disturbed sites. Birds distribute the seeds widely by eating the fruit and then discarding the seeds either by spitting them out or by passing them intact through their digestive tracts.

The common associates of elderberry are an assortment of marsh plants such as sawgrass, arrowroot, pickerelweed, maidencane, and Virginia chain fern, along with a few shrubs and vines such as laurel greenbriar, buttonbush, Virginia-willow, and primrose-willow, and trees such as coastal plain willow, red maple, and bald-cypress.

Elderberry has a high wildlife value. Its flowers attract many insect pollinators, and the abundant fruit produced from mid-May to late fall is an important food resource for more than 50 species of birds (Martin et al. 1951; Miller and Miller 1999). Pileated woodpeckers seem to be particularly fond of elderberry fruit (Martin et al. 1951; wildlife ecologist Katie Greenberg, personal communication). In addition, elderberry thickets isolated in open marshes are ideal nesting sites for herons, egrets, white ibis, and anhingas. The rookery at Bird Island in Alachua County's Orange Lake is the first established and oldest Audubon sanctuary in the world. This formerly successful heron and egret rookery was purchased by the Audubon Society in 1910 and was the only major heron-egret rookery successfully defended against plume hunters in the early twentieth century. It consisted of a nearly pure stand of elderberry surrounded by spadderdock marsh and open water.

Elderberry has no great commercial value, but the fruit is harvested by people for making jam and wine. The straight young stems, with their large pith chambers, were sometimes used in the past to make wooden whistles and pea shooters, but this was not a good idea. The plant is toxic, and children have gotten sick using the stems this way (retired University of Florida botanist Dana Griffin, personal communication).

Elderberry's showy flowers, useful fruit, and high wildlife value justify some landscape use of this species, especially for the edges of stormwater retention areas and along small streams and ditches. Unfortunately, rather than planting this species *for* wildlife, the opposite is what normally happens. Wet areas along streams and ditches, on the edges of ponds, and around retention areas are usually cleared of elderberry and other shrubs in an effort to make them appear more open and orderly and to prevent their harboring wildlife, particularly insects and snakes.

Viburnums—*Viburnum* spp. (5 native species)

Viburnaceae, Adroxaceae, Caprifoliaceae

Rusty Black-Haw—*Viburnum rufidulum*

This large shrub or small tree is often tree shaped with a single trunk. The leathery, deciduous, oblong to roundish, 2-inch-long leaves are arranged opposite each other on the stem and are shiny dark green on top with a variable amount of dark, rusty-colored pubescence below. The dark hairs also appear on the stems and are especially notable on the buds, which appear dark rusty-brown as a result. The trunk, if large and old enough, has a rather thick, checkered, knobby bark. The white, bisexual flowers, which appear in the spring after the leaves form, occur in showy clusters at the ends of the twigs.

Normally 5 to 15 feet tall with trunk diameters from 2 to 4 inches, rusty black-haw can grow 25 feet tall with a trunk diameter of 15 inches (Godfrey 1988). Rusty black-haw grows throughout the southeastern United States from Virginia to southeastern Kansas and from the central Florida peninsula into eastern Texas (Kartesz 2015), reaching as far south on the Florida peninsula as Hernando County (Wunderlin et al. 2020). It often grows on the upper edge of upland hardwood forests where the forest transitions to a fire-adapted upland pine forest. In this ecotone, rusty black-haw is subject to occasional fire to which it is somewhat resistant because of its thick bark. It is also drought and somewhat shade tolerant.

The wildlife value of rusty black-haw is good, primarily because of its fruits, which are borne in clusters at the ends of twigs and branches as oblong drupes about half an inch long and dark blue to purple in color at maturity. The fruit is eaten by gray squirrels and fox squirrels and various birds such as wild turkey, brown thrasher, cardinal, robin, pileated woodpecker, and cedar waxwing; white-tailed deer also browse on the twigs and foliage to some extent (Martin et al. 1951). The insect-pollinated flowers also contribute to rusty black-haw's wildlife value by supplying bees, butterflies, and other insects with nectar and pollen.

Rusty black-haw can be an attractive shrub or small tree when used ornamentally, but this is rarely done. It would be an attractive alternative to crape myrtle and dogwood in locations too shady for these often-overused ornamentals to thrive. It is carefree and drought tolerant. However, in my yard, it seems only to live for about 20 years.

Walter's Viburnum—*Viburnum obovatum*

Also called "small viburnum" and "small-leaved viburnum," this shrub or small tree is sometimes tree shaped and sometimes multistemmed and shrubby, especially when forming thickets by spreading from underground runners. When growing as a small tree, it can grow 32 feet tall with a 7-inch trunk diameter (American Forests 2004). I measured one at Owen's Spring on the Suwannee River in Lafayette County that was 32 feet tall with a trunk diameter of 22 inches on September 1, 2000. The leaves are about 1 inch long, oblong, shiny green on top, and semievergreen. Small white, bisexual flowers are produced in early spring in showy clusters at the ends of the branches, and clusters of small fruits (drupes) that start out red and then turn black are produced in the fall.

Walter's viburnum is very adaptable, growing in hydric hammocks, floodplains, along streams, and on dry, sandy uplands. Its native range includes eastern South Carolina, southern Georgia, and all of Florida except for the westernmost panhandle and the Florida Keys (Kartesz 2015). It is both flood and drought tolerant. Somewhat shade tolerant, it can grow in the understory of hydric hammocks and stream floodplains under trees such as live oak, laurel oak, sweetgum, red maple, and loblolly pine. In such situations it is usually a large bush or small tree, often with a single trunk. It also can grow in the sunny understory of upland longleaf pine forests, where it is usually a shrubby cluster of root sprouts.

The wildlife value of Walter's viburnum is good. Deer browse the foliage, and gray squirrels and birds such as wild turkeys, brown thrashers, cardinals, robins, pileated woodpeckers, and cedar waxwings eat the fruit (Martin et al. 1951). The flowers are visited by bees and other insects.

Walter's viburnum is used to some extent as an ornamental, with several selected cultivars available for sale in nurseries. It is an attractive, tough, carefree plant that is beneficial to bees and butterflies in the spring and to birds in the fall.

Possum-Haw (Smooth Witherod)—*Viburnum nudum*

This is a wetland shrub with deciduous elliptic leaves from 2 to 6 inches long and about half as wide. The buds are covered with rusty-colored scales; the spring flower clusters at the ends of the branches are white; and the fruits in the fall start out pink and then turn blue, often with a thin covering of whitish wax. It is often 6 to 10 feet tall with slender stems. It ranges from Maine and the upper peninsula of Michigan south into the south-central Florida peninsula and west from the Florida Panhandle into eastern Texas (Kartesz 2015).

In Florida, this plant is uncommon but is found in acidic wetlands such as bayheads, bogs, and seepage slopes, often within pine flatwoods landscapes such as the Green Swamp in Central Florida and the Osceola and Apalachicola National Forests in North Florida. It is not found on well-drained uplands and does not appear to be drought tolerant. It tolerates continuously wet conditions but perhaps not deep or prolonged flooding. It is moderately shade tolerant but grows best with abundant sunlight.

The wildlife value of this viburnum is similar to that of Walter's viburnum.

Over the wide range of this plant, which includes most of the eastern United States, it is sometimes used as an ornamental, with several cultivars available for sale from plant nurseries.

Arrow-Wood—*Viburnum dentatum* var. *scabrellum* and Maple-Leaved Viburnum—*Viburnum acerifolium*

These two species of small, deciduous viburnums are found as occasional understory shrubs in hardwood forests in northern Florida. *Viburnum dentatum* has ovate leaves 2 to 4 inches long with rough surfaces and strongly serrated margins. It is found throughout much of the eastern United States, reaching as far south as the central Florida peninsula (Kartesz 2015). *Viburnum acerifolium*, also known as "maple-leaved viburnum," has leaves 2 to 3 inches long and wide with three lobes somewhat similar to those on maple leaves. It is found in Florida only in the western panhandle (Godfrey 1988) but occurs throughout much of the rest of the eastern United States (Kartesz 2015). These two species are typically multistemmed shrubs growing to about 6 feet tall. They both produce clusters of white, bisexual flowers at the ends of branches in spring and clusters of dark blue fruits at the ends of branches in the fall. These shrubs grow in the shade of hardwood forests or mixed pine and hardwood forests on uplands and along streams on well-drained soils. They are shade tolerant but do well in full sun; they are also drought tolerant. *Viburnum acerifolium* is often found in clonal patches that spread by underground runners (Godfrey 1988).

The wildlife value of these shrubs is similar to that of Walter's viburnum.

These shrubs are sometimes used as ornamentals. They are drought tolerant and largely trouble free, with the possible exception of needing occasional discretionary trimming. *Viburnum dentatum* var. *scabrellum* is the species most often used ornamentally, with a number of cultivars commercially available. One in my yard lived for 50 years. Clones of *Viburnum acerifolium* could probably live much longer.

Pepper Vine—*Nekemias* (*Ampelopsis*) *arborea* and Heartleaf Pepper Vine (Raccoon-Grape)—*Ampelopsis cordata*

Vitaceae

Pepper vine is a high-climbing vine in the grape family that is quite distinctive. Its deciduous leaves are twice pinnately compound. Older stems are light gray and slightly roughened (but without thickened, scaly, or ridged bark) and display what appear to be joints (rings of slightly thicker stem) at regular intervals about 1 foot apart. The leaves on some plants are red while growing, which is during most of the spring and summer for the growing tips of the foliage, and they turn red in the fall. The small, yellow-green, mostly bisexual flowers occur in clusters and are followed by fruits similar in appearance to small clusters of black grapes, each about half an inch in diameter. The growing vines have tendrils similar to those of grape vines for attaching to other plants. Pepper vine occurs throughout most of Florida and the rest of the southeastern United States (Kartesz 2015).

Heartleaf pepper vine is a relative of pepper vine with simple, heart-shaped leaves and brown, shredded bark. It occurs from southern Ohio and the Carolinas westward and southward into Missouri, Nebraska, Oklahoma, Alabama, Louisiana, and eastern Texas, reaching into Florida in the central panhandle region (Kartesz 2015). Heartleaf pepper vine's habitat preferences, growth characteristics, and wildlife value are similar to those of pepper vine.

Pepper vine is not rare, but it is less common than wild grape vines, Virginia creeper, and trumpet creeper. It occurs in the wild on moist to wet fertile soil in swamps, floodplain forests, hammocks, and shrub thickets. It will grow in rather infertile, moist to wet soil but is rare on deep, infertile sandy soils. Its common associates include sweetgum, sugarberry, laurel oak, live oak, bald-cypress, swamp tupelo, waxmyrtle, Virginia creeper, wild grape, and supplejack.

Pepper vine is not especially shade tolerant, growing best in full sun. It usually thrives on the edges of forests such as along rivers, road rights-of-way, lake shores, or the edges of fields. It also does well in pioneer situations, growing on top of shrubs and on the sides of the crowns of young trees.

Pepper vine has a moderately high wildlife value because of its fruit, which ripens from mid-August through mid-October, with some fruit still hanging on in mid-November. The fruits, similar to small grapes, are marginally edible by humans, although they can produce a burning sensation in the mouth and throat because they contain calcium oxalate crystals. Produced in abundance in small clusters, pepper vine fruit is highly edible by birds and some other wild animals. Most of the birds and mammals that eat wild grapes and Virginia creeper fruit also eat pepper vine fruit. These include wild turkey, bobwhite quail, wood duck, mockingbird, cardinal, brown thrasher, great-crested flycatcher, thrush, vireo, woodpecker, blue jay, yellow-rumped warbler, gray squirrel, deer, opossum, black bear, and raccoon. I have seen several of these creatures eating the fruits and have also watched a Tennessee warbler eat one. The flowers bloom early summer to midsummer, attracting native bees.

Pepper vine has occasionally been used as a cover for walls and fences, usually by accident. However, it serves this purpose admirably, being quite attractive for most of the year. It is bare for two to three months in winter.

Pepper vine is one among a number of Florida's native vines being killed by the cutting of vines on public lands by persons unknown for no apparent reason. This vandalism renders the lands on which it occurs less valuable as wildlife habitat.

Virginia Creeper—*Parthenocissus quinquefolia*

Vitaceae

Virginia creeper is a woody vine with palmately compound leaves, each of which normally has five leaflets. The leaves are deciduous, turning bright red before dropping in November and December. The leaves are somewhat similar in appearance to poison ivy leaves, which, by contrast, have three leaflets per leaf. Clusters of small, bisexual flowers are produced in late May into early June. Virginia creeper is related to wild grape and produces clusters of black, grape-like fruits in late summer. It is often part of the ground cover beneath hardwood trees, and it also climbs tree trunks and spreads out on the branches in the upper crowns of even the tallest trees. It has specialized tendrils that form sticky pads for clinging to the trunks of trees or to any other surface such as the walls of buildings. The trunks of large old vines have thick dark brown bark with prominent ridges that are neither flaky nor scaly.

Virginia creeper is native to most of the eastern half of the United States and occurs throughout Florida (Wunderlin et al. 2020).

Virginia creeper does not normally occur in scrub, sandhill, or swamp forests but is common to abundant in most hardwood forests on uplands and floodplains that are seldom underwater for long periods. It is also common along fencerows, on trees or shrubs in agricultural areas, and in urban and suburban areas on trees, shrubs, fences, and buildings. It is a rapid invader because its seeds are so readily and widely distributed by the birds that eat its fruit. It is better at climbing into fully grown trees than most other native vines because of its shade tolerance and its ability to attach to the bark of tree trunks. Once in a tree, it does not normally harm the tree, as Virginia creeper does not tend to cover the crown. Instead, it tends to cover the trunk and main branches and to grow long, thin shoots that hang straight down as festoons. It will sometimes grow over and cover shrubs and tree saplings, especially in association with other vines.

Virginia creeper has a high wildlife value, mainly because of the abundant fruit it produces from mid-August to mid-October. If not eaten right away, some fruit may stay on the vine until late winter. Birds that regularly eat the fruit include mockingbird, brown thrasher, eastern bluebird, great-crested flycatcher, flicker, pileated woodpecker, red-bellied woodpecker, yellow-breasted sapsucker, robin, red-eyed and white-eyed vireos, and various thrushes (Martin et al. 1951). Other species that consume the fruit include gray squirrel, gray catbird, Carolina chickadee, and tree swallow (Miller and Miller 1999). The high value of the berries for birds in northern Florida is confirmed by the observations of Rex Rowan, coauthor of "A Birdwatchers Guide to Alachua County, Florida" (personal communication) and others, including me. Former Alachua Audubon Society president Michael Meisenburg has observed eastern kingbirds and summer tanagers feeding on the fruits in northern Florida (personal communication), and I have watched catbirds and Tennessee warblers eating them. Virginia creeper leaves are eaten by a variety of insects, which in turn feed insectivorous birds. I have seen vireos and warblers feeding on caterpillars they plucked from the foliage. The flowers of Virginia creeper are visited by native bees and other pollinators.

Virginia creeper is one of the vine species being cut and killed on public lands for no apparent reason. This seemingly careless action unfortunately renders these lands much less valuable as wildlife habitat.

The leaves of Virginia creeper turn bright crimson in late autumn. For this reason, and because it can densely cover a wall, this vine is sometimes used in landscaping as a wall cover. Several cultivated color forms, some variegated, are available from plant nurseries (professional botanist David Hall, personal communication). It more often invades by itself and may provide considerable wildlife value without the knowledge or consent of the landowner. It is certainly preferable to use Virginia creeper for surface coverage than it is to use English ivy, cat's claw vine, or ficus vine, which provide no wildlife value and can become problem invaders that disrupt native ecosystems. Where Virginia creeper already exists in a large tree in a landscaped situation, it can be left there to provide wildlife benefits and fall color without doing any significant harm to the tree.

Muscadine Grape (Bullace Grape, Scuppernong Grape)—*Vitis rotundifolia*

Vitaceae

Muscadine grape is the most distinctive among the several kinds of wild grape vines that are a common and natural component of the hardwood forests of northern Florida. This species can be distinguished from the others by the tan-colored bark that is smooth on young stems but splits into plates on older stems; by its unbranched tendrils; and by the half-inch-diameter grapes that occur in small, open clusters from mid-August to mid-September. It also often has numerous long roots that hang down in the open air from the stems, some of which eventually reach the ground to form additional roots for supporting the vine. Muscadine grape grows throughout Florida and most of the rest of the southeastern United States (Kartesz 2015).

Muscadine vines can grow to the tops of trees that are sometimes over 100 feet tall, and the main stem can be 6 inches or more in diameter, although they are more commonly 1 to 3 inches in diameter. They climb with the use of very strong tendrils that wrap around the stems of other plants, but they are not able to attach to large tree trunks or to the walls of buildings. The larger vines, often as old as the trees they are attached to, are sometimes over 100 years old. Many of these old vines got started after a disturbance of some kind—a hurricane or a logging operation, perhaps—opened up the forest canopy or after a large tree fell over, creating a treefall gap. Once a vine has gotten established in the forest canopy, it often spreads from one tree crown to another, making use of more than one tree and binding them to one another to some extent. Large live oaks and southern magnolia trees are particularly good at providing a structure for grape vines.

Muscadine grape prefers well-drained, fertile soil. It is not fire, shade, or particularly flood tolerant but is common in upland hardwood forests, along streams, and along the edges of forests. It also occurs along fencerows and in other altered landscapes. Various cultivars are planted for the purpose of producing grape crops. It was particularly common in some of the larger hardwood forests such as Gulf Hammock and San Felasco Hammock, but, unfortunately, Gulf Hammock has largely been destroyed, and the vines in San Felasco Hammock have recently been largely exterminated in an intentional, massive, and very unfortunate vine-cutting campaign.

The intentional killing of vines in San Felasco Hammock Preserve State Park is an example of an image problem vines have in general. Because some vines grow high into trees and compete for light, sometimes damaging the trees, people often dislike the vines enough to eradicate them by cutting the stems. In a private, landscaped situation, this can be acceptable, though not particularly wise if the person doing it likes wildlife. On public lands preserved for their conservation values, such cutting campaigns are unacceptable because of the damage such removal does to the ecosystem and particularly to its wildlife populations. The fact that the native vines in our area have been growing here in association with the forest trees for at least thousands of years without significantly harming tree populations should be proof that the vines are not a threat to the health of the forest as a whole. Indeed, they are an important part of the forest ecology. Unfortunately, there are a significant number of introduced invasive exotic vines such as cat's claw vine, skunk vine, oriental bittersweet, kudzu, air potato, confederate jasmine, Japanese clematis, creeping fig, climbing fern, English ivy, Japanese honeysuckle, and Chinese wisteria that are serious threats to our native forests. One of the bad things about cutting native vines in a forest is that it opens up more room for invasive vines to move in and makes the forest even more susceptible to being overrun by these and other destructive exotic pest plants.

The wildlife value of native grape vines, muscadine vines in particular, is very high (Miller and Miller 1999). Produced mid-August to mid-September, the grapes are eaten by many species including wild turkeys, songbirds, woodpeckers, black bears, raccoons, striped skunk, gray fox, and squirrels; the tangle of vine stems and foliage in tree crowns provides habitat for songbirds; the bark is used by birds for nest material; and some of the grapes shrivel and remain on the vine into winter, providing a continuing food source (Martin et al. 1951). Grape vines and tree foliage tangled together in the upper canopy of the forest creates unique habitat for various species of birds, insects, spiders, and reptiles. Green anoles and rough green snakes are particularly well adapted to living there and are often hunted on the upper surface of this habitat by swallow-tailed kites in spring and summer (personal observation). Closer to the ground, vine tangles are a preferred nesting habitat for the golden mouse. The clusters of small flowers produced in late spring provide nectar and pollen for native insect pollinators, especially sweat (halicid) bees (Sampson et al. 2001).

The commercial and landscape value of muscadine grape cultivars is high. Farm homesteads of the past usually had a grape arbor of muscadine vines, and this tradition survives today on some farms and in rural subdivisions. Various cultivated varieties are offered for sale by plant nurseries. Owning a grape arbor provides an annual supply of grapes and an excellent way to observe the wildlife value of the grapes. Indeed, it can be difficult to keep the deer, birds, squirrels, and raccoons away from the grapes long enough to get some for oneself. Even in the middle of the city of Gainesville, where two of my neighbors established small grape arbors, the mockingbirds, brown thrashers, cardinals, and red-bellied woodpeckers provide stiff competition for the grapes.

Wild Grape Vines (other than Muscadine)—*Vitis* spp. (3 native species)

Vitaceae

In our area, this group of several closely related species primarily includes summer grape (*Vitis aestivalis*), fox or frost grape (*Vitis vulpina*), and Simpson's grape (*Vitis cinerea* var. *floridana*). Fox grape and Simpson's grape are more common on floodplains, beside streams, and on fertile soil in hardwood forests, whereas summer grape is more common on well-drained sandy soils. All of these vines climb with the use of very strong, branched tendrils that wrap around the stems of other plants; they are not adapted to attaching to the walls of man-made structures or the bark of trees, however. The leaves of muscadine grape are more rounded and less likely to be lobed than the leaves of summer, fox, and Simpson's grape, and the tendrils of muscadine are unbranched, whereas the tendrils of these other grape species are bifurcated. Summer, fox, and Simpson's grape are also easily distinguished from muscadine grape by the bark on their older stems, which is dark brown and finely divided into shreds, whereas the tan-colored bark of muscadine grape adheres tightly to the stem until the stem is large enough for the bark to split into plates. Finally, the clusters of grapes on muscadine vines are very open with a few large grapes, whereas the clusters of grapes on these other grape species hang down in dense clusters and contain a dozen or more smaller grapes.

Wild grape vines are rapidly growing, sun-loving plants that cannot thrive in the dense shade under the closed canopy of hardwood forests. Nonetheless, they are common as mature vines in hardwood forests. The way this works is that the grape vines get their start in sunny openings called "treefall gaps." These are created by the death of a large canopy tree or after a major disturbance such as a hurricane or logging operation removes part of the tree canopy, letting in more sunlight. Once a grape vine gets a start in a sunny spot, it grows up with the young trees also growing up in the same sunny area, eventually reaching the upper tree canopy. Wild grape vines are often abundant on forest edges, within forest treefall gaps, and along tree-lined fencerows.

Summer, fox, and Simpson's grape vines grow into the canopies of trees, especially in hardwood forests on well-drained soil, and they often grow in association with muscadine grape. Many of the hardwood forests in northern Florida, such as San Felasco Hammock and Gulf Hammock, were rather like giant grape arbors until just a few years ago, with many large grape vines in the tree canopy supplying grapes in late summer and fall to many species of wildlife. The grape vines often grow to the tops of trees as tall as 100 feet and do compete to some extent with the host trees for light. However, most tree species are well adapted to living in harmony with grape vines, as these native species have been growing together in our native forests for many thousands of years.

The wildlife value of these native grape species is high (Miller and Miller 1999). The grapes they produce provide food for a wide variety of wildlife including wild turkeys, songbirds, woodpeckers, bears, raccoons, and squirrels; the bark provides nesting material for birds; and the tangle of grape vines and foliage in the canopy provides a unique habitat for songbirds and other animals (Martin et al. 1951). This unique habitat in the upper canopy of hardwood forests with its associated insects and spiders is particularly beneficial to green anoles and rough green snakes, which, in turn, provide a food source for swallow-tailed kites (personal observation) and probably other birds of prey. Closer to the ground, vine tangles are a preferred nesting habitat for the golden mouse. Bird species particularly fond of eating grapes include wild turkey, mockingbird, brown thrasher, cardinal, red-bellied woodpecker, and pileated woodpecker (Martin et al. 1951).

Although summer, fox, and Simpson's grapes are also edible by people, they are not as desirable as muscadine grapes because of their grapes' small size and variable flavor. They are therefore not normally planted in grape arbors or given other landscape uses and are usually treated as weeds when they do appear. Summer grape can grow very rapidly and can be weedy in some situations, such as in pine plantations and along fencerows in well-drained, sandy, sunny locations.

The eradication of grape and other native vines in parks, nature preserves, and stream floodplains that has recently been happening on a large scale in Alachua County (see species account for muscadine grape above) is shortsighted and destructive. It greatly reduces the wildlife habitat value of these places, often the only remaining areas where many of our native wildlife species have a chance to live in sustainable populations.

Tallow Wood (Hog-Plum, Tallow-Plum)—*Ximenia americana*

Ximeniaceae, Olacaceae

Tallow Wood is a large, dense, very thorny, sprawling shrub with glabrous, shiny, evergreen leaves. It somewhat resembles a small orange tree, although the leaves are only about half as large, being 2 to 3 inches long, half as wide, and often having a slight notch at the leaf tip. The petiole (leaf stalk) is short, averaging about one-quarter of an inch. The straight, stiff, sharp thorns are typically half an inch to 1 inch long. The sweetly aromatic, bisexual flowers bloom prolifically in dense clusters along the previous season's stems. Blooming can occur at any time in the spring, summer, or fall but is densest in midsummer. Tallow wood's flowers are small, yellowish-white, four petaled, and fuzzy. The fruits (drupes) are round to oblong, smooth skinned, yellow when ripe, and about three-quarters of an inch to 1 inch long by half to three-quarters of an inch wide, with sour, edible flesh and a single stone. The leaves and the stone contain cyanide. The bush gets from 5 to 25 feet tall with a similar spread and a trunk from 1 to 5 inches in diameter. Tallow wood usually has multiple trunks. Its main roots are thick, weak, and brittle like carrots.

Tallow wood is a tropical plant that reaches the northern limit of its native range near the northern end of the Florida peninsula at Goldhead Branch State Park in Clay County (personal observation), at Palm Point Hill in central Alachua

County (personal observation), and in coastal Duval and Levy Counties (Wunderlin et al. 2020). It is similar in cold hardiness to an orange tree and resprouts vigorously after being killed to the ground either by cold or by fire. The thick roots appear to be adapted for storing energy.

The worldwide native range of this species is huge. It inhabits the tropics and subtropics of both the New World and Old World, including Mexico, Central America, South America, the West Indies, sub-Saharan Africa, southern India, Thailand, Vietnam, Indonesia, Malaysia, the Philippines, Borneo, New Guinea, and Australia (Royal Botanical Society and Kew Science 2021).

The natural habitat of this plant in the inland northern part of its range on the Florida peninsula is primarily as an uncommon understory shrub in xeric to dry mesic upland hardwood forests. Farther south and along the coast, it occurs in many habitats, most typically scrub, xeric hammocks, and coastal maritime hammocks. It gets larger in the tropical hammocks of South Florida and the Florida Keys than it does farther north. One strange thing about tallow wood is that it is somewhat parasitic on the roots of other plants (Godfrey 1988). This parasitism occurs when tallow wood root attaches itself to the root of another plant by growing a special attachment called a "haustorium" on the surface of the other plant's roots through which it can draw water and nutrients. Oak trees are one of the reported host plants.

Tallow wood appears to be moderately shade tolerant, growing well, blooming, and setting fruit in the shade of a mature oak–hickory forest with an open understory or in the understory of sand pine scrub forests. It does even better in full sun and appears to be quite heat tolerant. It is very drought tolerant and does not appear to have many insect or disease problems. It is also salt tolerant, doing well along both coasts and in the Florida Keys.

The wildlife value of tallow wood is not well documented. Because they are eaten by humans, the fruits are probably also consumed by many other mammals and by birds as well. However, in my yard, neither squirrels nor birds seem to be interested in them. The dense, evergreen, thorny crown of this shrub would appear to be an ideal site for some birds to build nests. The rather strange, hairy, white flowers are strongly and pleasantly fragrant and are very attractive to honey bees and to native bees and wasps. Honey bees and 13 species of native bees and wasps were recorded visiting tallow wood flowers at Archbold Biological Station on the south-central Florida peninsula (Deyrup and Deyrup 2015).

In other parts of the world, this plant serves a wide range of human purposes. Tallow wood products are used in human and livestock food, as an ingredient in a wide assortment of medicines, and the hard, dense, strong wood is used to make various implements such as tool handles (Feyssa et al. 2012). It is also an excellent nectar and pollen plant for honey bees (Djonwangwe et al. 2011).

I have never seen this plant used for landscaping except in my own yard. It has the distinct drawbacks of not being completely cold hardy and of having many sharp thorns. Nonetheless, it is a pretty evergreen shrub with dense foliage that produces edible fruit. It begins flowering and producing fruit at about four years of age from seed. Unfortunately, the flavor of the fruit is a bit too sour, in the opinion of most people, to make it a best seller. One possible use for tallow wood in the landscape would be as an impenetrable hedge. Its low, densely packed, very thorny, intertwining twigs and branches are excellent for this purpose. Even though it would freeze back on occasion, the dead stems would remain impenetrable, and the plants would sprout back rapidly. The plants in my yard in Gainesville have not frozen back at any time during the past 30 years, so it would appear that they are cold hardy down to about 25 degrees Fahrenheit. Tallow wood has high value as a nectar and pollen plant for honey bees, with the total sugar content of the nectar averaging about 41 percent (Djonwangwe et al. 2011). Therefore, another possible use for this species would be to install it, along with a diversity of other good nectar-producing plants, specifically to provide nectar and pollen in an area where honey bees are desired or where they are kept in the summertime.

APPENDIX

Common and Scientific Names of Plants and Animals

The following list identifies the common names and Latin binomials for plant and animal species cited, but not detailed, in this book. The woody plant species that comprise the central focus of the text are discussed at length in its individual species accounts and are not listed here.

Acadian Flycatcher—*Empidonax virescens*
Agave Snout Weevil—*Scyphophorus acupunctatus*
Air Potato—*Dioscorea bulbifera*
American Crow—*Corvus brachyrhynchos*
American Kestrel—*Falco sparverius*
American Snout Butterfly—*Libytheana carinenta*
Anhinga—*Anhinga anhinga*
Ants—Family: Formicidae
Armadillo (Nine-banded Armadillo)—*Dasypus novemcinctus*
Asian Woolly Hackberry Aphid—*Shivaphis celti*
Azalea Caterpillar—*Datana major*
Azalea Miner (Bee)—*Andrena cornelli*
Bachman's Sparrow—*Peucaea aestivalis* (*Aimophila aestivalis*)
Bald Eagle—*Haliaeetus leucocephalus*
Ball Moss—*Tillandsia recurvata*
Baltimore Oriole—*Icterus galbula*
Barn Owl—*Tyto alba*
Barred Owl—*Strix varia*
Bartram's Airplant—*Tillandsia bartramii*
Bats—Order: Chiroptera
Bear (American Black Bear)—*Ursus americanus*
Beaver (American Beaver)—*Castor canadensis*
Big Cypress Fox Squirrel—*Sciurus niger avicennia*
Black Turpentine Beetle—*Dendroctonous terebrans*

Black Twig Borer—*Xylosandrus compactus*

Bladdernut Plant Bug—*Lopidea staphyleae*

Blazing Star—*Liatris* spp.

Blueberry Digger Bee (Southeastern Blueberry Bee)—*Habropoda laboriosa*

Bluebird (Eastern Bluebird)—*Sialia sialis*

Blue Calamintha Bee—*Osmia calaminthae*

Bluegray Gnatcatcher—*Polioptila caerulea*

Blue Grosbeak—*Passerina caerulea* (*Guiraca caerulea*)

Blue Jay—*Cyanocitta cristata*

Bluestar (Fringed Bluestar)—*Amsonia ciliata*

Bluestem Grasses—*Andropogon* spp.

Boat-Tailed Grackle—*Quiscalus major*

Bobwhite Quail (Northern Bobwhite)—*Colinus virginianus*

Bracken Fern—*Pteridium aquilinum*

Broomsedge Grasses—Genus: *Andropogon*

Brown Creeper—*Certhia americana*

Brown-Headed Nuthatch—*Sitta pusilla*

Brown Thrasher—*Toxostoma rufum*

Bumblebees—*Bombus* spp.

Butterfly Milkweed—*Asclepias tuberosa*

Butterfly Pea (Spurred Butterfly Pea)—*Centrosema virginianum*

Canary Island Date Palm—*Phoenix canariensis*

Caracara (Northern Crested Caracara)—*Caracara cheriway*

Cardinal (Northern Cardinal)—*Cardinalis cardinalis*

Carolina Chickadee—*Poecile carolinensis*

Carolina Wren—*Thryothorus ludovicianus*

Catalpa Sphinx Moth (Catalpa Worm)—*Ceratomia catalpae*

Catbird (Gray Catbird)—*Dumetella carolinensis*

Cattle—*Bos taurus*

Cedar-Apple Rust—*Gymnosporangium juniperi-virginianae*

Cedar Waxwing—*Bombycilla cedrorum*

Chimney Swift—*Chaetura pelagica*

Chinese Fringetree—*Chionanthus retusus*

Christmas Fern—*Polystichum acrostichoides*

Chuck-Will's-Widow—*Antrostomus carolinensis* (*Caprimulgus carolinensis*)

Climbing Fern (Japanese Climbing Fern)—*Lygodium japonicum*

Cochineal Insect—*Dactylopius coccus*

Cofaqui Giant Skipper—*Megathymus cofaqui*

Common Buckthorn—*Rhamnus cathartica*

Common Grackle—*Quiscalus quiscula*

Confederate Jasmine (Star Jasmine)—*Trachelospermum jasminoides*

Coral Hairstreak Butterfly—*Satyrium titus*
Cotton Mouse—*Peromyscus gossypinus*
Cotton Rat—*Sigmodon hispidus*
Cottonwood Leaf Beetle—*Chrysomela scripta*
Cottonwood Leaf Curl Mite—*Tetra lobulifera*
Coyote—*Canis latrans*
Creeping Fig—*Ficus pumila*
Deer (White-tailed Deer)—*Odocoileus virginianus*
Deer's-Tongue (Vanillaleaf)—*Carphephorus odoratissimus*
Dogwood Anthracnose—*Discula destructiva*
Downy Woodpecker—*Dryobates pubescens* (*Picoides pubescens*)
Dry Wood Termites—*Cryptotermes* spp.
Eastern Bumelia Borer—*Plinthocoelium suaveolens*
Eastern Cottontail Rabbit—*Sylvilagus floridanus*
Eastern Kingbird—*Tyrannus tyrannus*
Eastern Palearctic Seed Beetle—*Bruchidius terrenus*
Eastern Phoebe—*Sayornis phoebe*
Eastern Pygmy Blue Butterfly—*Blephidium pseudofea* (*B. isophthalma*)
Eastern Tent Caterpillar—*Malacosoma americanum*
Eastern Tiger Swallowtail Butterfly—*Papilio glaucus*
Eastern Towhee—*Pipilo erythrophthalmus*
Echo Moth—*Seirarctia echo*
Egrets—Family: Ardeidae
Emerald Ash Borer—*Agrilus planipennis*
English Ivy—*Hedera helix*
Erythrina Leafroller—*Agathodes designalis*
Erythrina Stem Borer—*Terastia meticulosalis*
European Starling—*Sturnus vulgaris*
Faithful Beauty Moth—*Composia fidelissima*
Fall Webworm—*Hyphantria cunea*
Fish Crow—*Corvus ossifragus*
Florida Beargrass—*Nolina atopocarpa*
Florida Bonamia (Florida Lady's Nightcap)—*Bonamia grandiflora*
Florida Leaf-Footed Bug—*Acanthocephala femorata*
Florida Mouse—*Podomys floridanus*
Florida Pink Scavenger Moth—*Anatrachyntis badia*
Florida Red Scale—*Chrysomphalus aonidum*
Florida Scrub Jay—*Aphelocoma coerulescens*
Flying Squirrel (Southern Flying Squirrel)—*Glaucomys volans*
Fox Squirrel—*Sciurus niger*
Giant Bee Fly—*Bombylius major*

Giant Milkweed Bug—*Sephina gundlachi*

Giant Swallowtail Butterfly (Orange Swallowtail Butterfly)—Papilio cresphontes

Golden Mouse—*Ochrotomys nuttalli*

Golden Polypody (Goldfoot Fern, Rabbit's Foot Fern)—*Phlebodium aureum*

Goldfinch (American Goldfinch)—*Spinus tristis* (*Carduelis tristis*)

Gopher Tortoise—*Gopherus Polyphemus*

Gray Catbird—*Dumetella carolinensis*

Gray Fox—*Urocyon cinereoargenteus*

Gray Hairstreak Butterfly—*Strymon melinus*

Gray Squirrel (Eastern Gray Squirrel)—*Sciurus carolinensis*

Great Crested Flycatcher—*Myiarchus crinitus*

Great Horned Owl—*Bubo virginianus*

Great Southern White Butterfly—*Ascia monuste*

Great Purple Hairstreak—*Atlides halesus*

Green Anole—*Anolis carolinensis*

Greenfly Orchid—*Epidendrum conopseum* (*E. magnoliae*)

Ground Dove (Common Ground Dove)—*Columbina passerina*

Hackberry Emperor—*Asterocampa celtis*

Hairy Woodpecker—*Leuconotopicus villosus*

Hand Fern—*Ophioglossum palmatum* (*Cheiroglossa palmata*)

Hemispherical Scale—*Saissetia coffeae*

Hermit Thrush—*Catharus guttatus*

Herons—Family: Ardeidae

Hessel's Hairstreak Butterfly—*Callophrys hesseli*

Hickory Bark Beetle—*Scolytus quadrispinosus*

Hog (Wild Hog)—*Sus scrofa*

Hooded Owlet Moths—Genus: *Cucullia*

House Finch—*Haemorhous mexicanus* (*Carpodacus mexicanus*)

Hydrangea Sphinx Moth—*Darapsa versicolor*

Insect—Class: Insecta or Hexapoda

Io Moth—*Automeris io*

King's Hairstreak Butterfly—*Satyrium kingi*

Kudzu Beetle—*Megacopta cribraria*

Laurel (Palamedes) Swallowtail Butterfly—*Papilio palamedes*

Leafy Liverworts—Order: Jungermanniales

Little Underwing Moth—*Catocala minuta*

Lizard's Tail—*Saururus cernuus*

Loggerhead Shrike—*Lanius ludovicianus*

Long-Tailed Mealybug—*Pseudococcus longispinus*

Lopsided Indian Grass—*Sorghastrum secundum*

Maidencane—*Panicum hemitomon* (*Hymenachne hemitomon*)

Mimosa Webworm—*Homadaula anisocentra*

Mimosa Wilt—*Fusarium oxysporum* forma specialis *Perniciosum*

Mockingbird (Northern Mockingbird)—*Mimus polyglottos*

Mosses—Division: Bryophyta

Mournful Thyris Moth—*Pseudothyris sepulchralis*

Mourning Dove—*Zenaida macroura*

Nighthawk (Common Nighthawk)—*Chordeiles minor*

Northern Flicker—*Colaptes auratus*

Oldfield Mouse—*Peromyscus polionotus*

Opossum (Virginia Opossum)—*Didelphis virginiana*

Orchard Oriole—*Icterus spurius*

Oriental Bittersweet—*Celastrus orbiculatus*

Osprey—*Pandion haliaetus*

Palmetto Weevil—*Rhynchophorus cruentatus*

Partridge Pea—*Chamaecrista fasciculata*

Pileated Woodpecker—*Dryocopus pileatus*

Pine Engraver Beetles—*Ips* spp. (*Ips avulsus, Ips calligraphus*, and *Ips grandicollis* in Florida)

Pineland Wild Indigo—*Baptisia lecontei*

Pine Siskin—*Spinus pinus* (*Carduelis pinus*)

Pine Warbler—*Setophaga pinus*

Poplar Tentmaker—*Clostera inclusa* (*Ichthyura inclusa*)

Poppy Mallow—*Callirhoe involucrata*

Powderpost Beetles—Subfamily: Lyctinae

Promethea Moth—*Callosamia promethea*

Prothonotary Warbler—*Protonotaria citrea*

Purple Finch—*Haemorhous purpureus* (*Carpodacus purpureus*)

Pyracantha (Firethorn)—*Pyracantha* spp.

Question Mark Butterfly—*Polygonia interrogationis*

Raccoon—*Procyon lotor*

Rat Snakes (Eastern Rat Snake, Gray Rat Snake, Corn Snake)—Genus: *Pantherophis* (*Elaphe*)

Red Bay Ambrosia Beetle—*Xyleborus glabratus*

Red-Bellied Woodpecker—*Melanerpes carolinus*

Red-Cockaded Woodpecker—*Picoides borealis*

Red-Eyed Vireo—*Vireo olivaceus*

Red-Headed Woodpecker—*Melanerpes erythrocephalus*

Red-Shouldered Bug—*Jadera haematoloma*

Red Spider Mites—Family: Tetranychidae

Red-Spotted Purple Butterfly—*Limenitis arthemis astyanax*

Red-Tailed Hawk—*Buteo jamaicensis*

Red Widow Spider—*Latrodectus bishopi*

Red-Winged Blackbird—*Agelaius phoeniceus*

Regal Moth (Royal Walnut Moth, Hickory Horned Devil Caterpillar)—
    Citheronia regalis

Resurrection Fern—*Pleopeltis polypodioides*

Ring-Billed Gull—*Larus delawarensis*

Robin (American Robin)—*Turdus migratorius*

Root Collar Borer Moth—*Euzophera ostricolorella*

Rose-Breasted Grosbeak—*Pheucticus ludovicianus*

Rosemary Grasshopper—*Schistocerca ceratiola*

Rough Green Snake—*Opheodrys aestivus*

Ruby-Crowned Kinglet—*Regulus calendula*

Ruby-Throated Hummingbird—*Archilochus colubris*

Rusty Blackbird—*Euphagus carolinus*

Sago Palm—*Cycas revoluta*

Sandhill Crane (Florida Sandhill Crane)—*Antigone canadensis* (*pratensis*)

Sandhill Dropseed (Pineywoods Dropseed)—*Sporobolus junceus*

Sandyfield Beaksedge (Scrub Beakrush)—*Rhynchospora megalocarpa*

Scarlet Tanager—*Piranga olivacea*

Scorpion—Order: Scorpiones

Screech Owl (Eastern Screech Owl)—*Otus asio*

Scrub Jay (Florida Scrub Jay)—*Aphelocoma coerulescens*

Sea Oats—*Uniola paniculata*

Shoestring Fern—*Vittaria lineata*

Silk-Grass (Grassyleaf Golden-Aster)—*Pityopsis graminifolia*

Silkworm—*Bombyx mori* moth (larva)

Silver Croton—*Croton argyranthemus*

Silver-Spotted Skipper—*Epargyreus clarus*

Snail Kite—*Rostrhamus sociabilis*

Southern Dogface Butterfly—*Zerene cesonia*

Southern Pine Beetle—*Dendroctonus frontalis*

Southern Yellowjacket—*Vespula squamosa*

Soybean Rust Fungus—*Phakopsora pachyrhizi*

Spanish Moss—*Tillandsia usneoides*

Spicebush Swallowtail Butterfly—*Papilio troilus*

Spider—Order: Araneae

Splitbeard Bluestem—*Andropogon ternarius*

Striped Hairstreak Butterfly—*Satyrium liparops liparops*

Striped Skunk—*Mephitis mephitis*

Subterranean Termite (Eastern Subterranean Termite)—*Reticulitermes flavipes*

Summer Farewell—*Dalea pinnata*
Summer Tanager—*Piranga rubra*
Swainson's Thrush—*Catharus ustulatus*
Swallow-Tailed Kite—*Elanoides forficatus*
Swamp Rabbit—*Sylvilagus aquaticus*
Sweat Bee (Halicid Bees)—Order: Hymenoptera; Family: Halictidae
Sweetbay Silkmoth—*Callosamia securifera*
Tawny Emperor—*Asterocampa clyton*
Towhee (Eastern Towhee)—*Pipilo erythrophthalmus*
Tree Swallow—*Tachycineta bicolor*
Tufted Titmouse—*Baeolophus bicolor*
Tulip-Tree Beauty Moth—*Epimecis hortaria*
Tulip-Tree Silkmoth—*Callosamia angulifera*
Veery—*Catharus fuscescens*
Viceroy Butterfly—*Limenitis archippus*
Virginia Chain Fern—*Woodwardia virginica*
Walnut Caterpillar—*Datana integerrima*
Warblers (Wood Warblers)—Family: Parulidae
White-Crowned Sparrow—*Zonotrichia leucophrys*
White-Eyed Vireo—*Vireo griseus*
White Ibis—*Eudocimus albus*
White Peach Scale—*Pseudaulacaspis pentagona*
White-Tailed Deer—*Odocoileus virginianus*
White Wild Indigo—*Baptisia alba*
Wild Turkey—*Meleagris gallopavo*
Wire Grass—*Aristida stricta (A. beyrichiana)*
Wood Duck—*Aix sponsa*
Woodrat (Eastern Woodrat)—*Neotoma floridana*
Wood Stork—*Mycteria americana*
Wood Thrush—*Hylocichla mustelina*
Yellow-Bellied Sapsucker—*Sphyrapicus varius*
Yellow-Billed Cuckoo—*Coccyzus americanus*
Yellow-Rumped Warbler (Myrtle Warbler)—*Setophaga coronata (Dendroica coronata)*
Yellow-Throated Vireo—*Vireo flavifrons*
Yucca Giant Skipper—*Megathymus yuccae*
Yucca Moth—*Tegeticula yuccasella (Pronuba yuccasella)*
Zebra Swallowtail Butterfly—*Eurytides marcellus*

REFERENCES

Abrahamson, Warren G. 1995. "Habitat Distribution and Competitive Neighborhoods of Two Florida Palmettos." *Bulletin of the Torrey Botanical Club* 122, no. 1: 1–14.

Adams, Robert P. 1986. "Geographic Variation in *Juniperus silicicola* and *J. virginiana* of the Southeastern United States: Multivariant Analysis of Morphology and Terpenoids." *Taxon* 35: 61–75.

Adams, Susan B., Paul B. Hamel, Kristina Conner, Bryce Burke, Emile S. Gardiner, and David Wise. 2007. "Potential Roles of Fish, Birds, and Water in Swamp Privet (*Forestiera acuminata*) Seed Dispersal." *Southeastern Naturalist* 6, no. 4: 669–82.

Alabama Herbarium Consortium. 2018. "*Wisteria frutescens.*" Alabama Plant Atlas (database). University of West Alabama. June 12, 2018. www.floraofalabama.org/plant.aspx?id=2106.

Alderman, Derek H., and Donna G'Segner Alderman. 2001. "Kudzu: A Tale of Two Vines." *Southern Cultures* 7, no. 3: 49–64.

Alward, Ashley, Candice A. Corriher, Michelle H. Barton, Debra C. Sellon, Anthony T. Blikslager, and Samuel L. Jones. 2006. "Red Maple (*Acer rubrum*) Leaf Toxicosis in Horses: A Retrospective Study of 32 Cases." *Journal of Veterinary Internal Medicine* 20, no. 5: 1197–201.

American Forests. 1986. "National Register of Big Trees." *American Forests* 92, no. 4: 21–35.

———. 1994. "The 1994 National Register of Big Trees." *American Forests* 100, nos. 1 & 2: 14–41.

———. 2000. "The National Register of Big Trees 2000–2001." *American Forests* 106, no. 1: 22–64.

———. 2004. "The National Register of Big Trees 2004–2005." *American Forests* 110, no. 1: 14–43.

———. 2008. "The National Register of Big Trees 2008." *American Forests* 114, no. 1: 14–41.

———. 2015. American Forests Champion Trees National Register (database). July 18, 2015. http://www.americanforests.org/bigtrees/bigtrees-search/.

———. 2018. American Forests Champion Trees National Register (database). September 17, 2018. http://www.americanforests.org/bigtrees/bigtrees-search/.

———. 2019. American Forests Champion Trees National Register (database). September 27, 2019; October 31, 2019. http://www.americanforests.org/bigtrees/bigtrees-search/.

———. 2020. American Forests Champion Trees National Register (database). June 12, 2020. http://www.americanforests.org/bigtrees/bigtrees-search/.

Anderson, Robert L., John L. Knighton, Mark Windham, Keith Langdon, Floyd Hendrix, and Ron Roncadori. 1993. *Dogwood Anthracnose and Its Spread in the South*. Forest Health Protection Report R8—PR-26. Atlanta, Ga.: USDA (U.S. Department of Agriculture) Forest Service, Southern Region.

Anich, Nicholas M., Thomas J. Benson, Jeremy D. Brown, Carolina Roa, James C. Bednarz, Raymond E. Brown, and J. G. Dickson. 2010. "Swainson's Warbler (*Limnothlypis swainsonii*)." In *The Birds of North America*, Version 2.0., edited by Alan F. Poole. Ithaca, N.Y.: Cornell Lab of Ornithology: Birds of the World (database). https://doi.org/10.2173/bna.126.

Appleton, Bonnie, Roger Berrier, Roger Harris, Dawn Alleman, and Lynette Swanson. 2015. "The Walnut Tree: Allelopathic Effects and Tolerant Plants." Virginia Cooperative Extension pub. no. 430–021. Blacksburg: Virginia Polytechnic Institute and State University.

Ayers, George. 2016. "The Rhamnaceae, the Buckthorn Family." *American Bee Journal*. October 14, 2016. https://americanbeejournal.com/rhamnaceae-buckthorn-family/.

Baker, James B., and O. Gordon Langdon. 1990. "*Pinus taeda* L.—Loblolly Pine." In *Silvics of North America*, vol. 1, *Conifers*, edited by Russell M. Burns and Barbara H. Honkala, 497–512. USDA Forest Service Agriculture Handbook 654. Washington, D.C.: Government Printing Office.

Barta, S.W.T. 1985. Red Maple (*Acer rubrum* L.), an Important Early Spring Food Resource for Honey Bees and Other Insects. *Journal of the Kansas Entomological Society* 58, no. 1: 169–72.

Bartram, John. (1766) 1942. *Diary of a Journey through the Carolinas, Georgia and Florida: From July 1, 1765 to April 10, 1766*. Edited and annotated by Francis Harper. Transactions of the American Philosophical Society, vol. 33, pt. l. Philadelphia, Pa.: American Philosophical Society.

Bartram, William. (1791) 1928. *Travels Through North and South Carolina, Georgia, East and West Florida* [. . .]. Edited by Mark Van Doren. New York: Dover.

Beal, Foster Ellenborough Lascelles. 1911. *Food of the Woodpeckers of the United States*. USDA Biological Survey Bulletin 37. Washington, D.C.: Government Printing Office.

Beck, Donald E. 1990. "*Liriodendron tulipifera* L.—Loblolly Pine." In *Silvics of North America*, vol 2 *Hardwoods*, edited by Russell M. Burns and Barbara H. Honkala, 408–16. USDA Forest Service Agriculture Handbook 654. Washington, D.C.: Government Printing Office.

Beckman, Emily. 2017. "Bluff Oak—*Quercus austrina*." IUCN (International Union for Conservation of Nature and Natural Resources) Red List of Threatened Species (database). https://dx.doi.org/10.2305/IUCN.UK.2017-2.RLTS.T194067A2296028.en.

Belanger, Roger P. 1990. "*Quercus falcata* Michx.—Southern Red Oak." In *Silvics of North America*, vol. 2, *Hardwoods*, edited by Russell M. Burns and Barbara H. Honkala, 640–44. USDA Forest Service Agriculture Handbook 654. Washington, D.C.: Government Printing Office.

Bell, Karen L., Haripriya Ragan, Manuel Fernandes, Christian A. Kull, and Daniel J. Murphy. 2017. "Chance Long-Distance or Human Mediated Dispersal? How *Acacia s.l. farnesiana* Attained Its Pan-Tropical Distribution." *Royal Society Open Science Publishing* (website). April 4, 2017. https://royalsocietypublishing.org/doi/10.1098/rsos.170105.

Benning, John W. 2015. "Odd For an Ericad: Nocturnal Pollination of *Lyonia lucida* (Ericaceae)." *American Midland Naturalist* 174: 204–17.

Bernheim Arboretum and Research Forest. 2019. "Downy Serviceberry." Trees and Plants: Select Urban Trees (database). https://bernheim.org/learn/trees-plants/bernheim-select-urban-trees/downy-serviceberry/.

Blair, Robert M. 1990. "*Gleditsia triancanthos* L.—Honeylocust." In *Silvics of North America*, vol. 2, *Hardwoods*, edited by Russell M. Burns and Barbara H. Honkala, 358–64. USDA Forest Service Agriculture Handbook 654. Washington, D.C.: Government Printing Office.

Bohall, Petra G. 1984. "Habitat Selection, Seasonal Abundance, and Foraging Ecology of American Kestrel Subspecies in North Florida." Master's thesis, University of Florida, Gainesville.

Bond, G. 1956. "Evidence for Fixation of Nitrogen by the Root Nodules of Alder (*Alnus*) under Field Conditions." *New Phytologist* 55, no. 2: 147–53.

Bonner, Franklin T. 1974. *Liquidambar styraciflua* L.—Sweetgum. In *Seeds of Woody Plants of the United States*, edited by Clifford Scharff Schopmeyer, 505–7. USDA Forest Service Agriculture Handbook 450. Washington, D.C.: Government Printing Office.

Boyer, William D. 1990. "*Pinus palustris* Mill.—Longleaf Pine." In *Silvics of North America*, vol. 1, *Conifers*, edited by Russell M. Burns and Barbara H. Honkala, 405–12. USDA Forest Service Agriculture Handbook 654. Washington, D.C.: Government Printing Office.

Braman, Kris, Bodie Pennisi, Elizabeth Benton, and Kim Toal. 2017. "Selecting Trees and Shrubs as Resources for Pollinators." University of Georgia Cooperative Extension Bulletin 1483. Athens, Ga.: University of Georgia.

Broschat, Timothy K. (1993) 2013. "*Sabal Palmetto*: Sabal or Cabbage Palm." Pub. no. ENH-733. Gainesville: University of Florida Environmental Horticulture Department and Institute of Food and Agricultural Sciences (IFAS).

Broun, Maurice. 1941. "Gulls Eating Fruit of Cabbage Palmetto." *Auk* 58, no. 4: 579.

Brown, Harry P., Alexis J. Panshin, and Carl C. Forsaith. 1949. *Textbook of Wood Technology.* Vol. 1, *Structure, Identification, Defects, and Uses of the Commercial Woods of the United States.* New York: McGraw-Hill.

Brown, Stephen H., and Kim Cooprider. 2012. "*Yucca aloifolia.*" Plant fact sheet. Fort Myers: University of Florida IFAS Extension, Lee County. https://drive.google.com/file/d/1UuhsfaiexU3CPI1USnI28Sl2-UbFVHdE/view.

Brunswig, Norman L., Stephen G. Winton, and Paul B. Hamel. 1983. "A Dietary Overlap of Evening Grosbeaks and Carolina Parakeets." *Wilson Bulletin* 95, no. 3: 452.

Burnett, Leroy. 2016. "The Origins and History of Yaupon Holly." Teamuse (website). https://www.teamuse.com/article_161116.html.

Burtchaell, Peter Edward. 1949. "Economic Change and Population at Cedar Key." Master's thesis, University of Florida, Gainesville.

Burton, James D. 1990. "*Maclura pomifera* (Raf.) Schneid., Osage-Orange." In *Silvics of North America*, vol. 2, *Hardwoods*, edited by Russell M. Burns and Barbara H. Honkala, 426–32. USDA Forest Service Agriculture Handbook 654. Washington, D.C.: Government Printing Office.

Camus, Aimée. 1936–1954. *Les Chenes: Monographie du Genre "Quercus."* 3 vols. Paris: Paul Lechevalier et Fils.

Cane, James H., and Jerry A Payne. 1988. "Foraging Ecology of the Bee *Habropoda laboriosa* (Hymenoptera: Anthrophoridae), an Oligolege of Blueberries (Ericaceae: Vaccinium) in the Southeastern United States. *Annals of the Entomological Society of America* 81, no. 3: 419–27.

Carey, Jennifer H. 1992. "*Quercus hemisphaerica, Q. laurifolia.*" Fire Effects Information System (database). USDA Forest Service, Rocky Mountain Research Station, Fire Science Laboratory, Fort Collins, Colorado. October 3, 2016. https://www.fs.fed.us/database/feis/plants/tree/quesppl/all.html.

———. 1992. "*Quercus stellata.*" Fire Effects Information System (database). USDA Forest Service, Rocky Mountain Research Station, Fire Science Laboratory, Fort Collins, Colorado. October 10, 2016. https://www.fs.fed.us/database/feis/plants/tree/queste/all.html.

Catling, Paul M., L. Dumouchel, and Vivian R. Brownell. 1998. "Pollination of the Miccosukee Gooseberry (*Ribes echinellum*)." *Castanea* 63, no. 4: 402–7.

Chafin, Linda G. (2007) 2020. "*Illicium floridanum* Ellis: Florida Anise." Georgia Biodiversity Portal: Plants (database). Georgia Department of Natural Resources, Wildlife Resources Division. https://georgiabiodiversity.org/natels/profile?es_id=15861.

———. (2008) 2020. "*Sideroxylon macrocarpum* (Nutt.) J. R. Allison—Ohoopee Bumelia." Georgia Biodiversity Portal: Plants (database). Georgia Department of Natural Resources, Wildlife Resources Division. https://georgiabiodiversity.org/natels/profile?es_id=20327.

———. (2008) 2020. "*Morella inodora* (Bartr.) Small: Odorless Bayberry." Georgia Biodiversity Portal: Plants (database). Georgia Department of Natural Resources, Wildlife Resources Division. https://georgiabiodiversity.org/natels/profile?es_id=16011.

———. (2009) 2020. "*Crataegus triflora* Chapman: Three-Flower Hawthorn." Georgia Biodiversity Portal: Plants (database). Georgia Department of Natural Resources, Wildlife Resources Division. https://georgiabiodiversity.org/natels/profile?es_id=21985.

Chiver, Ioana, Leslie J. Evans Ogden, and B. J. Stutchbury. 2011. "Hooded Warbler (*Setophaga citrina*)." In *The Birds of North America*, Version 2.0, edited by Alan F. Poole. Ithaca, N.Y.: Cornell Lab of Ornithology: Birds of the World (database). https://birdsoftheworld.org/bow/historic/bna/hoowar/2.0/introduction.

Christman, Steve. (1997) Updated 2003. "*Illicium floridanum.*" *Floridata Plant Encyclopedia.* https://floridata.com/plant/284.

———. (1997) Updated 2016. "*Lagerstroemia indica.*" *Floridata Plant Encyclopedia.* https://floridata.com/plant/193.

———. (1997) Updated 2008. "*Lyonia lucida.*" *Floridata Plant Encyclopedia.* https://floridata.com/plant/1017.

———. (1997) Updated 2012. "*Yucca filamentosa.*" *Floridata Plant Encyclopedia.* https://floridata.com/plant/266.

———. (1999) Updated 2003. "*Hamelia patens.*" *Floridata Plant Encyclopedia.* https://floridata.com/plant/174.

———. 2003. "*Illicium parviflorum.*" *Floridata Plant Encyclopedia.* https://floridata.com/plant/969.

———. 2008. "*Chrysoma pauciflosculosa.*" *Floridata Plant Encyclopedia.* July 21, 2018. https://floridata.com/plant/1099.

———. 2017. "*Amorpha fruticosa.*" *Floridata Plant Encyclopedia.* May 15, 2018. https://floridata.com/plant/1283.

Clewell, Andre. 1985. *Guide to the Vascular Plants of the Florida Panhandle.* Tallahasee: University Presses of Florida.

Coder, Kim D. 2016. *American Mistletoe: Tree Infection, Damage and Assessment Manual.* Warnell Outreach pub. no. 36. Athens: Warnell School of Forestry and Natural Resources, University of Georgia.

Colorado State University. 2019. "Clematis." *Guide to Poisonous Plants.* Fort Collins: Colorado State University, James L. Voss Veterinary Teaching Hospital. August 8, 2019. https://csuvth.colostate.edu/poisonous_plants/Plants/Details/136.

Cooley, John H., and J. W. Van Sambeek. 1990. "*Ulmus rubra* Muhl.—Slippery Elm." In *Silvics of North America*, vol. 2, *Hardwoods*, edited by Russell M. Burns and Barbara H. Honkala, 812–16. USDA Forest Service Agriculture Handbook 654. Washington, D.C.: Government Printing Office.

Cooper, D. T. 1990. "*Populus deltoides* Bartr. ex Marsh. var. *deltoides*—Eastern Cottonwood." In *Silvics of North America*, vol. 2, *Hardwoods*, edited by Russell M. Burns and Barbara H. Honkala, 530–35. USDA Forest Service Agriculture Handbook 654. Washington, D.C.: Government Printing Office.

Corogin, Paul T., and Walter S. Judd. 2014. "New Geographic and Morphological Data for *Sideroxylon reclinatum* Subspecies *austrofloridense* (Sapotaceae), a Taxon Endemic to Southwestern Peninsular Florida, U.S.A." *Journal of the Botanical Research Institute of Texas* 8, no. 2: 413–17.

Creech, Dave. 2017. *Magnolia pyramidata*–Bigger and Better in Texas. *Dave Creech–Life on the Green Side* (blog). https://dcreechsite.com/2017/08/15/magnolia-pyramidata-bigger-and-better-in-texas/.

Culbert, Daniel F. (1995) 2010. "Florida Coonties and Atala Butterflies." Pub. no. ENH-117. Gainesville: University of Florida Environmental Horticulture Department and IFAS. https://edis.ifas.ufl.edu/mg347.

Darley-Hill, Susan, and W. Carter Johnson. 1981. "Acorn Dispersal by the Blue Jay (*Cyanocitta cristata*)." *Oecologia* 50, no. 2: 231–32.

Dekle, G. W., and Thomas R. Fasulo. 2000. "Azalea Caterpillar—*Datana major* [. . .]." Featured Creatures (website). Gainesville: University of Florida IFAS. http://entnemdept.ufl.edu/creatures/orn/azalea_caterpillar.htm.

de la Vega, Garcilasco. (1605) 1996. *The Florida of the Inca*. Translated by John and Jeannette Varner. Austin: University of Texas Press.

Deyrup, Mark A., and D. N. Deyrup. 2015. *Database of Observations of Hymenoptera Visitations to Flowers of Plants on Archbold Biological Station, Florida, USA*. Venus, Fla.: Archbold Biological Station.

Deyrup, Mark A., and Leif D. Deyrup. 2011. "*Colletes francesae,* a New Species of Colletid Bee (Hymenoptera: Colletidae) Associated with *Sideroxylon tenax* (Sapotaceae) in Florida Scrub Habitat." *Florida Entomologist* 94, no. 4: 897–901.

Deyrup, Mark, Jayanthi Edirisinghe, and Beth Norden. 2002. "The Diversity and Floral Hosts of Bees at the Archbold Biological Station, Florida." *Insecta Mundi* 16, nos. 1–3: 87–120.

Dickerson, James G. 1990. "*Cercis canadensis* L.—Eastern Redbud." In *Silvics of North America*, vol. 2, *Hardwoods*, edited by Russell M. Burns and Barbara H. Honkala, 266–69. USDA Forest Service Agriculture Handbook 654. Washington, D.C.: Government Printing Office.

Djonwangwe, Denis, Fernand-Nestor Tchuenguem Fohouo, and Jean Messi. 2011. "Foraging and Activities of *Apis millifera adamsonii* Latreille (Hymenoptera: Apidae) on *Ximenia americana* (Olacaceae) Flowers at Ngaoundere (Cameroon)." *International Research Journal of Plant Science* 2, no. 6: 170–78. June 22, 2018. http://www.interesjournals.org/IRJPS.

Dolan, R. W. 2004. "Conservation Assessment for Trailing Arbutus (*Epigaea repens* L.)." Milwaukee, Wisc.: USDA Forest Service, Eastern Region. https://www.researchgate.net/publication/237383859_Conservation_Assessment_for_Trailing_arbutus_Epigaea_repens_L.

Doty, Sharon L., Brian Oakley, Gang Xin, Jun Won Kang, Glenda Singleton, Zareen Khan, Azra Vajzovic, and James T. Staley. 2009. "Diazotrophic Endophytes of Native Black Cottonwood and Willow." *Symbiosis* 47, No. 1:23–33.

Duke, James A. 1983. *Handbook of Energy Crops*. West Lafayette, Ind.: Purdue University Center for New Crops and Plant Products. September 21, 2018. https://www.hort.purdue.edu/newcrop/duke_energy/dukeindex.html.

Dziergowski, Annie. 2009. "Species Account/Biologue: Etonia Rosemary (*Conradina etonia*)." U.S. Fish and Wildlife Service, North Florida Ecological Services Office, Jacksonville. https://www.fws.gov/northflorida/Species-Accounts/Etonia-Rosemary-2005.htm.

Eaton, Stephen W. 1992. "Wild Turkey." In *The Birds of North America*, edited by Alan F. Poole, Peter Stettenheim, and Frank B. Gill. No. 22 of 716 issues, 18 vols. Philadelphia, Pa., and Washington, D.C.: Academy of Natural Sciences of Philadelphia and American Ornithologists Union.

Eckert, Christopher G., Marcel E. Dorken, and Stacy A. Mitchell. 1999. "Loss of Sex in Clonal Populations of a Flowering Plant, *Decodon verticillatus* (Lythraceae)." *Evolution* 53, no. 4: 1079–92.

Edwards, Christine E., Walter S. Judd, Gretchen M. Ionta, and Brenda Herring. 2009. "Using Population Genetic Data as a Tool to Identify New Species: *Conradina cygniflora* (Lamiaceae), a New Endangered Species from Florida." *Systematic Botany* 34, no. 4: 747–56.

Edwards, M. B. 1990. "*Quercus michauxii* Nutt.—Swamp Chestnut Oak." In *Silvics of North America*, vol. 2, *Hardwoods*, edited by Russell M. Burns and Barbara H. Honkala, 693–96. USDA Forest Service Agriculture Handbook 654. Washington, D.C.: Government Printing Office.

———. 1990. "*Quercus shumardii* Buckl.—Shumard Oak." In *Silvics of North America*, vol. 2, *Hardwoods*, edited by Russell M. Burns and Barbara H. Honkala, 734–37. USDA Forest Service Agriculture Handbook 654. Washington, D.C.: Government Printing Office.

Emerald Ash Borer Information Network. 2020. Network regularly updated. USDA Forest Service and Michigan State University. http://www.emeraldashborer.info/index.php.

Encyclopedia Britannica. 2020. "Constitution (Old Ironsides)." June 19, 2020. https://www.britannica.com/topic/constitution-ship.

Engstrom, R. Todd, and Tom Radzio. 2014. "What's Eating the Fruit of the Miccosukee Gooseberry." *Castanea* 79, no. 1: 27–31.

Feyssa, Debela H., Jesse T. Njoka, Zemede Astaw, and M. M. Nyangito. 2012. "Uses and Management of *Ximenia Americana, Olacaceae* in Semi-Arid East Shewa, Ethiopia." *Pakistan Journal of Botany* 44, no. 4: 1177–184.

Flores, Alfredo. 2003. "Scouring the World for a Skunkvine Control." *USDA Agricultural Research Magazine* 51, no. 10. February 29, 2016. https://agresearchmag.ars.usda.gov/2003/Oct/skunk.

Florida Champion Trees (database). 2020. Florida Division of Forestry. Tallahassee, Fla. https://www.fdacs.gov/Forest-Wildfire/Our-Forests/Florida-Champion-Trees.

Florida Department of Environmental Protection (Florida DEP). 1995. Division of Resource Management. *Resource Management Notes* 7, no. 3.

Florida Exotic Pest Plant Council (FLEPPC). 2013. *Florida Exotic Pest Plant Council Invasive Plant Lists*. Athens: University of Georgia Center for Invasive Species and Ecosystem Health. http://www.fleppc.org/list/list.htm.

Florida Natural Areas Inventory. 2000. "Apalachicola Rosemary." August 17, 2018. https://www.fnai.org/FieldGuide/pdf/Conradina_glabra.pdf.

———. 2000. "Miccosukee Gooseberry." https://www.fnai.org/FieldGuide/pdf/Ribes_echinellum.pdf.

———. 2000. "Mock Pennyroyal." August 17, 2018. https://www.fnai.org/FieldGuide/pdf/Stachydeoma_graveolens.pdf.

———. 2019. "Florida Willow." July 11, 2019. https://www.fnai.org/FieldGuide/pdf/Salix_floridana.pdf.

Fordham, Alfred J. 1961. "Liquidambar." *Arnoldia* 21, no. 10: 59–66.

Fowells, Harry A., compiler. 1965. *Silvics of Forest Trees of the United States*. USDA Agriculture Handbook 271. Washington, D.C.: United States Division of Timber Management Research and USDA Forest Service.

Francis, John K. 1990. "*Carya aquatica* (Michx. f.) Nutt.—Water Hickory." In *Silvics of North America*, vol. 2, *Hardwoods*, edited by Russell M. Burns and Barbara H. Honkala, 186–89. USDA Forest Service Agriculture Handbook 654. Washington, D.C.: Government Printing Office.

Gabriel, William J. 1990. "*Acer saccharinum* L.—Silver Maple." In *Silvics of North America*, vol. 2, *Hardwoods*, edited by Russell M. Burns and Barbara H. Honkala, 70–77. USDA Forest Service Agriculture Handbook 654. Washington, D.C.: Government Printing Office.

Giblin-Davis, Robin M., and F. W. Howard. 1989. "Vulnerability of Stressed Palms to Attack by *Rhynchophorus cruentatus* (Fabricius) (Coleoptera: Curculionidae) and Insecticidal Control of the Pest." *Journal of Economic Entomology* 82, no. 4: 1185–190.

Gilman, Edward F., and Dennis G. Watson. 1993. "*Osmanthus americanus:* Devilwood." Fact sheet ST-424. Gainesville: University of Florida Environmental Horticulture Department, IFAS Cooperative Extension Service, and IFAS. https://edis.ifas.ufl.edu/st424.

———. 1993. "*Pinckneya pubens:* Pinckneya." Publication ENH-614. Gainesville: University of Florida Environmental Horticulture Department, IFAS Cooperative Extension Service, and IFAS. https://edis.ifas.ufl.edu/st455.

Glasgow, Leslie L. 1977. "Common Persimmon—*Diospyros virginiana* L." In *Southern Fruit-Producing Woody Plants Used by Wildlife*, edited by Lowell K. Halls, 103. General Technical Report SO-16. New Orleans, La.: USDA Forest Service, Southern Forest Experiment Station.

Glassberg, Jeffrey, Marc C. Minno, and John V. Calhoun. 2000. *Butterflies through Binoculars: A Field, Finding, and Gardening Guide to Butterflies in Florida*. New York: Oxford University Press.

Global Invasive Species Database. 2020. "*Raffaelea lauricola*." July 21, 2016. http://www.iucngisd. org/gisd/speciesname/Raffaelea+lauricola.

———. 2020. "*Sesbania punicea*." August 11, 2018. http://193.206.192.138/gisd/speciesname/ Sesbania+punicea.

Godfrey, Robert K. 1988. *Trees, Shrubs, and Woody Vines of Northern Florida and Adjacent Georgia and Alabama*. Athens: University of Georgia Press.

Godfrey, Robert K., and Daniel B. Ward. 1979. "Hand Fern." In *Rare and Endangered Biota of Florida*, vol. 5, *Plants*, edited by Daniel B. Ward, 44–46. Gainesville: University Presses of Florida.

Greenberg, Cathryn H., and Robert W. Simons. 2000. "Age and Stand Structure of Old-Growth Oak in Florida High Pine." In *Fire and Forest Ecology: Innovative Silvaculture and Vegetation Management*, edited by W. Keith Moser and Cynthia F. Moser, 30. Tall Timbers Fire Ecology Conference Proceedings no. 21. Tallahassee, Fla.: Tall Timbers Research Station.

Grelen, H. E. 1990. "*Betula Nigra* L.—River Birch." In *Silvics of North America*, vol. 2, *Hardwoods*, edited by Russell M. Burns and Barbara H. Honkala, 153–57. USDA Forest Service Agriculture Handbook 654. Washington, D.C.: Government Printing Office.

———. 1990. "*Ilex opaca* Ait.—American Holly." In *Silvics of North America*, vol. 2, *Hardwoods*, edited by Russell M. Burns and Barbara H. Honkala, 379–85. USDA Forest Service Agriculture Handbook 654. Washington, D.C.: Government Printing Office.

Gresham, Charles A., and Donald J. Lipscomb. 1990. "*Gordonia lasianthus* (L.) Ellis—Loblolly-Bay." In *Silvics of North America*, vol. 2, *Hardwoods*, edited by Russell M. Burns and Barbara H. Honkala, 365–69. USDA Forest Service Agriculture Handbook 654. Washington, D.C.: Government Printing Office.

Griggs, Margene M. 1990. "*Sassafras albidum* (Nutt.) Nees—Sassafras." In *Silvics of North America*, vol. 2, *Hardwoods*, edited by Russell M. Burns and Barbara H. Honkala, 773–77. USDA Forest Service Agriculture Handbook 654. Washington, D.C.: Government Printing Office.

Halls, Lowell K. 1990. "*Diospyros virginiana* L.—Common Persimmon." In *Silvics of North America*, vol. 2, *Hardwoods*, edited by Russell M. Burns and Barbara H. Honkala, 294–98. USDA Forest Service Agriculture Handbook 654. Washington, D.C.: Government Printing Office.

Halls, Lowell K., ed. 1977. *Southern Fruit Producing Woody Plants Used by Wildlife*. USDA Forest Service General Technical Report SO-16. New Orleans, La.: USDA Forest Service, Southern Forest Experiment Station.

Hammer, Roger L. 2018. *Complete Guide to Florida Wildflowers*. Guilford, Conn. and Helena, Mont.: Falcon Guides.

Harlow, Richard F. 1990. "*Quercus laevis* Walt.—Turkey Oak." In *Silvics of North America*, vol. 2, *Hardwoods*, edited by Russell M. Burns and Barbara H. Honkala, 672–76. USDA Forest Service Agriculture Handbook 654. Washington, D.C.: Government Printing Office.

Harlow, William M., and Ellwood S. Harrar. 1958. *Textbook of Dendrology*, 4th ed. New York: McGraw-Hill.

Harms, William R. 1990. "*Fraxinus profunda* (Bush) Bush—Pumpkin Ash." In *Silvics of North America*, vol. 2, *Hardwoods*, edited by Russell M. Burns and Barbara H. Honkala, 355–57. USDA Forest Service Agriculture Handbook 654. Washington, D.C.: Government Printing Office.

Harper, Roland M. 1911. "The Relation of Climax Vegetation to Islands and Peninsulas." *Bulletin of the Torrey Botanical Club* 38, no. 11: 515–25.

———. 1914. "Geography and Vegetation of Northern Florida." *Florida Geological Survey*, 6th Annual Report, 163–451. Tallahassee, Fla.: State Geological Survey.

———. 1915. "Vegetation Types." In "Natural Resources Survey of an Area in Central Florida" by E. H. Sellards, Roland M. Harper, Charles N. Mooney, W. J. Latimer, Herman Gunter, and

Emil Gunter, 135–88. *Florida Geological Survey*, 7th Annual Report. Tallahassee, Fla.: State Geological Survey.

Harrison, Nigel A., and Monica Lynn Elliott. (2007) 2016. "Texas Phoenix Palm Decline." Plant Pathology publication PP243. Gainesville: University of Florida IFAS, Cooperative Extension Service.

Hickman, Kennedy. (2017) 2020. "Hundred Years' War: English Longbow." ThoughtCo, Lifelong Learning (website). https://www.thoughtco.com/hundred-years-war-english-longbow-2361241.

Hicks, David J., Robert Wyatt, and Thomas R. Meagher. 1985. "Reproductive Biology of Distylous Partridgeberry, *Mitchella repens*." *American Journal of Botany* 72, no. 10: 1503–514.

Hill, Geoffrey E. 1993. "House Finch (*Carpodacus mexicanus*)." In *The Birds of North America*, edited by Alan F. Poole and Frank B. Gill. No. 46 of 716 issues, 18 vols. Philadelphia, Pa., and Washington, D.C.: Academy of Natural Sciences of Philadelphia and American Ornithologists Union.

Hill, K. 2009. "Mangrove Habitats." Indian River Lagoon Species Inventory (database). Fort Pierce, Fla.: Smithsonian Marine Station at Fort Pierce. https://naturalhistory2.si.edu/smsfp/irlspec/Mangroves.htm.

Hill, Steven R. 2002. "Conservation Assessment for Deerberry (*Vaccinium stamineum*)." Illinois Natural History Survey (INHS) Technical Report, prepared for USDA Forest Service, Vienna Ranger District, Shawnee National Forest.

Hilty, John. 2018. "False Indigo (*Amorpha fruticosa*)." *Illinois Wildflowers* (website). https://www.illinoiswildflowers.info/trees/plants/false_indigo.htm.

———. 2018. "Swamp Loosestrife (*Decodon verticillatus*)." *Illinois Wildflowers* (website). https://www.illinoiswildflowers.info/wetland/plants/sw_loosestrife.html.

———. 2019. "Black Willow (*Salix nigra*)." *Illinois Wildflowers* (website). https://www.illinoiswildflowers.info/trees/plants/bl_willow.htm.

———. 2019. "Bladdernut (*Stephylea trifolia*)." *Illinois Wildflowers* (website). https://www.illinoiswildflowers.info/trees/plants/bladdernut.htm.

———. 2019. "Downy Serviceberry (*Amelanchier arborea*)." *Illinois Wildflowers* (website). https://www.illinoiswildflowers.info/trees/plants/dwn_service.html.

———. 2019. "New Jersey Tea (*Ceanothus americanus*)." *Illinois Wildflowers* (website). https://www.illinoiswildflowers.info/prairie/plantx/nj_teax.htm.

———. 2019. "Ninebark (*Physocarpus opulifolius*)." *Illinois Wildflowers* (website). https://www.illinoiswildflowers.info/trees/plants/ninebark.htm.

———. 2019. "Prairie Willow (*Salix humilis humilis*)." *Illinois Wildflowers* (website). https://www.illinoiswildflowers.info/prairie/plantx/pr_willowx.htm.

———. 2019. "Swamp Rose (*Rosa palustris*)." *Illinois Wildflowers* (website). https://www.illinoiswildflowers.info/wetland/plants/sw_rose.html.

———. 2019. "Wild Black Cherry (*Prunus serotine*)." *Illinois Wildflowers* (website). https://www.illinoiswildflowers.info/trees/plants/wb_cherry.htm.

———. 2019. "Wild Crab Apple (*Malus coronaria*)." *Illinois Wildflowers* (website). https://www.illinoiswildflowers.info/trees/plants/wild_crab.htm.

———. 2019. "Wild Hydrangea (*Hydrangea arborescens*)." *Illinois Wildflowers* (website). https://www.illinoiswildflowers.info/woodland/plants/hydrangea.htm.

Hoffmann, John H., and V. Cliff Moran. 1989. "Novel Graphs for Depicting Herbivore Damage on Plants: The Biocontrol of *Sesbania punicea* (Fabaceae) by an Introduced Weevil." *Journal of Applied Ecology* 26, no. 1: 353–60.

Houston, David R. 1987. "Beech Bark Diseases." In *Exotic Pests of Eastern Forests Conference Proceedings, April 8–10, 1997*, edited by Kerry O. Britton, 29–41. Nashville, Tenn.: USDA Forest Service and Tennessee Exotic Pest Plant Council.

Hua, Catherine, Shayla Salzman, and Naomi Pierce. 2018. "A First Record of *Anatrachyntis badia* (Hodges 1962) (Lepidoptera: Cosmopterigidae) on *Zamia integrifolia* (Zamiaceae)." *Florida Entomologist* 101, no. 2: 335–38. May 17, 2020. https://doi.org/10.1653/024.101.0230.

Huegel, Craig N. 2009. "Apalachicola Rosemary–*Conradina glabra*." *Native Florida Wildflowers* (blog). October 31. http://hawthornhillwildflowers.blogspot.com/search?q=Apalachicola+Rosemary.

———. 2009. "Coastalplain Balm–*Dicerandra linearifolia*." *Native Florida Wildflowers* (blog). November 1. http://hawthornhillwildflowers.blogspot.com/2009/11/coastalplain-balm-dicerandra.html.

———. 2009. "Georgia Calamint–*Calamintha georgiana*." *Native Florida Wildflowers* (blog). October 31. http://hawthornhillwildflowers.blogspot.com/search?q=Georgia+Calamint.

———. 2009. "Large-Flowered Rosemary–*Conradina grandiflora*" *Native Florida Wildflowers* (blog). October 31. http://hawthornhillwildflowers.blogspot.com/search?q=Large-flowered+rosemary.

———. 2009. "Longspur Balm–*Dicerandra cornutissima*." *Native Florida Wildflowers* (blog). November 1. http://hawthornhillwildflowers.blogspot.com/2009/11/longspur-balm-dicerandra-cornutissima.html.

———. 2009. "Toothed Savory–*Calamintha dentata*." *Native Florida Wildflowers* (blog). October 31. http://hawthornhillwildflowers.blogspot.com/search?q=Toothed+Savory.

———. 2009. "Wild Rosemary–*Conradina canescens*." *Native Florida Wildflowers* (blog). October 31. http://hawthornhillwildflowers.blogspot.com/2009/10/wild-rosemary-conradina-canescens.html.

———. 2010. *Native Plant Landscaping for Florida Wildlife*. Gainesville: University Press of Florida.

Humphrey, Stephen R., and Patrick G. R. Jodice. 1992. "Big Cypress Fox Squirrel *Sciurus niger avicennia*." In *Rare and Endangered Biota of Florida*, vol. 1, *Mammals*, edited by Stephen R. Humphrey, 224–33. Gainesville: University Press of Florida.

Hunt, April. 2001. "Family Sues Resort over Toxic Berries." *Orlando Sentinel*. May 5.

Hutchinson, Mark. 2013. "*Psychotria nervosa*–Wild Coffee." *Native Plant Owner's Manual*. Florida Native Plant Society. https:www.fnps.org/assets/pdf/pubs/sychotria_nervosa_wildcoffee.pdf.

Hutto, Joe. 1995. *Illumination in the Flatwoods: A Season with the Wild Turkey*. New York: Lyons Press.

Iowa State University. (2003–2020). "Nymphs on *Sebastiania fruticosa*—*Orsilochides guttata*." Bug Guide: Identification, Images, and Information for Insects, Spiders and Their Kin for the United States and Canada (database). Iowa State University Department of Entomology. https://bugguide.net/node/view/321173/bgimage; https://bugguide.net/node/view/321172/bgimage; https://bugguide.net/node/view/321171/bgimage.

Johnson, Ann. 1982. "Some Demographic Characteristics of the Florida Rosemary *Ceratiola ericoides* Michaux." *American Midland Naturalist* 108, no. 1: 170–74.

Johnson, A. Sydney, Philip E. Hale, William M. Ford, James M. Wentworth, Jeffrey R. French, Owen F. Anderson, and Gerald B. Pullen. 1995. "White-Tailed Deer Foraging in Relation to Successional Stage, Overstory Type, and Management of Southern Appalachian Forests." *American Midland Naturalist* 133, no. 1: 18–35.

Johnson, Robert L. 1990. "*Populus heterophylla* L.—Swamp Cottonwood." In *Silvics of North America*, vol. 2, *Hardwoods*, edited by Russell M. Burns and Barbara H. Honkala, 551–54. USDA Forest Service Agriculture Handbook 654. Washington, D.C.: Government Printing Office.

Jones, Earle P., Jr. 1990. "*Acer barbatum* Michx.—Florida Maple." In *Silvics of North America*, vol. 2, *Hardwoods*, edited by Russell M. Burns and Barbara H. Honkala, 29–32. USDA Forest Service Agriculture Handbook 654. Washington, D.C.: Government Printing Office.

Kartesz, John T. 2015. *Floristic Synthesis of North America*, Version 1.0. (2014 CD-Rom distributed separately and unavailable online.) Biota of North America Program (BONAP). Taxonomic Data Center. Chapel Hill, N.C. http://www.bonap.net/tdc.

Kellam, John O., Deborah K. Jansen, Annette T. Johnson, Ralph W. Arwood, Melissa J. Merrick, and John L. Koprowski. 2016. "Big Cypress Fox Squirrel (*Sciurus niger avicennia*) Ecology and Habitat Use in a Cypress Dome–Pine Forest Mosaic." *Journal of Mammology* 97, no. 1: 200–210.

Kennedy, Harvey E., Jr. 1990. "*Celtis laevigata* Willd.—Sugarberry." In *Silvics of North America*, vol. 2, *Hardwoods*, edited by Russell M. Burns and Barbara H. Honkala, 258–61. USDA Forest Service Agriculture Handbook 654. Washington, D.C.: Government Printing Office.

———. 1990. "*Fraxinus pennsylvanica* Marsh.—Green Ash." In *Silvics of North America*, vol. 2, *Hardwoods*, edited by Russell M. Burns and Barbara H. Honkala, 348–54. USDA Forest Service Agriculture Handbook 654. Washington, D.C.: Government Printing Office.

Kilham, Lawrence. 1989. *The American Crow and the Common Raven*. College Station: Texas A&M University.

Knopf, Jim, Sally Wasowski, John Kadel Boring, Glenn Keator, Jane Scott, and Erica Glasener. 1995. *Natural Gardening (A Nature Company Guide)*. Edited by R. G. Turner Jr. Berkeley, Calif.: Nature Company.

Kormanik, Paul P. 1990. "*Liquidambar styraciflua* L.—Sweetgum." In *Silvics of North America*, vol. 2, *Hardwoods*, edited by Russell M. Burns and Barbara H. Honkala, 400–405. USDA Forest Service Agriculture Handbook 654. Washington, D.C.: Government Printing Office.

Kossuth, Susan V., and J. L Michael. 1990. "*Pinus glabra* Walt.—Spruce Pine." In *Silvics of North America*, vol. 1, *Conifers*, edited by Russell M. Burns and Barbara H. Honkala, 355–58. USDA Forest Service Agriculture Handbook 654. Washington, D.C.: Government Printing Office.

Kossuth, Susan V., and Robert L. Scheer. 1990. "*Nyssa ogeche* Bartr. ex Marsh.—Ogeechee Tupelo." In *Silvics of North America*, vol. 2, *Hardwoods*, edited by Russell M. Burns and Barbara H. Honkala, 479–81. USDA Forest Service Agriculture Handbook 654. Washington, D.C.: Government Printing Office.

Krochmal, Connie. 2017. "The Eastern Huckleberries as Bee Plants." *Bee Culture: The Magazine of American Beekeeping*. November 15. https://www.beeculture.com/eastern-huckleberries-bee-plants/.

Kubes, Amanda J. 2009. "Modeling Species Distributions of Three Endemic Florida Panhandle Mints under Climate Change: Comparing Plant Pollinator Distribution Shifts under Future Conditions." Master's thesis, Florida State University College of Arts and Sciences, Tallahassee.

Kurten, Erin L., Carolyn P. Snyder, Terri Iwata, and Peter M. Vitousek. 2008. "*Morella cerifera* Invasion and Nitrogen Cycling on a Lowland Hawaiian Lava Flow." *Biological Invasions* 10, no. 1: 19–24.

Lamson, Neil I. 1990. "*Morus rubra* L.—Red mulberry." In *Silvics of North America*, vol. 2, *Hardwoods*, edited by Russell M. Burns and Barbara H. Honkala, 470–73. USDA Forest Service Agriculture Handbook 654. Washington, D.C.: Government Printing Office.

Lance, Ron. 2004. *Woody Plants of the Southeastern United States: A Winter Guide*. Athens: University of Georgia Press.

Laqueur, Gert L., and Maria Spatz. 1968. "Toxicology of Cycesin." *Cancer Research* 28: 2262–267.

Lawson, Edwin R. 1990. "*Pinus echinata* Mill.—Shortleaf Pine." In *Silvics of North America*, vol. 1, *Conifers*, edited by Russell M. Burns and Barbara H. Honkala, 316–26. USDA Forest Service Agriculture Handbook 654. Washington, D.C.: Government Printing Office.

Layne, James N., and Warren G. Abrahamson. 2006. "Scrub Hickory: A Florida Endemic." *Palmetto* 23, no. 2: 4–7, 13.

Leckie, Seabrooke, and David Beadle. 2018. *Peterson Field Guide to Moths of Southeastern North America*. New York and Boston: Houghton Mifflin Harcourt.

Lee, Jisun, and Madeleine M. Joullie. 2018. "Total Synthesis of the Reported Structure of Ceanothine D via a Novel Microcyclization Strategy." *Chemical Science* 2018, no. 9: 2432–436.

Legare, Donna. 2014. "Meet the Southeastern Blueberry Bee—The Main Pollinator of Blueberry Flowers." *Tallahassee Democrat*, May 30.

Little, Silas, and Peter W. Garrett. 1990. "*Chamaecyparis thyoides* (L.) B.S.P.—Atlantic White Cedar." In *Silvics of North America*, vol. 1, *Conifers*, edited by Russell M. Burns and Barbara H. Honkala, 103–8. USDA Forest Service Agriculture Handbook 654. Washington, D.C.: Government Printing Office.

Lohrey, Richard E., and Susan V. Kossuth. 1990. "*Pinus elliottii* Engelm.—Slash Pine." In *Silvics of North America*, vol. 1, *Conifers*, edited by Russell M. Burns and Barbara H. Honkala, 338–47. USDA Forest Service Agriculture Handbook 654. Washington, D.C.: Government Printing Office.

Louisiana State University (LSU) Agricultural Center. 2021. Louisiana Ecosystems and Plant Identification (database). Baton Rouge: LSU Agricultural Center and LSU College of Agriculture. https://www.lsuagcenter.com/portals/our_offices/departments/renewable-natural-resources/features/ecosystems_plant_id.

Martin, Alexander C., Herbert S. Zim, and Arnold L. Nelson. 1951. *American Wildlife and Plants*. New York: McGraw-Hill.

McCormack, Jeffrey Holt. 1975. "Beetle Pollination of *Calycanthus floridanus* L.: Pollination Behavior as a Function of the Volatile Oils." PhD diss., University of Connecticut, Storrs.

Mead, Frank W. 1985. "*Jadera* Scentless Plant Bugs in Florida (Hemiptera: Rhopalidea)." Entomology Circular no. 277. Gainesville: Florida Department of Agriculture and Consumer Services, Division of Plant Industry.

Menges, Eric S. 2014. "*Ceratiola ericoides*—Species Account." Venus, Fla.: Archbold Biological Station. www.archbold-station.org/html/research/plant/cererisppacc.html.

Metalmark Web and Data. 2019. "Mournful Thyris (*Thyris sepulchralis*)." Butterflies and Moths of North America (website). August 8, 2019. https://www.butterfliesandmoths.org/species/thyris-sepulchralis.

Metzger, F. T. 1990. "*Carpinus caroliniana* Walt.—American Hornbeam." In *Silvics of North America*, vol. 2, *Hardwoods*, edited by Russell M. Burns and Barbara H. Honkala, 179–85. USDA Forest Service Agriculture Handbook 654. Washington, D.C.: Government Printing Office.

Michaux, François André. 1810–13. *Histoire des arbres forestiers de l'Amérique septentrionale*. Paris: L. Haussmann & Hautel; Philadelphia: Samuel Bradford and Inskeep.

Milbrandt, Elliott C., and Melissa N. Tinsley. 2006. "The Role of Saltbush (*Batis maritima* L.) in Regeneration of Degraded Mangrove Forests." *Hydrobiologia* 568, no. 1: 369–77.

Miller, James H., and Karl V. Miller. 1999. *Forest Plants of the Southeast and Their Wildlife Uses*. Champaign, Ill.: Southern Weed Science Society.

Moler, Paul E. 1985. "Home Range and Seasonal Activity of the Eastern Indigo Snake, *Drymarchon corais couperi*, in Northern Florida." Master's thesis, University of Florida, Gainesville.

Montz, Glen N., and Arthur Cherubini. 1973. "An Ecological Study of a Baldcypress Swamp in St. Charles Parish, Louisiana." *Castanea* 38, no. 4: 378–86.

Moore, Dwight M., and William P. Thomas. 1977. "Red Mulberry/*Morus rubra* L." In *Southern Fruit Producing Woody Plants Used by Wildlife*, edited by Lowell K. Halls, 55. USDA Forest Service General Technical Report SO-16. New Orleans, La.: USDA Forest Service, Southern Forest Experiment Station.

Morris, R. C., T. H. Filer, J. D. Solomon, Francis I. McCracken, N. A. Overgaard, and M. J. Weiss.

1975. *Insects and Diseases of Cottonwood*. USDA Forest Service General Technical Report SO-8. New Orleans, La.: USDA Forest Service, Southern Forest Experiment Station.

Moyroud, Richard. 1996. "Cabbage Palms: Can We Continue to Transplant from the Wild?" *Palmetto* 16, no. 3: 13–15.

Myers, Ronald L. 1990. "Scrub and High Pine." In *Ecosystems of Florida*, edited by Ronald L. Myers and John J. Ewel, 150–93. Orlando: University of Central Florida Press.

Nash, George V. 1895. "Notes on Some Florida Plants." *Bulletin of the Torrey Botanical Club* 22, no. 4: 141–61.

North Carolina Cooperative Extension. n.d. "*Agarista populifolia*." North Carolina Extension Gardener Plant Toolbox (database). North Carolina State University, North Carolina Agricultural and Technical State University, and North Carolina Cooperative Extension. https://plants.ces.ncsu.edu/plants/all/agarista-populifolia/.

———. 2018. "*Sassafras albidum*." North Carolina Extension Gardener Plant Toolbox (database). North Carolina State University, North Carolina Agricultural and Technical State University, and North Carolina Cooperative Extension. https://plants.ces.ncsu.edu/plants/all/sassafras-albidum/.

Nelson, Gil. 1994. *The Trees of Florida: A Reference and Field Guide*. Sarasota, Fla.: Pineapple Press.

———. 1996. *Shrubs and Woody Vines of Florida*. Sarasota, Fla.: Pineapple Press.

Nooney, Jill. 1994. "A Very Valuable Shrub: *Xanthorhiza simplicissima*." *Arnoldia* 54, no. 2: 31–35.

Noss, Reed F. 2018. *Fire Ecology of Florida and the Southeastern Coastal Plain*. Gainesville: University Press of Florida.

Oertel, Everett. 1967. "Nectar and Pollen Plants." In *Bee Keeping in the United States*, 10–16. USDA Agriculture Handbook 335. Agricultural Research Service USDA. Washington, D.C.: Government Printing Office. https://naldc.nal.usda.gov/download/CAT87208707/PDF.

Ornduff, Robert. 1970. "The Systematics and Breeding System of *Gelsemium* (Loganiaceae)." *Journal of the Arnold Arboretum* 51, no. 1: 1–17.

Oswalt, Christopher M., Jason A. Cooper, Dale G. Brockway, Horace W. Brooks, Joan L. Walker, Kristina F. Connor, Sonja N. Oswalt, and Roger C. Conner. 2012. *History and Current Condition of Longleaf Pine in the Southern United States*. USDA Forest Service General Technical Report SRS-166. Asheville, N.C.: USDA Forest Service, Southern Research Station.

Outcalt, Kenneth W. 1990. "*Magnolia grandiflora* L.-Southern Magnolia." In *Silvics of North America*, vol. 2, *Hardwoods*, edited by Russell M. Burns and Barbara H. Honkala, 445–48. USDA Forest Service Agriculture Handbook 654. Washington, D.C.: Government Printing Office.

Overton, Ronald P. 1990. "*Acer negundo* L.—Boxelder." In *Silvics of North America*, vol. 2, *Hardwoods*, edited by Russell M. Burns and Barbara H. Honkala, 41–45. USDA Forest Service Agriculture Handbook 654. Washington, D.C.: Government Printing Office.

Peles, John D., Christopher K. Williams, and Gary W. Barrett. 1995. "Bioenergetics of Golden Mice: The Importance of Food Quality." *American Midland Naturalist*, 133, no. 2: 373–76.

Perkins, Kent D., and Willard W. Payne. 1978. *Guide to the Poisonous and Irritant Plants of Florida*. Circular 441. Gainesville: University of Florida IFAS, Cooperative Extension Service.

Peterson, J. K. 1990. "*Carya illinoensis* (Wangenh.) K. Koch.—Pecan." In *Silvics of North America*, vol. 2, *Hardwoods*, edited by Russel M. Burns and Barbara H. Honkala, 205–10. USDA Forest Service Agricultural Handbook 654. Washington, D.C.: Government Printing Office.

Phipps, James B., and Kenneth A. Dvorsky. 2008. "A Taxonomic Revision of *Crataegus* Series *Lacrimatae* (Rosaceae)." *Journal of the Botanical Research Institute of Texas* 2, no. 2: 1101–162.

Pitcher, J. A., and J. S. McKnight. 1990. "*Salix nigra* Marsh.—Black Willow." In *Silvics of North America*, vol. 2, *Hardwoods*, edited by Russell M. Burns and Barbara H. Honkala, 768–72. USDA Forest Service Agriculture Handbook 654. Washington, D.C.: Government Printing Office.

Platt, William J., and Mark W. Schwartz. 1990. "Temperate Hardwood Forests." In *Ecosystems of Florida*, edited by Ronald L. Myers and John J. Ewel, 194–229. Orlando: University of Central Florida Press.

Pomeroy, K. B. and, Dorothy Dixon. 1966. "These Are the Champs." *American Forests* 72, no. 5: 15–35.

Priester, David S. 1990. "*Magnolia virginiana* L.—Sweetbay." In *Silvics of North America*, vol. 2, *Hardwoods*, edited by Russell M. Burns and Barbara H. Honkala, 449–54. USDA Forest Service Agriculture Handbook 654. Washington, D.C.: Government Printing Office.

Pyle, Robert Michael. 1981. *The Audubon Society Field Guide to North American Butterflies*. New York: Knopf.

Randolph, KaDonna C. 2017. "Status of *Sassafras albidum* (Nutt.) Nees in the Presence of Laurel Wilt Disease and throughout the Eastern United States." *Southeastern Naturalist* 16, no. 1: 37–58.

Rightmyer, Molly G., Mark Deyrup, John S. Ascher, and Terry Griswold. 2011. "*Osmia* Species (Hymenoptera, Megachilidae) from the Southeastern United States with Modified Facial Hairs: Taxonomy, Host Plants, and Conservation Status." *ZooKeys* 148: 257–78. https://zookeys.pensoft.net/article/2987/.

Rogers, George. 2011. "Dogs Beware of the Leafless Swallow-wort!" *Treasure Coast Natives* (blog). https://treasurecoastnatives.wordpress.com/category/leafless-swallowwort/.

———. 2015. "Corkwood." *Treasure Coast Natives* (blog). https://treasurecoastnatives.wordpress.com/2015/11/29/corkwood/.

Rogers, Robert. 1990. "*Quercus alba* L.—White Oak." In *Silvics of North America*, vol. 2, *Hardwoods*, edited by Russell M. Burns and Barbara H. Honkala, 605–13. USDA Forest Service Agriculture Handbook 654. Washington, D.C.: Government Printing Office.

Romans, Bernard. (1775) 1961. *A Concise Natural History of East and West Florida*. New Orleans, La.: Pelican.

Royal Botanical Society and Kew Science. 2021. "*Ximenia americana* L." Plants of the World Online (database). http://www.plantsoftheworldonline.org/taxon/urn:lsid:ipni.org:names:316341-2.

Sampson, Blair, Steve Noffsinger, Creighton Gupton, and James McGee. 2001. "Pollination Biology of the Muscadine Grape." *HortScience* 36, no. 1: 120–24.

Sander, Ivan L. 1990. "*Quercus muehlenbergii* Engelm.—Chinkapin Oak." In *Silvics of North America*, vol. 2, *Hardwoods*, edited by Russell M. Burns and Barbara H. Honkala, 697–700. USDA Forest Service Agriculture Handbook 654. Washington, D.C.: Government Printing Office.

Schlesinger, Richard C. 1990. "*Fraxinus americana* L.—White Ash." In *Silvics of North America*, vol. 2, *Hardwoods*, edited by Russell M. Burns and Barbara H. Honkala, 333–38. USDA Forest Service Agriculture Handbook 654. Washington, D.C.: Government Printing Office.

Sharma, Jyotsna, James A. Schrader, and William R. Graves. 2008. "Ecology and Phenotypic Variation of *Leitneria floridana* (Leitneriaceae) in Disjunct Native Habitats." *Castanea* 73, no. 2: 94–105.

Short, Henry L., Robert M. Blair, and E. A. Epps Jr. 1975. "Composition and Digestibility of Deer Browse in Southern Forests. USDA Forest Service Research Paper SO-111. New Orleans, La.: USDA Forest Service, Southern Forest Experiment Station.

Sikora, Edward J. 2014 "Kudzu: Invasive Weed Supports the Soybean Rust Pathogen through Winter Months in Southeastern United States." *Outlooks on Pest Management* 25, no. 2: 175–79. https://www.researchgate.net/publication/262577502_Kudzu_Invasive_Weed_Supports_the_Soybean_Rust_Pathogen_Through_Winter_Months_in_Southeastern_United_States.

Simons, Robert W. 1983. *Recovery Plan for Chapman's Rhododendron, Rhododendron chapmanii A. Gray*. Atlanta, Ga.: U.S. Fish and Wildlife Service. https://ecos.fws.gov/docs/recovery_plan/chapmans%20rhododendron%20rp.pdf.

———. 1984. "The Native Habitat of Chapman's Rhododendron." *American Rhododendron Society Indiana Chapter Newsletter* 4, no. 1: 1–3. https://scholar.lib.vt.edu/ejournals/JARS/v38n2/v38n2-simons.htm.

Small, John Kunkel. (1933) 1972. *Manual of the Southeastern Flora.* 2 vols. New York: Hafner.

Smalley, G. W. 1990. "*Carya glabra*—Pignut Hickory." In *Silvics of North America*, vol. 2, *Hardwoods*, edited by Russell M. Burns and Barbara H. Honkala, 198–204. USDA Forest Service Agriculture Handbook 654. Washington, D.C.: Government Printing Office.

Smith, H. Clay 1990. "*Carya cordiformis*—Bitternut Hickory." In *Silvics of North America*, vol. 2, *Hardwoods*, edited by Russell M. Burns and Barbara H. Honkala, 190–97. USDA Forest Service Agriculture Handbook 654. Washington, D.C.: Government Printing Office.

———. 1990. "*Carya tomentosa*—Mockernut Hickory." In *Silvics of North America*, vol. 2, *Hardwoods*, edited by Russell M. Burns and Barbara H. Honkala, 226–33. USDA Forest Service Agriculture Handbook 654. Washington, D.C.: Government Printing Office.

———. 1990. "*Magnolia acuminata* L.—Cucumbertree." In *Silvics of North America*, vol. 2, *Hardwoods*, edited by Russell M. Burns and Barbara H. Honkala, 433–38. USDA Forest Service Agriculture Handbook 654. Washington, D.C.: Government Printing Office.

Smith, Jason A., Kerry O'Donnell, Lacey L. Mount, Keumchul Shin, Kelly Peacock, Aaron Trulock, Tova Spector, Jenny Cruse-Sanders, and Ron Determann. 2011. "A Novel *Fuserium* Species Causes a Canker Disease of the Critically Endangered Conifer, *Torreya taxifolia.*" *Plant Disease* 95, no. 6: 633–39.

Snow, G. A. 1990. "*Ulmus alata*—Winged elm." In *Silvics of North America*, vol. 2, *Hardwoods*, edited by Russell M. Burns and Barbara H. Honkala, 797–800. USDA Forest Service Agriculture Handbook 654. Washington, D.C.: Government Printing Office.

Solomon, J. D. 1990. "*Quercus lyrata* Walt.—Overcup Oak." In *Silvics of North America*, vol. 2, *Hardwoods*, edited by Russell M. Burns and Barbara H. Honkala, 681–85. USDA Forest Service Agriculture Handbook 654. Washington, D.C.: Government Printing Office.

Sourakov, Andrei. 2011. "Niche Partitioning, Co-evolution and Life Histories of *Erythrina* Moths, *Terastia meticulosalis* and *Agathodes designalis* (Lepidoptera: Crombidae)." *Tropical Lepidoptera Research* 21, no. 2: 84–94.

Spence, Don, Marc Hughes, and Jason Smith. 2013. "Laurel Wilt: An Exceptionally Damaging Exotic Disease that Threatens Florida's Forests." *Journal of Florida Studies* 1, no. 2. Open access. http://www.journaloffloridastudies.org/0102laurelwilt.html.

Sprunt, Alexander, Jr. 1954. *Florida Bird Life.* New York: Coward-McCann.

Stahle, David W., Malcolm K. Cleaveland, and John G. Hehr. 1988. "North Carolina Climate Changes Reconstructed from Tree Rings: A.D. 372 to 1985." *Science* 240, no. 4858: 1517–519.

Stahle, David W., Jesse R. Edmondson, Ian M. Howard, Charles R. Robbins, R. Daniel Griffin, A. Carl, C. B. Hall, Daniel K. Stahle, and Max C. A. Torbenson. 2019. "Longevity, Climate Sensitivity, and Conservation Status of Wetland Trees at Black River, North Carolina." *Environmental Research Communications* 1, no. 4. Open access. https://iopscience.iop.org/article/10.1088/2515-7620/ab0c4a.

Stiling, Peter D. 1989. *Florida's Butterflies and Other Insects.* Sarasota, Fla.: Pineapple Press.

Stransky, John J., and Sylvia M. Bierschenk. 1990. "*Ulmus crassifolia* Nutt.—Cedar Elm." In *Silvics of North America*, vol. 2, *Hardwoods*, edited by Russell M. Burns and Barbara H. Honkala, 808–11. USDA Forest Service Agriculture Handbook 654. Washington, D.C.: Government Printing Office.

Sullivan, Janet. 1993. "*Taxus floridana.*" Fire Effects Information System (database). USDA Forest Service, Rocky Mountain Research Station, Fire Science Laboratory, Fort Collins, Colorado. March 18, 2019. http://www.fs.fed.us/database/feis/plants/tree/taxflo/all.html.

———. 1994. "*Platanus occidentalis.*" Fire Effects Information System (database). USDA Forest

Service, Rocky Mountain Research Station, Fire Science Laboratory, Fort Collins, Colorado. March 18, 2019. https://www.fs.fed.us/database/feis/plants/tree/plaocc/all.html.

Surdick, Jim, and Amy Jenkins. 2009. "Pondspice (*Litsea aestivalis*) Population Status and Response to Laurel Wilt Disease in Northeast Florida." Final Report for the Division of Forestry, Florida Department of Agriculture and Consumer Services Contract no. 025665. Florida Natural Areas Inventory. Tallahassee. https://sjrda.stuchalk.domains.unf.edu/files/content/sjrda_717.pdf.

———. 2010. "Population Surveys of Rare Lauraceae Species to Assess the Effects of Laurel Wilt Disease in Florida." Final Report for the Division of Forestry, Florida Department of Agriculture and Consumer Services Contract no. 027356. Florida Natural Areas Inventory, Tallahassee.

Tang, William. 1987. "Insect Pollination in the Cycad *Zamia pumila* (Zamiaceae)." *American Journal of Botany* 74, no. 1: 90–99.

Takahashi, Mizuki K., Liana M. Horner, Toshiro Kubota, Nathan A. Keller, and Warren G. Abrahamson. 2011. "Extensive Clonal Spread and Extreme Longevity in Saw Palmetto, a Foundation Plant." *Molecular Ecology* 20, no. 18: 3730–742.

Takayuk, Aoki, Jason A. Smith, Lacey L. Mount, David M. Geiser, and Kerry O'Donnell. 2013. "*Fuserium Torreyae* sp. Nov., a Pathogen Causing Canker Disease of Florida Torreya (*Torreya taxifolia*), a Critically Endangered Conifer Restricted to Northern Florida and Southeastern Georgia." *Mycologia* 105, no. 2: 312–19.

Terres. John K. 1980. *Audubon Society Encyclopedia of North American Birds*. New York: Knopf.

Tesky, Julie L. 1993. "*Sciurus niger*." *Fire Effects Information System* (database). USDA Forest Service, Rocky Mountain Research Station, Fire Science Laboratory, Fort Collins, Colorado. www.fs.fed.us/database/feis/animals/mammal/scni/all.html.

Texas Invasive Species Institute. 2014. "Silk Tree/Mimosa." Inventory: Plants (database). Texas State University System. http://www.tsusinvasives.org/home/database/albizia-julibrissin.

Thomas, P. 2011. "West Himalayan Yew—*Taxus contorta*." *IUCN* (International Union for Conservation of Nature and Natural Resources) *Red List of Threatened Species* (database). https://dx.doi.org/10.2305/IUCN.UK.2011-2.RLTS.T39147A10170545.en.

Thomas, P., N. Li, and T. Christian. 2013. "Chinese Yew—*Taxus chinensis*." IUCN Red List of Threatened Species (database). https://dx.doi.org/10.2305/IUCN.UK.2013-1.RLTS.T42548A2987120.en.

Tillman, Glynn P. 2015. "First Record of *Sesbania punicea* (Fabales: Fabaceae) as a Host Plant for *Chinavia hilaris* (Hemiptera: Pentatomidae)." *Florida Entomoligist* 98, no. 3: 989–90.

Tirmenstein, D. 1990. "*Vaccinium myrsinites*." Fire Effects Information System (database). USDA Forest Service, Rocky Mountain Research Station, Fire Science Laboratory, Fort Collins, Colorado. https://www.fs.fed.us/database/feis/plants/shrub/vacmys/all.html.

Trelease, William 1924. *The American Oaks*. Vol. 20 of *Memoirs of the National Academy of Sciences*. Washington, D.C.: National Academy of Sciences.

Turgeon, Jean J., K. Kamijo, and Gary L. DeBarr. 1997. "A New Species of *Megastigmus dalman* (Hymenoptera: Torymidae) Reared from Seeds of Atlantic White Cedar (Cupressaceae) with Notes on Infestation Rates." *Proceedings of the Entomological Society of Washington* 99, no. 4: 605–13.

United States Fish and Wildlife Service. 2018. "Peruvian Primrose (*Ludwigia peruviana*) Ecological Risk Screening Summary." https://www.fws.gov/fisheries/ANS/erss/highrisk/ERSS-Ludwigia-peruviana-FINAL.pdf.

Vignoles, Charles Blacker. 1823. *Observations upon the Floridas*. New York: E. Bliss and E. White.

Vozzo, J. A. 1990. "*Quercus nigra* L.—Water Oak." In *Silvics of North America*, vol. 2, *Hardwoods*, edited by Russell M. Burns and Barbara H. Honkala, 701–3. Forest Service Agriculture Handbook 654. Washington, D.C.: Government Printing Office.

Wahlenberg, William G. 1946. *Longleaf Pine: Its Use, Ecology, Regeneration, Protection, Growth, and Management*. Washington, D.C.: Charles Lathrop Pack Forestry Foundation and USDA Forest Service.

Walters, Russell S., and Harry W. Yawney. 1990. "*Acer rubrum* L.—Red Maple." In *Silvics of North America*, vol. 2, *Hardwoods*, edited by Russell M. Burns and Barbara H. Honkala, 60–69. USDA Forest Service Agriculture Handbook 654. Washington, D.C.: Government Printing Office.

Ward, Carlton, Jr. 2015. "Bombing Range Is National Example for Wildlife Conservation." *National Geographic Society Newsroom* (blog). April 1. https://blog.nationalgeographic.org/2015/04/01/bombing-range-is-national-example-for-wildlife-conservation.

Ward, Daniel B. 2012. "*Yucca filamentosa* and *Yucca flaccida* (Agavaceae) Are Two Distinct Taxa in Their Type Localities." *Castanea* 77, no. 3: 273–79.

Ward, Daniel B., and Robert T. Ing. 1997. *Big Trees: The Florida Register*. Orlando: Florida Native Plant Society.

Watkins, John V., Thomas J. Sheehan, and Robert J. Black. 1975. *Florida Landscape Plants, Native and Exotic*. 2nd revised ed. Gainesville: University Presses of Florida.

Weidensaul, Scott, Tara Rodden Robinson, Robert R. Sargent, Martha B. Sargent, and Theodore J. Zenzal. 2019. "Ruby-Throated Hummingbird (*Archilochus colubris*), Version 2.1." In *The Birds of North America*, edited by Paul G. Rodenwald. Ithaca, N.Y.: Cornell Lab of Ornithology: Birds of the World (database). https://doi.org/10.2173/bna.rthhum.02.1.

Weissling, Tom J., and Robin M. Giblin-Davis. (1997) 2013. "Palmetto Weevil, *Rhynchophorus cruentatus* Fabricius (Insecta: Coleoptera: Curculionidae)" Pub. no. EENY013. Gainesville: University of Florida IFAS Extension. https://edis.ifas.ufl.edu/in139.

Wells, O. O., and R. C. Schmidtling. 1990. "*Platanus occidentalis* L.—Sycamore." In *Silvics of North America*, vol. 2, *Hardwoods*, edited by Russell M. Burns and Barbara H. Honkala, 511–17. USDA Forest Service Agriculture Handbook 654. Washington, D.C.: Government Printing Office.

West, Erdman, and Lillian E. Arnold. 1950. *The Native Trees of Florida*. Gainesville: University of Florida Press.

Whittemore, Alan T., Julian J. Campbell, Zheng-Lian Xia, Craig H. Carlson, David Atha, and Richard T. Olsen. 2018. "Ploidy Variation in *Fraxinus* L. (Oleaceae) of Eastern North America." *International Journal of Plant Sciences* 179, no. 5: 377–89.

Wikipedia. Nov. 12, 2020 (last modified). "Florida Mangroves." https://en.wikipedia.org/wiki/Florida_mangroves.

———. Nov. 10, 2020 (last modified). "Kudzu: Ecological Damage and Roles." https://en.wikipedia.org/wiki/Kudzu#Ecological_damage_and_roles.

———. Nov. 6, 2020 (last modified). "*Megacopta cribraria* [Kudzu Bug]." https://en.wikipedia.org/wiki/Megacopta_cribraria.

———. Oct. 26, 2020 (last modified). "Sassafras Albidum: History of Commercial Use of the Sassafras Albidum Plant." https://en.wikipedia.org/wiki/Sassafras_albidum#History_of_commercial_use_of_the_Sassafras_albidum_plant.

———. Oct. 26, 2020 (last modified). "Witch-Hazel: Folk Medicine." https://en.wikipedia.org/wiki/Witch-hazel#Folk_medicine.

Wilhite, L. P. 1990. "*Juniperus silicicola* (Small) Bailey—Southern Redcedar." In *Silvics of North America*, vol. 1, *Conifers*, edited by Russell M. Burns and Barbara H. Honkala, 127–30. USDA Forest Service Agriculture Handbook 654. Washington, D.C.: Government Printing Office.

Wilhite, L. P., and J. R. Toliver. 1990. "*Taxodium distichum* (L.) Rich.—Baldcypress." In *Silvics of North America*, vol. 1, *Conifers*, edited by Russell M. Burns and Barbara H. Honkala, 563–72. USDA Forest Service Agriculture Handbook 654. Washington, D.C.: Government Printing Office.

Williams, Robert D. 1990. "*Juglans nigra* L.—Black Walnut." In *Silvics of North America*, vol. 2, *Hardwoods*, edited by Russell M. Burns and Barbara H. Honkala, 391–99. USDA Forest Service Agriculture Handbook 654. Washington, D.C.: Government Printing Office.

Witmer, Mark C., D. Jim Mountjoy, and Lang Elliot. 1997. "Cedar Waxwing (*Bombycilla cedrorum*)." In *The Birds of North America*, edited by Alan F. Poole, Peter Stettenheim, and Frank B. Gill. No. 309 of 716 issues, 18 vols. Philadelphia, Pa., and Washington, D.C.: Academy of Natural Sciences of Philadelphia and American Ornithologists Union.

Wood, Don A. 1994. *Official Lists of Endangered and Potentially Endangered Fauna and Flora in Florida*. Tallahassee: Florida Game and Fresh Water Fish Commission.

Woolfenden, Glen E. 1973. "Nesting and Survival in a Population of Florida Scrub Jays." *Living Bird* 12: 25–49.

Wormley, Theo. G. 1870. "A Contribution to Our Knowledge of the Chemical Composition of *Gelsemium sempervirens*." *American Journal of Pharmacy* 42 (January): 1–16.

Wright, Jonathan W. 1957. "New Chromosome Counts in *Acer* and *Fraxinus*." *Morris Arboretum Bulletin* 8: 33–34.

———. 1965. "Green Ash (*Fraxinus pennsylvanica* Marsh)." In *Silvics of Forest Trees of the United States*, compiled and revised by Harry A. Fowells, 185–90. USDA Agriculture Handbook 271. Washington, D.C.: USDA Forest Service.

Wunderlin, Richard P. 1998. *Guide to the Vascular Plants of Florida*. Gainesville: University Press of Florida.

Wunderlin, Richard P., Bruce F. Hansen, Alan R. Frank, and Fred B. Essig. 2020. Atlas of Florida Plants (database). University of South Florida Institute for Systematic Botany (ISB), Tampa. http://florida.plantatlas.usf.edu/.

Youssef, Hany. 2008. "Toxicology Brief: Cycad Toxicosis in Dogs." *DMV360* (e-zine). May 1. https://www.dvm360.com/view/toxicology-brief-cycad-toxicosis-dogs.

Zhang, Yanzhuo, James L. Hanula, and Scott Horn. 2012. "The Biology and Preliminary Host Range of *Megacopta cribraria* (Heteroptera: Plataspidae) and Its Impact on Kudzu Growth." *Environmental Entomology* 41, no. 1: 40–50.

Zielinski, Sarah. 2013. "Fewer Freezes Let Florida's Mangroves Move North." *Smithsonian Magazine*. December 30. https://www.smithsonianmag.com/science-nature/fewer-freezes-let-floridas-mangroves-move-north-180948075/.

Zimmermann, Helmuth G., V. Cliff Moran, and John H. Hoffmann. 2001. "The Renowned Cactus Moth, *Cactoblastis cactorum* (Lepidoptera: Pyralidae): Its Natural History and Threat to the Native *Opuntia* Floras in Mexico and the United States of America." *Florida Entomologist* 84, no. 4: 543–51.

INDEX

ROBERT W. SIMONS is a forest management and ecological consultant for public agencies and private landowners. He served for periods ranging from 23 to 26 years on the UF School of Forest Resources and Conservation Advisory Board, the Florida Forest Stewardship Coordinating Committee, the Florida Silvaculture Best Management Practices Technical Advisory Committee, and the Gainesville Tree Advisory Board.

Printed in the United States
by Baker & Taylor Publisher Services